LASER MACHINING AND WELDING

LASER MACHINING AND WELDING

N. RYKALIN, A. UGLOV and A. KOKORA

Translated from the Russian by
O. Glebov

MIR PUBLISHERS
MOSCOW

PERGAMON PRESS
OXFORD · NEW YORK · TORONTO
SYDNEY · PARIS · FRANKFURT

U. K.	Pergamon Press Ltd., Headington Hill Hall, Oxford OX3 0BW, England
U. S. A.	Pergamon Press Inc., Maxwell House, Fairview Park, Elmsford, New York 10523, U. S. A.
CANADA	Pergamon of Canada Ltd., 75 The East Mall, Toronto, Ontario, Canada
AUSTRALIA	Pergamon Press (Aust.) Pty. Ltd., 19a Boundary Street, Rushcutters Bay, N S. W. 2011, Australia
FRANCE	Pergamon Press SARL, 24 rue des Ecoles, 75240 Paris, Cedex 05, France
FEDERAL REPUBLIC OF GERMANY	Pergamon Press GmbH, 6242 Kronberg-Taunus, Pferdstrasse 1, Federal Republic of Germany

First edition 1978

British Library Cataloging in Publication Data

RYKALIN, N.
Laser Machining and Welding
1. Lasers
I. Title II. Uglov, A. III. Kokora, A.
IV. Glebov, O.
621.36'6 TA1675 78-40166

ISBN 0 08 022724 4

Translated from 'Lazernaya obrabotka materialov'
published by Mashinostroyenie Publishers, Moscow, 1975

CONTENTS

PREFACE

Machining with the focussed laser beam is a new research and technological field which has developed since 1960 when powerful pulse generators of monochromatic radiation were built.

Laser machining is one of the techniques for machining with concentrated energy beams (light, electron and ion beams, plasma jet, etc.).

The theory and practice of laser machining are still being developed. However, considerable advances have been made in theoretical analysis of the effects of powerful light beams on materials and in practical applications of these effects.

This book summarizes the available theoretical and experimental results on various aspects of laser machining. The first chapter deals with the basic physical effects of laser radiation on opaque mediums, primarily, metals. The second chapter describes the techniques for analyzing the laser effects including the measurement of parameters of the laser beam and studies of the laser-affected zones in materials. The third chapter is concerned with the laser heat treatment and welding; various thermophysical problems related to the laser effects are considered there. The laser beams discussed in this chapter have the radiation power density up to about 10^6 W/cm^2. In the fourth chapter the reader will find the data on drilling of materials, analyzing the motion of the liquid phase during formation of a hole, considering the factors which affect drilling, and precision and reproducibility of laser machining. The fifth chapter covers the use of the continuous-wave lasers, mainly CO_2 lasers, in technology. The sixth chapter discusses the changes in metal structure and properties in the zones affected by laser radiation, including the effects of the Q-switched laser radiation on opaque materials.

1
BASIC PHYSICAL EFFECTS OF LASER RADIATION ON OPAQUE MEDIUMS

1.1. PROPERTIES OF RADIATION, LASER TYPES AND POWER CHARACTERISTICS OF LASER BEAM

Laser radiation has a number of unique properties — high intensity (power) of electromagnetic energy flux, high monochromaticity and high spatial and temporal coherence. Hence, laser radiation differs from other types of electromagnetic radiation in that it travels as a very narrow beam.

Lasers of various types emit radiation in a very wide wavelength range — from the ultraviolet to the far infrared region (approximately 0.1-70 μm), and this range is still expanding. We shall discuss here mostly the laser radiation in the wavelength range from 0.4 to 0.6 μm where sufficiently high energy parameters (power output for the continuous-wave and pulsed operation, etc.) can be obtained making possbile efficient application of lasers in technology.

Let us discuss briefly the basic properties of laser radiation. A fundamental feature of laser radiation is its monochromaticity which determines the width of the frequency or wavelength range occupied by the laser radiation, that is, the laser linewidth. The degree of monochromaticity μ for a spectral line of the wavelength λ_0 (frequency ω_0) is defined in the following way [88]:

$$\mu = \Delta\lambda/\lambda_0 = \Delta\omega/\omega_0 \qquad (1)$$

where $\Delta\lambda$ is the linewidth of laser radiation.

Typically, lasers are assumed to emit monochromatic radiation because their spectral lines are narrow enough to be described by a single frequency or wavelength.

Since laser radiation is highly monochromatic lasers can be used in atmospheric studies, spectroscopy and studies of photoluminescence and photoeffect, as frequency standards, to control chemical reactions and so on. Monochromaticity of laser radiation is closely related to its other fundamental property, namely, its coherence, both spatial and temporal. Coherence is a correlation between the parameters of the radiation fields produced by two spatially separated sources simultaneously (spatial coherence) or by the same source at different moments (temporal coherence).

When the sources are coherent the intensity J resulting from superposition of the fields at a given point Q can have any value between $(\sqrt{J_1} - \sqrt{J_2})^2$ and $(\sqrt{J_1} + \sqrt{J_2})^2$ depending on the phase difference. When the sources are not coherent the intensity J is the sum of intensities J_1 and J_2.

The relationship between the temporal and spatial coherence is given by the following equation

$$l_{coh} = c\tau_{coh} \tag{2}$$

Here c is the velocity of light. Equation (2) means that if the path difference between the beams J_1 and J_2 is greater than the parameter l_{coh} known as the coherence length there is no correlation between the parameters of the electromagnetic field at various moments. The parameter τ_{coh} is the coherence time during which the phase difference between electromagnetic waves from the sources does not change by more than π; τ_{coh} is proportional to the width of the line, that is, to the degree of monochromaticity of radiation:

$$\tau_{coh} \sim \Delta\lambda \sim \mu \tag{3}$$

The coherence time for laser radiation can be as long as 10^{-2}-10^{-1} s while for conventional light sources it is 10^{-8} s.

The coherence properties of radiation are described usually by the coherence function which, by definition, is the time average of the product of two components of the electromagnetic field at points r_1 and r_2:

$$\Gamma(r_1, r_2, \tau) = \lim_{T\to\infty} \frac{1}{2T} \int_{-T}^{T} V(r_1, t+\tau)\, V^*(r_2, t)\, dt \tag{4}$$

$$V(r, t) = A(r, t) \cos[\varphi(r, t) - \bar{\omega}t] \tag{5}$$

where $A(r, t)$ and $\varphi(r, t)$ are the amplitude and phase of the wave and $\bar{\omega}$ is the mean frequency.

The magnitude of the coherence function determines the extent of correlation of waves at various points of the field as a function of time. The interference pattern at the point P yields data on the correlation of waves at the points P_1 and P_2 at the same moment ($\tau = t_2 - t_1 = 0$). In this case the spatial coherence is present, which is described by the function $\Gamma(r_1, r_2, 0)$. Correlation of the waves at the point P_1 at various moments τ describes the temporal coherence which is characterized by the function $\Gamma(r_1, r_2, \tau)$.

High temporal coherence of laser radiation makes it useful for various research and technological applications which involve interference, measurements of length, linear and angular veloci-

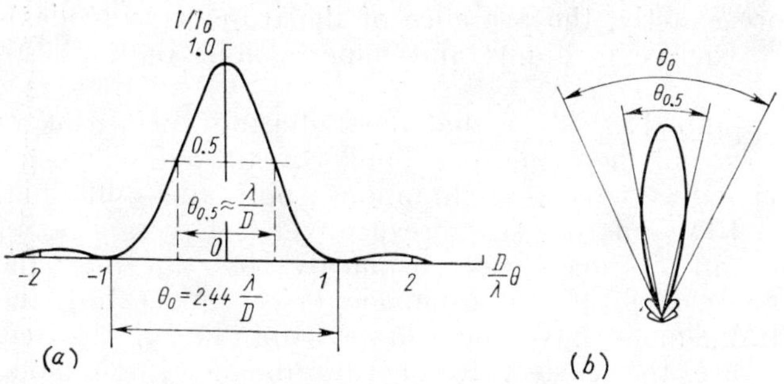

FIG. 1. Radiation patterns for a circular emitting area
(*a*) cartesian system; (*b*) polar coordinate system

ties, and small displacements, for data transmission at optical frequencies, etc.

The spatial coherence of laser radiation makes it highly directional so that it can be focussed upon small areas. The directivity of radiation is described by the solid angle in which the bulk of the radiation energy is emitted. If this solid angle is a cone laser radiation is characterized by the plane angle of spread (the divergence angle). If the solid angle is not a cone two angles in the vertical and horizontal planes are usually specified as parameters.

Figure 1 presents a theoretical radiation pattern of a circular emitting area.

The bulk of the radiated energy is concentrated in the main lobe of the pattern.

The peak field strength of the side lobes is not more than 2% of the peak intensity I_0 of the radiation field.

If the amplitude and the phase are assumed to be uniform in the cross section of the radiation flux then the diffraction-limited divergence of the beam is the angle (at the half-power level)

$$\theta_{0.5} = 1.22\lambda/D \tag{6}$$

where D is the diameter of the beam.

For a ruby laser with $\lambda = 0.69$ μm and $D = 1$ cm $\theta_{0.5} \approx \approx 0.85 \times 10^{-4}$ rad (about 14″) while for a CO_2 gas laser with $\lambda = 10.6$ μm and the same diameter of the radiation flux $\theta_{0.5} = = 1.21 \times 10^{-3}$ rad (about 3′20″). In fact, the divergence of the laser radiation is considerably higher amounting to a few or a few tens of angular minutes. The real lasers have considerably higher divergences than the predicted value owing to many factors — nonuniform distribution of amplitude and phase of the radiation field within the radiating area, multimode nature of generation (more exactly, the presence of the transverse modes), inhomogeneity of active mediums and imperfections of the resonator elements.

References [168, 173] show that the radiating surface of an active medium is not homogeneous and constitutes a mosaic structure comprising complexes of luminous spots and individual spots 850 and 100 μm in size, respectively. Divergence of laser radiation can be decreased, primarily, by selecting the higher transverse modes. The single-mode lasers generating the lowest-order TEM_{00} mode have the smallest angular divergence. The gas lasers are the easiest to operate in the single-mode regime.

The experimental results show that the above properties of radiation are not significant by themselves in the effects of laser radiation on the opaque solids when the energy flux densities are so high that the solids disintegrate. Development of heating zones, which determines the character of laser machining, is determined, primarily, by the energy parameters — energy, power, energy density, pulse duration, spatial and temporal structure of radiation, spatial distribution of the radiation power density over the focal spot, focussing conditions and the physical properties of the material (reflectivity, thermal properties, melting point, etc.).

Let us discuss the main types of lasers and the parameters of their radiation. Solids, liquids and gases are used as active mediums in lasers. The active mediums in solid-state lasers are crystalline or amorphous materials with admixtures of certain elements. Many solids are known to be suitable for lasers but only a few are widely used for building lasers for machining, such as, Al_2O_3 with chromic oxide admixture (ruby); glass, yttrium aluminium garnet $Y_3Al_5O_{12}$ and calcium tungstate $CaWO_4$ doped with neodymium. The lasers with these active mediums have higher output powers and efficiencies. The ruby lasers are the most popular out of the solid-state lasers. Though ruby has a number of advantages compared with other active mediums its drawback is a high excitation threshold.

In most active mediums radiation is generated according to the four-level system [112]. The following solids with rare-earth admixtures have relatively low excitation energies and are the most suitable for continuous operation: yttrium aluminium garnet and calcium tungstate doped with ions Nd^{+3}, and calcium fluoride CaF_2 doped with ions Dy^{+2} or U^{+3}.

The neodymium-doped glasses are the most widely used active mediums since they are easily manufactured in any shape or size, have low losses, high uniformity and other advantages.

The solid-state lasers are similar to the liquid lasers which use liquid dielectrics with dissolved impurities as active mediums.

The generation processes in the solid-state and liquid lasers are essentially similar for the electron energy level structures of impurities in some solutions are close to such structures of solid dielectrics. The liquid lasers can be classified in two groups — inorganic-liquid lasers and organic dye lasers.

The inorganic-liquid lasers, as a rule, make use of phosphorus or selenium oxychloride with an admixture of tin tetrachloride or other metal halides in which neodymium oxide Nd_2O_3 is dissolved to a concentration of a few per cent (the active substance). Such a liquid laser is an analogue of the solid-state neodymium laser but the liquid laser has a more narrow bandwidth than the neodymium-glass laser owing to the more uniform structure of the former.

The radiation in a wide spectral range from the visible to the infrared regions of the spectrum can be produced by using the organic dyes with wide absorption bands (polymethine dyes, rhodamines, pyronine, trypaflavine, stilbenyloxazoles, etc. which are dissolved in alcohols, glycerine, acids or water).

The pumping sources for the dye lasers are, typically, the ruby and neodymium-glass lasers. Since the lifetimes of the excited states are short, a high power is required for pumping and the pumping pulse must have a steep front.

When a pulsed solid-state or liquid laser emits free-generation radiation with the total pulse duration of about 10^{-3} s the pulse consists of a series of successive pulses (spikes) having the duration of about 10^{-6} s with the time interval between them varying from 1 to 5×10^{-6} s. Reference [31] describes the experimentally observed fine structure of the spikes which consist of individual light spots that are diffraction maxima. For the ruby laser with the active medium size 7×120 mm and the pumping power exceeding the threshold power by 40% the energy of the maxima varies from 6×10^{-4} to 1.5×10^{-3} J. The mean duration of micropulses is about 0.2 μs.

Studies in laser resonator systems have resulted in generation [88, 57] of spikes of approximately the same amplitude and the constant duty ratio (ordered generation). Generation of another type is known as the quasistationary generation [40, 47, 134, 135] for the pulse in this case practically does not contain spikes if its fine temporal structure is neglected.

The *Q*-switched laser can generate pulses about 10^{-7}-10^{-9} s long [112]. In this case the pulse practically does not have a spike structure *, that is, the giant pulse is a monopulse with an almost delta shape.

* Reference [164] reports a discrete structure of a giant pulse.

Techniques and equipment have been developed for recording picosecond laser pulses with the duration less than 10^{-9} s. Generation of ultrashort light pulses about 4×10^{-12} s long is described in [154].

The pulse energy of the solid-state and liquid lasers varies from a few tenths to 10^3 J or more (for the free generation) and to tens and hundreds of joules for the Q-switching operation.

The radiation power of pulsed lasers can vary from hundreds of kilowatts (free generation) to gigawatts (Q-switching operation). For the spike operation (unordered generation) there can be a two-order-of-magnitude difference between the integral mean power of a pulse and the power of an individual spike. This difference is somewhat smaller for a pulse with an ordered structure (regular pulse). For a quasistationary pulse the integral mean power differs only slightly from the power at a given moment. Therefore, the quasistationary operation is of practical interest for welding and machining of materials since it can be used for "soft" heating which decreases the loss of material from the affected zone [39].

The efficiency limit of lasers is determined mostly by internal losses in the active medium and by the efficiency of utilization of the pumping energy. For instance, for a ruby laser the practical efficiency is not more than 1% and for a neodymium-glass laser it is 2%.

The active medium in the gas lasers is a gas, a mixture of gases or a mixture of a gas and metal vapour. The gas lasers include also the chemical lasers in which gaseous chemical mediums are used. In chemical lasers excitation of the active medium occurs owing to rapid chemical reactions. The active particles in the gas lasers can be neutral atoms, ions and gas molecules.

The neutral-atom lasers emit light mainly in the infrared spectral range and some of them in the red visible range.

The ion lasers emit mostly visible and ultraviolet radiation. The molecular gas lasers emit radiation with the wavelength of 10-100 μm (the infrared and submillimetre ranges). The neutral-atom (for instance, He-Ne) lasers have the continuous-operation power of not more than 10 mW, the power of the ion (argon) lasers can be as high as 500 W while the molecular lasers are the most powerful ones. For instance, a carbon dioxide laser has a continuous-operation output power of a few tens of kilowatts. The efficiency of the neutral-atom and ion lasers practically does not exceed 0.1% while the molecular lasers can have efficiencies as high as 20%. As most of the gas lasers operate in the single-mode regime the angular divergence is not more than 1-2′ while for the multimode operation it is not more than 10′.

Among the factors contributing to the laser radiation effects on materials there are the focussing conditions, that is, the mutual arrangement of the focal plane of the optical laser system and the irradiated surface. The minimum size of the irradiated area on the surface of the sample being illuminated is the radius of the central spot of the diffraction pattern given by the following equation:

$$r_f = \theta F \tag{7}$$

where F is the local distance of the lens.

For the half-power level $\theta = \theta_{0.5}$ we have

$$r_f = 1.22 \frac{\lambda F}{D} \tag{8}$$

For instance, at $D/F = 0.3$, r_f for a ruby laser is about 3 μm. If the laser power is 10^6 W the flux density at the spot centre is about 10^{13} W/cm^2.

Focussing of the coherent laser radiation yields the following distribution of intensity in the focal plane of the lens [88]:

$$q(r) = q_0 \left[\frac{2J_1(\mathrm{B}r)}{(\mathrm{B}r)}\right]^2 \tag{9}$$

Here $J_1(u)$ is the Bessel function of first kind and of order 1

$$\mathrm{B} = \pi D/\lambda F$$

The intensity in the spot centre ($r = 0$) is

$$q_0 = \frac{\pi D^2}{4\lambda^2 F^2} P_0 \tag{10}$$

where P_0 is the emitted power.

The diffraction distribution of the flux density [Eq. (9)], typically, does not describe the radiation of the solid-state lasers. Owing to inhomogeneities of the laser system the distribution is more complicated and frequently asymmetric. The angular distribution of radiation intensity varies with the distance from the emitting surface. Thus, there are the near-field, Fresnel and far-field (Fraunhofer) intensity distribution zones with respect to the emitter. The angular distribution in the far-field zone almost does not depend on the distance to the emitting surface of the laser. The border of this region is at the distance from $D^2/2\lambda$ to $2D^2/\lambda$. Since the diffraction pattern in this region is known as the Fraunhofer diffraction this region is called the Fraunhofer zone. The region between the border of the far-field zone and another border at the distance of $(D/2)/(D/\lambda)^{1/3}$ from the emitter is known as the Fresnel zone.

The region between the inner border of the Fresnel zone and the emitting surface is referred to as the near-field zone. No approximations can be used in this zone for describing the field strength distribution. The radiation fields in this region can be very complex and rapidly varying and have sharp shadow borders.

Table 1 [153] describes the zones of spatial field strength distribution for a coherent radiation source with diffraction effects.

TABLE 1. Field Strength as a Function of Distance to Emitter

Zone	Distance to emitter	Phase difference at the distance L from points on surface of diameter D
Near-field	$L = D$	No difference
Fresnel	$L \approx (D/2)/(D/\lambda)^{1/3}$	$\Delta\varphi \approx \lambda/16$
Rayleigh	$L \approx D^2/2\lambda$	$\Delta\varphi \approx \lambda/4$; parallel beam
Fraunhofer (far-field)	$L \approx 2D^2/\lambda$	$\Delta\varphi \approx \lambda/16$; diverging beam

1.2. ENERGY TRANSFER TO IRRADIATED MATERIALS

A part of the light incident on the surface of the material is reflected, and the remaining light flux is absorbed in the bulk of the material. This process is equivalent to the existence of a heat source with a certain temporal and spatial distribution on the surface and inside the body. In most cases of practical interest the absorbed radiation power density varies in the bulk of a solid according to Bouguer's law

$$q_v(z) = q_{v_0}(1 - R)\exp(-\alpha z) \tag{11}$$

where $q_v(z)$ and q_{v_0} are the absorbed radiation powers per unit volume (in W/cm^3) at the distance z from the surface of the irradiated body and at its surface, respectively, and $1 - R$ and α are the light absorptivity and absorption coefficient in cm^{-1}, respectively.

Absorption of light and subsequent energy transfers in the body occur via different processes in metals, insulators and semiconductors.

Let us consider the physical processes at the initial stage of laser radiation effect on metals. The optical properties of metals can be described in the framework of the classical free-electron model [8] according to which the light flux minus the reflected part is fully absorbed in interaction with the conduction electrons in the surface layer 10^{-4}-10^{-5} cm thick. The conduction electrons * absorb photons and emit or absorb phonons or collide with each other, etc.

The electron energy is increased by light absorption. The electrons transfer part of the absorbed energy to the crystal lattice but the efficiency of such transfer is low owing to the large mass difference between electrons and ions. Therefore, the electron gas becomes highly overheated in comparison to the lattice. The heating of the electron gas and the energy transfer to the lattice are analyzed in detail in [58] proceeding from the kinetic equation for the electron energy distribution function.

The electron gas and the lattice in metal are two weakly interacting subsystems. The electron and ion relaxation frequencies ν_{ee} and ν_{ii} are considerably higher than the relaxation frequency ν_{ei} for the energy exchange between them. If we have

$$\nu_{ee} \gg \nu_{ef} \tag{12}$$

and

$$\nu_{ii} \gg \nu_{ei} \tag{13}$$

where ν_{ee} and ν_{ii} are the electron — electron and ion — ion collision frequencies, ν_{ef} is the electron — photon collision frequency and ν_{ei} is the electron — ion collision frequency then the electron gas and the lattice can be described by different temperatures — the electron temperature T_e and the lattice temperature T_i. Equation (12) means that the absorbed energy is rapidly distributed among the free electrons. Similarly, Eq. (13) means that the energy transferred by the electrons to the lattice is rapidly distributed between the ions.

Let us consider the relaxation frequencies ν_{ef}, ν_{ee}, ν_{ei} and ν_{ii} which determine the radiation-absorption processes. The frequency of the electron-photon collisions ν_{ef} is proportional to the power density q_s and can be written as [43]

$$\nu_{ef} = \frac{\alpha q_s}{h\nu n'} \tag{14}$$

where q_s is the surface power density, erg/cm^2 s, $h\nu$ is the photon energy, ergs and n' is the density of photon-absorbing electrons, cm^{-3}.

In the visible spectral range $h\nu$ is much higher than kT_e where the electron temperature T_e is about 10^3 K. Hence

$$n' \approx \frac{h\nu}{\varepsilon_F} n$$

* If the photon energy $h\nu$ is higher than the energy gap the photons can be also absorbed by the valence electrons which are excited to the conduction band.

where $n \approx 5 \times 10^{22}$ cm^{-3} is the conduction electron density and ε_F is the Fermi energy;

$$\varepsilon_F = \frac{h}{8m\pi^2}(3\pi^2)^{1/3} n^{2/3}$$

where h is Planck's constant, m is the electron mass and $\varepsilon_F \approx$ $\approx$ 5-10 eV.

For the laser with the ruby active medium ($h\nu = 1.67$ eV) $n' \approx (1\text{-}2) \times 10^{22}$ cm^{-3} and $\nu_{ef} \approx (1.5\text{-}3) \times 10^{-5}\ q_0$.

The frequency ν_{ee} of the electron — electron collisions in metals is determined primarily by the electron density in the Fermi surface vicinity of width kT_e; it is given by the following relationship [43]:

$$\nu_{ee} = v_F \sigma_{ee} n \left(\frac{kT_e}{\varepsilon_F}\right)^2 \tag{15}$$

Here v_F is the electron velocity at the Fermi surface (about 10^8 cm/s) and σ_{ee} is the Rutherford cross section of the electron — electron interaction ($\sigma_{ee} \approx 5 \times 10^{-16}$ cm^2). When $T_e \approx 10^3$ K we have $\nu_{ee} \approx 10^{14}$ s^{-1} and the time τ_{ee} of relaxation to the equilibrium distribution of the electron gas is $\nu_{ee}^{-1} \approx 10^{-14}$ s. The frequency ν_{ei} can be found from the data reported in [58]. The rate of energy transfer between the electron gas and the lattice and the temperature are determined by the heat source and the heat transfer coefficient α_1 for the heat transfer from the electrons to the lattice.

Analysis of the energy exchange between electrons and lattice [58] shows that in the framework of quantum mechanics the generation of phonons corresponds to the Čerenkov radiation of the sound waves by supersonic electrons ($v_F \approx 10^8$ cm/s, the sound velocity s is about 10^5 cm/s) in the classical treatment of the problem.

In this case for $T_i > \theta_D$ (θ_D is the Debye temperature) when $\tau'_{ei} \approx 1/T$ (τ'_{ei} is the mean free path time of the electron between collisions with phonons) the energy transferred from the electron gas to the lattice per unit volume in unit time is

$$\bar{u} = \alpha_1 (T_e - T_i); \quad \alpha_1 = \frac{\pi^2 m^* s^2 n}{\sigma \theta_D \tau'_{ei}(\theta_D)} \tag{16}$$

where m^* is the effective mass of the electron.

According to [58] $\theta_D \tau'_{ei}(\theta_D)$=const so that for metals of medium atomic weight $\theta_D \tau'_{ei}(\theta_D) \approx 10^{-11}$ s K. Then the coefficient of heat transfer between electrons and lattice α_1 is about 10^{17} erg/cm^3 s K.

The electron — ion relaxation frequency ν_{ei} can be expressed in terms of the heat transfer coefficient:

$$\nu_{ei}=\frac{\alpha_1}{\rho_i c_i} \quad \text{or} \quad \nu_{ei}=\frac{\pi^2}{30}\,\frac{m^* v_F n k s^2}{\rho_i c_i d_0 \varepsilon_F} \tag{17}$$

where $\rho_i c_i$ is the lattice specific heat, erg/cm^3 K and d_0 is the lattice parameter (about 10^{-8} cm).

Substituting numerical values of parameters in Eq. (17) we find $\nu_{ei} \approx 10^{11}$ s^{-1} and $\tau_{ei} \approx 10^{-11}$ s.

The frequency ν_{ii} can be estimated from expression

$$\nu_{ii} \approx \frac{k\gamma_0^2 T_i}{d_0 \mu_i s} \tag{18}$$

where γ_0 is the Gruneisen constant, μ_i is the ion mass (about 10^{-22} g), $T_i \approx 10^3$ K, $\nu_{ii} \approx 10^{13}$ s^{-1} and $\tau_{ii} \approx 10^{-13}$ s.

A comparison of Eqs. (14) and (15) shows that when the radiation flux density $q_0 \leqslant 10^9$ W/cm^2 condition (12) is always satisfied, and the electron gas in metal is described by the temperature T_e. Similarly, condition (13) is also satisfied, and the thermal state of the lattice can be characterized by the lattice temperature T_i. When the time t is less than τ_{ei} ($\tau_{ei} \approx 10^{-11}$ s) only the electron gas is heated in the metal while the change in the ion temperature is small because the relaxation time of the electron gas is considerably less than the lattice ion relaxation time.

The difference $\Delta T = T_e - T_i$ is determined by the difference between the frequencies ν_{ef} and ν_{ei}. The higher the difference between the frequencies ν_{ei} and ν_{ef}, the lower the difference ΔT and when $\nu_{ef} \ll \nu_{ei}$ we have $T_e \approx T_i$. However, the electron gas temperature increases only until the amount of energy transferred to the lattice becomes equal to the energy received from the external source. The transfer of energy between the electron gas and the lattice intensifies when $t > \tau_{ei}$ at the highest difference $T_e - T_i$. Further on, the difference $T_e - T_i$ decreases tending to zero. Under radiation with the constant flux density a metal can be characterized by the single temperature T starting from the moment $t \gg 100\,\tau_{ei} \approx 10^{-9}$ s.

The energy absorbed in the zone irradiated by laser is transferred to the bulk of the material via various mechanisms of heat conduction, namely, electron, phonon and radiative conduction.

In the temperature range from hundreds to a few thousand degrees the prevalent energy transfer mechanism in metals is the electron heat conduction; in this temperature range the phonon heat conduction is small compared to the electron conduction though it is significant at lower temperatures. The radiative conduction makes an important contribution to the energy transfer at the temperatures above $\approx 10^4$ K.

The general problem of heating of a semi-infinite body for the moving mediums (or moving heat source) can be described by the following system of differential equations if we neglect the radiative heat conduction and the finiteness of the energy flow velocity:

$$c\rho\left(\frac{\partial T}{\partial \tau} + \vec{v}\,\mathrm{grad}\,T\right) + \mathrm{div}\,q_s = q_v \tag{19}$$

$$q_s = -\lambda_e\,\mathrm{grad}\,T \tag{20}$$

where $\vec{v}$ is the velocity vector and q_v is the heat released per unit volume in the body owing to the effect of the heat source.

The specific heat (per unit volume) $c\rho$ and the electron heat conductivity λ_e depend on the temperature that is why the system of equations (19) and (20) is nonlinear. To solve it we must make some simplifying assumptions. For instance, the temperature dependence of the thermal parameters is typically assumed to have no significant effect on the temperature distribution in metals. If we assume also that the medium is stationary and the radiation is absorbed in a thin surface layer about α^{-1} thick (α is the absorption coefficient) the system (19) and (20) is transformed into

$$\frac{1}{a}\frac{\partial T}{\partial \tau} = \Delta T + q_v\delta(z) \tag{21}$$

where $a = \lambda_e/c\rho$ is the coefficient of thermal diffusivity.

The boundary conditions for the problem are

$$\frac{\partial T}{\partial z} = 0 \text{ for } z = 0 \tag{22}$$

$$T = 0 \quad \text{for } z = \infty \tag{23}$$

$$\tau = 0$$

The system of equations (21), (22) and (23) describes the heating of a semi-infinite body starting from the time $\tau_i > 10^{-9}$ s. The boundary condition (22) is satisfied only when the density of the heat losses from the surface of the body by radiation (according to the Stefan — Boltzmann law), convection (according to the Newton law) or evaporation is small compared to the density of the incident laser radiation flux. If this condition is not satisfied the boundary condition (22) must be changed into $\partial T/\partial z > 0$ when the heat loss is by radiation or convection and $\partial T/\partial z \leqslant 0$ with evaporation heat loss. This problem is discussed in more detail in Chapters 3 and 4. The temperature T is a function of the spatial coordinates x, y and z and the time τ. Absorption of radiation in semiconductors depends significantly on the relation between the energy gap E_g and the laser photon energy $h\nu$, the density of the free charge carriers and the time of radiationless transitions. The following mechanisms of the laser radiation absorption are the most important for semiconductors [111].

Dielectric mechanism ($h\nu \ll E_g$). The multiphoton processes of the electron — hole pair generation are not significant for light absorption. The laser radiation is absorbed by its direct interaction with the lattice.

Metallic mechanism ($h\nu < E_g$). The multiphoton processes and the impurity absorption are not significant; there are free charge carriers. The light is absorbed by photon interaction with the free carriers.

Induced metallic mechanism ($h\nu > E_g$). The radiationless transition time is large. Radiation is mainly absorbed by the non-equilibrium free carriers generated by the radiation itself so that absorption by generated free carriers is higher than absorption by equilibrium carriers.

Semiconductor mechanism ($h\nu > E_g$). In the semiconductors with low impurity concentrations laser radiation generates electron — hole pairs whose recombination releases heat if radiationless transitions are allowed. This mechanism of the energy transfer from light to matter is significant if the radiationless transition time is small.

At the initial stage of light absorption in semiconductors when $h\nu > E_g$ the valence electrons are excited to the conduction band (the intrinsic photoeffect). The concentration of the free photoelectrons increases up to a certain level being limited by such effects as recombination (nonradiative recombination at the traps (the rate is proportional to n_e), radiative recombination between the bands (the rate is proportional to n_e^2) [51], the impact recombination (the rate is proportional to n_e^3)* and the ambipolar diffusion which is due to the gradients of concentrations of photoelectrons and holes in the region irradiated by laser.

The increase in concentration of the free photoelectrons results in variation in the electric and optical parameters of semiconductors, for instance, the reflectivity. The authors of [51] have analyzed the absorption and transfer of energy in semiconductors ($h\nu > E_g$) prior to disintegration of the material; they have shown that the light fluxes with the energy densities $q > 10^6$ W/cm^2 increase the concentration of free electrons up to 10^{20}-10^{21} cm^{-3} in time period of the order of 10^{-9}-10^{-8} s.

* Radiative recombination between bands and impact recombination can significantly limit the concentration of free photoelectrons only at high n_e owing to low probabilities of double and triple collisions.

Under these circumstances the optical properties of semiconductors are close to those of metals, and their reflectivity increases. For instance, the reflectivity of germanium increases by a factor of two when the radiation flux reaches approximately 10^7 W/cm^2, and the absorption coefficient increases to 10^4-10^5 cm^{-1}. At the same time the relaxation frequency ν_{ei} varies approaching the frequency for metals.

Since the lifetime of carriers before recombination at the traps (the rate proportional to n_e) is 10^{-7}-10^{-8} s [54] and in the time $\tau_{ei} \approx 10^{-10}$-$10^{-11}$ s the free photoelectrons transfer to the lattice only the energy $(h\nu - E_g)/h\nu$ (per electron), a low concentration of free photoelectrons and, hence, a low absorption of the incident light flux mean that the energy is transferred to the lattice much later than in metals, and an increase in the concentration n_e does not result in an increase in the lattice temperature T_i. At a certain concentration n_e' the temperature T_i starts to increase noticeably. Under such conditions the moment when the semiconductor lattice temperature starts to increase is determined by the time τ' needed to reach the concentration n_e' rather than by the time τ_{ei} of electron — phonon relaxation.

For instance, for the neodymium laser with the photon flux density $I \approx 10^{25}$ cm^{-2}s^{-1} the following estimates have been made: $n_e' \approx 5 \times 10^{20}$ cm^{-3} and $\tau' \approx 10^{-8}$-10^{-11} s [51].

Absorption of radiation by the conduction electrons in semiconductors occurs for $h\nu < E_g$, that is, below the intrinsic absorption band. This absorption is typically negligible since the concentration of free carriers is smaller by the factor $\exp(-E_g/2kT)$ than the concentration of electrons responsible for the intrinsic semiconductor absorption.

When radiation is absorbed by the free carriers they can be "heated"; the initially low carrier concentration can be increased owing to thermal ionization, that is, self-maintaining heating of the material can proceed with an increasing rate. Light absorption by impurities depends on the ratio between the photon energy $h\nu$ and the energy "distance" between the impurity level and the bottom of the conduction band or the top of the valence band; it also depends on the concentration of the impurities in the material.

The lattice absorption is the highest in the far infrared range (about 10-100 μm), and it is not very important in the visible and near infrared ranges. The exception is the effect of continuous radiation of metal by powerful CO_2 lasers in oxidizing environment, which produces semiconductor layers on the metal surface.

The absorbed energy is transferred from the surface layer to the semiconductor bulk via the thermal conduction mechanism [90]. At the initial stage of the process when the free electron concentration in the semiconductor is negligible the process is dominated by the phonon conduction. As the concentration n_e increases the increasing part of energy is transported by the conduction electrons which make a significant contribution to the total heat conduction at $n_e > n'_e$. In semiconductors energy can be also transported by recombination radiation [44].

In dielectrics radiation can be absorbed by direct interaction between photons and lattice. In transparent dielectrics with low absorption coefficients radiation absorption and subsequent disintegration can depend significantly on various absorption centres (impurity clusters, gas bubbles, microcracks, etc.) [109, 113].

1.3. HEATING EFFECTS

Conventional heat conductivity theory (linear or nonlinear) can be used for analyzing heating of materials by radiation from lasers with the pulse duration considerably larger than 10^{-9} s and the energy flux densities below 10^8-10^9 W/cm^2 in most practical cases. The time limitation is given by the pulse duration $\tau_i \approx 10^{-9}$ s. There are the following problems in analyzing the thermal effects of laser radiation on materials: to analyze the space — time characteristics of the heat sources generated by absorption of laser radiation in various materials; to formulate the boundary-value problem of heat conduction taking into account the typical features of laser radiation effect on materials; to develop techniques for exact and approximate analysis of the boundary-value problems and the numerical techniques for analysis of temperature fields; to derive analytical formulas suitable for evaluations and comparison with experimental data; to find the parameters of radiation pulses required for specific technological processes by analyzing the temperature field variation on the surface and in the bulk of the material.

To analyze the effects of laser radiation we have to know the energy parameters, namely, the absorbed part of the incident radiation flux, the highest flux energy density, the pulse duration, the wavelength, the spatial distribution of the energy density and the focussing conditions. These energy parameters are determined by the laser type, the active medium, the technique of pumping and design of the pumping system, resonator and the optical focussing system, on the one hand, and the optical properties of the irradiated material on the other hand.

To describe the heat sources generated by laser radiation with the known space — time parameters) in opaque materials we have to know their reflectances in order to estimate the absorbed radiation energy. The reflectance R at the flux densities

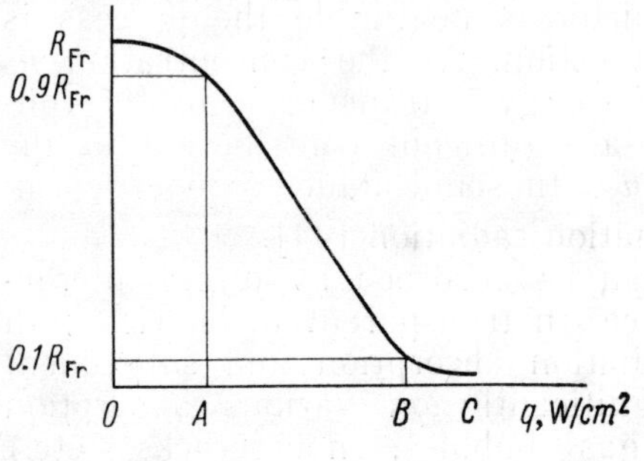

FIG. 2. Reflectance R_{Fr} as a function of incident power density q (qualitative estimate)

much lower than 10^6 W/cm² can be evaluated using the well-known Fresnel formulas [133]; to a significant extent it depends on the state of the surface (roughness of the surface, grade of finishing, the presence of an oxide film). When the radiation flux density increases up to 10^6-10^7 W/cm², for most materials reflectance decreases.

As shown by experiments R decreases sharply from the Fresnel value to 10% of it when the radiation flux density increases up to 10^{10} W/cm² [149] (Fig. 2). Since the laser pulse is noticeably nonuniform in time and space the experimentally measured reflectances correspond to an averaged radiation flux density.

Therefore, it is advisable to describe complicated spatial and temporal distributions of radiation flux density by simpler approximations. For instance, a useful approximation for spatial distribution of the radiation flux density is the normal distribution.

With this approximation we can make use of many results of the theory of heat sources [117] which has been sufficiently well developed for the processes of welding, cutting, etc. On the other hand, some experimental [11] and theoretical results suggest that the spatial distribution of the laser beam intensity is close to the normal (gaussian) distribution.

In the heat process theory two types of spatial distributions of the thermal flux intensity — gaussian distribution and uniform distribution over the heated spot of the radius r_f — are generally used. For the gaussian distribution we have the following formula:

$$q(r) = q_0 \exp(-kr^2), \quad r > 0 \tag{24}$$

where k is the concentration factor in cm^{-2} which determines the focussing property of the radiation source.

When the intensity is uniformly distributed over the focussed spot of the radius r_f we have

$$q'(r) = \begin{cases} q_0^* & r_f \geqslant r \geqslant 0 \\ 0 & r > r_f \end{cases} \tag{25}$$

The distributions (24) and (25) can be related through the concentration factor k [117] in the gaussian distribution:

$$r_f = B^* k^{-1/2} \tag{26}$$

Here B^* is a numerical factor which depends on definition of r_f for the gaussian distribution. If r_f is defined as the distance from the spot centre in which the intensity decreases by the factor e then $B^* = 1$. For the laser radiation with $q_0 \approx 10^6$ W/cm^2 the difference between these two distributions on the symmetry axis of the source is shown to be of the order of 10% and to become significant at $r \geqslant r_f$ [123].

A heat source equivalent to the effect of the laser beam can be surface or volume, and extended or discrete, according to the problem being solved, the calculation technique and the physical properties of the material.

When we consider the problems encountered in analyzing the heating of materials with laser pulses of a few milliseconds and with radiation intensity q_0 about 10^4-10^6 W/cm^2 we should bear in mind the following important features:

1. In most cases the heat losses from the heated surface by radiation and convection may be neglected.

2. In some cases the size and position of the affected zone is not much changed by taking into account the temperature dependence of the optical and thermal parameters of the material, and the equations are simplified by using constant, temperature-independent factors.

3. It is especially important to take into consideration the heats of phase transitions (melting, crystallization, evaporation) in those cases when evaporation and removal of gas and solid-state products are decisive factors in the energy balance. When the primary effect of radiation is melting as it is in laser welding it becomes important to take into account the heat of melting. Under these circumstances a modification of the Stefan problem [116] with the phase transition at a moving boundary should be solved.

4. The effects of laser radiation on materials can be described in the framework of the thermal process theory only at the laser intensities up to 10^8-10^9 W/cm^2. For higher radiation intensities the effects of laser beams on the materials should be analyzed using the theory of explosions accompanied by shock waves travelling through the bulk of the material. The thin surface layer in which the laser beam energy is absorbed acts here as an explosive. Under such conditions thin surface layers at the free surfaces can be split off if the pressure of the shock waves is higher than the tensile strength of the material [24]. Powerful lasers

with $q_0 \approx 10^{12}$-10^{14} W/cm^2 can produce in materials shock waves with amplitudes high enough to give rise to structural changes and phase transitions.

Let us consider the techniques for solving the thermal analysis problems. If the problems encountered in laser machining can be formulated linearly the techniques generally used are the source technique (Green's functions), Fourier methods, and finite and infinite integral transformations over spatial variables.

Now, let us discuss how to select a technique for solving a problem according to the specific results we have to obtain. The source technique is the most universal method for analyzing the three-dimensional problems of heating of semi-infinite bodies practically for any period of heating; in combination with the local effect principle [117] this technique yields the general equations for variation of the temperature field in the bulk of the material.

To analyze the temperature fields of composite bodies it is convenient to use the Laplace transformation over the time which yields solutions that are valid for short times of the source effect and for the steady-state temperature field.

The finite integral transformations over spatial variables are convenient to use when the joined or heated materials have significantly different thermal properties (for instance, for butt welding of sheet materials or for heating of two-layer finite slabs).

The classical Fourier methods in combination with integral transformations can be successfully used for analyzing heating problems when the surface temperature is close to the steady-state distribution [128, 129].

In many cases laser heating of materials should be analyzed taking into account the temperature dependences of the thermal parameters. There are no general analytical techniques for solving such problems so that a special approach has to be developed for each particular case. The techniques for solving the nonlinear heat conduction problems have been classified to a certain extent in [65, 82].

1.4. DISINTEGRATION OF MATERIALS

When the radiation intensity exceeds a certain critical level the irradiated material disintegrates after a certain period. The definition of the term "disintegration" under the effect of powerful radiation is an arbitrary one since radiation beams of practically any intensity give rise to physical processes in the bulk of the material, which result in irreversible changes, due to, for instance, diffusion or generation of structural imperfections.

Disintegration of opaque materials under the effect of laser radiation will be understood below, unless otherwise stated, as pitting of the material or development of cavities owing to vaporization processes. Melting of material and pushing out of the liquid phase by reactive pressure of vapour and the subsequent crystallization of the melt can also result in disintegration of the material giving rise to craters or through holes in thin plates. Brittle materials can "crack" and chip off under the effect of thermal stresses.

No general physical model has been developed to describe the typical features of radiation effects in a wide range of radiation intensities and for a wide variety of materials. There are two main models used for describing the disintegration processes — the thermal disintegration model and the gas dynamic model.

(a) THE THERMAL DISINTEGRATION THEORY

The theory of disintegration of materials under the effect of radiation with the power densities of 10^6-10^9 W/cm^2 has been developed by Anisimov, Ready and others [34, 55, 195]. This model assumes that the material is removed from the affected zone by evaporation from the surface. The process is described by the heat conduction equation for the condensed phase in the coordinate system fixed to the moving boundary at which the material is vaporized.

To simplify the mathematical treatment and to concentrate on the fundamental features of the evaporation process we shall not analyze the temporal structure of the laser pulse and ignore the fact that there is a liquid phase * on the surface of the material. Since the beam power density q_0 is more than 10^6 W/cm^2 the time of transition to the quasistationary process with the constant evaporation rate v_0 is small so that this transition stage of the process is neglected. If the inequality

$$r_f \gg \sqrt{a\tau_i} \qquad (27)$$

(where τ_i is the laser pulse duration) is satisfied the quasistationary motion of the evaporation boundary in the semi-infinite body can be analyzed as a one-dimensional problem:

$$\frac{\partial^2 T}{\partial z^2} + \frac{v_0}{a}\frac{\partial T}{\partial z} = 0$$

$$-\lambda \frac{\partial T}{\partial z} = q - \rho\,\Delta H v_0, \ (z = 0) \qquad (28)$$

$$T(z, 0) = T(\infty, t) = 0$$

* Batanov [55] has taken into account the presence of a liquid phase layer on the metal surface.

Here ΔH is the difference between the specific enthalpies of the solid and gaseous phases:

$$\Delta H = L_b - \frac{R_0 T}{2} \tag{29}$$

where L_b is the specific heat of evaporation and R_0 is the gas constant per mole.

In the travelling coordinate system the solution of the system (28) is

$$T = T^* \exp\left(-\frac{z v_0}{a}\right) \tag{30}$$

The following relationship links v_0 and T^*:

$$v_0 = \frac{q_0}{\rho (L_b + 2.5 R_0 T^*)} \tag{31}$$

To determine two parameters v_0 and T^* in Eq. (31) we have to write down an additional condition. In the thermal disintegration model this is the equation for evaporation kinetics [151]:

$$v_0 = s \exp(-L_b/R_0 T^*) \tag{32}$$

where s is close to the sound velocity in the material and determined by the lattice model being used.

For instance, for the Debye lattice model we have

$$s = \left[\frac{4\pi}{9}\left(s_l^{-3} + 2s_t^{-3}\right)\right]^{-1/3} \tag{33}$$

where s_l and s_t are the velocities of the longitudinal and transverse waves, respectively.

Solving Eqs. (31) and (32) simultaneously we can express v_0 and T^* via the physical parameters of the material. Then we can find the displacement of the evaporation front:

$$\Delta z = v_0 (t - t_0) \tag{34}$$

Here t_0 is the time of transition to the quasistationary state:

$$t_0 = \frac{a}{v_0^2} \frac{9\pi}{4(p + 2.5)^2} \tag{35}$$

where $p = L_b/R_0 T^*$.

Qualitative analysis shows that we can establish the optimum process of evaporation for the given material to provide for maximum vaporized mass and maximum displacement of the evaporation front. The optimum pulse duration is

$$t^* = \frac{0.15}{a}\left(\frac{Q}{\rho L_b}\right)^2 \frac{p^*}{p^* + 1/2} \tag{36}$$

where Q is the energy density, J/cm^2, p^* is the root of the transcendental equation:

$$\mathbf{e}^p = K\,\frac{2p+5}{2p+1} \tag{37}$$

and $K = 100\ s/9\pi a\rho L_b$.

The optimum laser beam power density is

$$q^* \approx \frac{2.8a}{Q}\,(\rho L_b)^2 \tag{38}$$

The displacement of the evaporation front in the optimum process is $\varepsilon\sqrt{at^*}$ where $\varepsilon \approx 1$. Thus, in the optimum evaporation process the rate of displacement of the evaporation front is close to the averaged (over the pulse duration) velocity of the heating front in the material.

The above treatment of the motion of the evaporation front does not take into account the temperature dependences of the thermal and optical properties of the material. Anisimov *et al* [22] have taken into consideration the reflection of light from the surface and the temperature dependence of the reflectance. The absorptivity $A(T)$ increases linearly with temperature. It has been shown that under these circumstances the optimum evaporation process requires higher beam power densities and smaller pulse durations. The authors of [22] have analyzed the one-dimensional problem.

The applicability of the above model is limited by the condition of low radiation absorption by the vaporized material, that is, by the lack of screening of the incident radiation. The analysis of evaporation of the material should take into account the kinetic phenomena in the thin layer adjacent to the surface, for instance, those that are due to the back flow of vapour as a result of atomic collisions.

The conditions near the evaporation front can be analyzed in more detail by solving the gas kinetic problem of vapour motion in the thin layer at the phase interface [1].

Anisimov has shown that the expression for the velocity of the evaporation front

$$v = \frac{q_0}{\rho\,(L_b + 2.2kT^*/m)} \tag{39}$$

which takes into account the kinetic phenomena differs from the respective expression which does not take into account them.

(b) THE GAS DYNAMIC EVAPORATION THEORY

This theory analyzes disintegration of material under radiation fluxes with intensities exceeding the critical value for beginning of evaporation using a system of gas dynamic equations. The gas dynamic disintegration model has been developed by Afanasyev and Krokhin [2] for a wide range of high power densities from 10^6 W/cm^2.

This approach to disintegration of material is based on the fact that the contribution of heat conduction in the condensed phase in the energy balance of the process decreases considerably with increasing laser beam power density while the increasingly important role is played by the effects due to the motion of the vaporized material and its interaction with the incident radiation.

The evaporation process depends significantly on the incident power density q. If the resulting temperature is low compared to ω/k (where ω is the bonding energy per atom and k is the Boltzmann constant) the main contribution to the energy balance is made by the evaporation heat. When the incident power density is high and the temperature exceeds ω/k the parameters of the condensed phase do not affect noticeably the disintegration process. When the incident power density is low and the maximum resulting temperature is low the vapour of the material will be transparent for the incident radiation. When the incident radiation power increases the temperature also increases and the radiation absorption by the vapour becomes increasingly significant; it is known as screening of radiation.

The motion of material under the effect of radiation incident from the point $x = \infty$ on the surface at the moment $t = 0$ is described by the following system of one-dimensional gas dynamic equations [2]:

$$\frac{\partial \rho}{\partial t} + \frac{\partial}{\partial x}(\rho v) = 0 \tag{40a}$$

$$\frac{\partial}{\partial t}(\rho v) + \frac{\partial}{\partial x}(p + \rho v^2) = 0 \tag{40b}$$

$$\frac{\partial}{\partial t}\left(\rho\varepsilon + \frac{\rho v^2}{2}\right) + \frac{\partial}{\partial x}\left[\rho v\left(\varepsilon + \frac{v^2}{2} + \frac{p}{\rho}\right)\right] - \frac{\partial q}{\partial x} = 0 \tag{40c}$$

The term $\frac{\partial q}{\partial x}$ in Eq. (40c) describes radiation absorption both in the thin surface layer and in the vaporized material. If the radiation power density is not high the term $\frac{\partial q}{\partial x}$ can be neglected everywhere with the exception of the thin layer on the evaporation surface. With low radiation power densities the evaporation boundary coincides with the absorption boundary, with high radiation power densities the absorption boundary shifts into the material vapour where its position is determined by the density at which the material is opaque. For comparatively low radiation power densities when the interface between the condensed phase and vapour coincides with the absorption boundary the gas dynamic disintegration theory treats the heating zone having the thickness of the order of a/v_b as a large-discontinuity region where the equations of conservation of mass, momentum and energy have the following form:

$$\begin{aligned} &\rho_1(v_0 - v_1) = \rho_0 v_0 \\ &p_1 - \rho_0 v_0 v_1 = p_0 \\ &-\rho_0 v_0\left(\varepsilon_1 + \frac{v_1^2}{2}\right) + p_1 v_1 = q_0 \end{aligned} \tag{41}$$

Here v_0 is the rate of evaporation, v_1, ρ_1 and p_1 are the velocity, density and pressure of the vapour at the vapour — condensed phase interface, respectively, p_0 and ρ_0 are the pressure and density of the condensed phase, respectively, q_0 is the incident radiation power density at the surface of the condensed phase and

$$\varepsilon_1 = p_1/(\varkappa - 1)\,\rho_1 + L_b$$

where L_b is the specific heat of vaporization and $\varkappa$ is the adiabatic exponent.

For the beam power density $q_c < q < q'$ where q' is the power density level above which there is no phase transition at the absorption boundary (the absorption boundary shifts into the region of the vaporized material) the vapour is transparent so that the kinetics of phase transition has to be analyzed.

In this power density range the bulk evaporation in the thin layer at the surface of the condensed phase is not taken into account.

The systems of equations (40) and (41) with $\partial q/\partial x = 0$ and the constant $q = q_0$ have a solution which is the limiting case of the centred rarefaction wave [46] corresponding to isentropic expansion of gas into vacuum. As the problem of the centred rarefaction wave is self-regulating it is convenient to introduce the dimensionless variable $\lambda = L_b^{-1/2}\,x/t$ and to express the functions v, ρ and p in the following form:

$$\begin{aligned} &v = L_b^{1/2} V(\lambda) \\ &\rho = q_0 L_b^{-3/2} R(\lambda) \\ &p = q_0 L_b^{-1/2} P(\lambda) \end{aligned} \tag{42}$$

where $V(\lambda)$, $R(\lambda)$ and $P(\lambda)$ are the dimensionless velocity, density and pressure, respectively.

Then the system (40) can be written in the dimensionless form:

$$(V-\lambda)\frac{dR}{d\lambda}=-R\frac{dV}{d\lambda}$$

$$(V-\lambda)\frac{dP}{d\lambda}=-\frac{1}{R}\frac{dP}{d\lambda} \qquad (43)$$

$$(V-\lambda)\frac{dV}{d\lambda}\left(\frac{P}{R}\right)=(\varkappa-1)\frac{P}{R}\frac{dV}{d\lambda}$$

The solution of the system (43) has the following form [2]:

$$R(\lambda)=R(\lambda_1)\left(1-\frac{\lambda-\lambda_1}{\lambda_2-\lambda_1}\right)^{2/(\varkappa-1)}$$

$$P(\lambda)=P(\lambda_1)\left(1-\frac{\lambda-\lambda_1}{\lambda_2-\lambda_1}\right)^{2\varkappa/(\varkappa-1)} \qquad (44)$$

$$V(\lambda)=V(\lambda_1)\left(1+\gamma\frac{\lambda-\lambda_1}{\lambda_2-\lambda_1}\right)$$

Here λ_1 and λ_2 are the values of the dimensionless variable λ corresponding to the interfaces between the vaporized material and the surface of the condensed phase and the vacuum ($\lambda_1 < 0$, $\lambda_2 > 0$) and γ is the constant determined by λ_1 and λ_2.

Then Eqs. (44) can be written as

$$R=R(\lambda_1)\left[1-\frac{\lambda-\lambda_1}{\lambda_2-\lambda_1}\right]^{2/(\varkappa-1)}$$

$$P=R(\lambda_1)(\varkappa-1)^2(\lambda_2-\lambda_1)^2\left[1-\frac{\lambda-\lambda_1}{\lambda_2-\lambda_1}\right]^{2\varkappa/(\varkappa-1)} \qquad (45)$$

$$V=\frac{2\lambda_1}{\varkappa+1}+\frac{\varkappa-1}{\varkappa+1}\lambda_2$$

At the vacuum interface ($\lambda=\lambda_2$) the solutions of Eqs. (45) satisfy the condition $P(\lambda_2)=R(\lambda_2)=T(\lambda_2)=0$ where T is the gas temperature. Equations (45) yield the condition

$$V(\lambda)-c(\lambda)=\lambda \qquad (46)$$

where $c(\lambda)=\left(\varkappa\frac{P}{R}\right)^{1/2}$ is the dimensionless sound velocity.

At the interface with the condensed phase ($\lambda=\lambda_1$) Eq. (46) determines the Jouguet condition which can be written in the dimensional form as

$$v_1-c_1=v_0 \qquad (47)$$

where $c_1=\left(\varkappa\frac{p_1}{\rho_1}\right)^{1/2}$ is the sound velocity in the gas at the interface with the condensed phase.

Equation (47), the boundary conditions (41) and the equation of the phase-equilibrium curve make up a system of five equations for six variables — p_0, p_1, v_1, v_0, ρ_1 and ρ_s (where ρ_s is the saturated vapour density). The sixth equation must relate the temperature of vapour to that of the condensed phase. The authors of [2] assume that these temperatures are equal at the evaporation surface. Anisimov [1] has shown that this assumption is not correct. The equality of temperatures yields the following condition:

$$\frac{p_0}{\rho_s} = \frac{p_1}{\rho_1} \tag{48}$$

The equation for the phase-equilibrium curve has the following form [151]:

$$\rho_s = B \exp\left(-\frac{L_b \rho_s}{p_0}\right) \tag{49}$$

where B is a constant.

Using the conditions (47) and (48) we can find the gas dynamic parameters in the evaporated material without solving Eqs. (44). However, using Eqs. (44) we can find the unknown parameters λ_1 and λ_2 and the quantity $R(\lambda_1)$ in Eqs. (44). A few simple algebraic transformations and the phase-equilibrium curve yield a system of two algebraic equations for λ_1 and λ_2.

This system of equations is given in [2]. If we assume that

$$\xi = \frac{|\lambda_1|}{\lambda_2} \ll 1 \tag{50}$$

that is, the velocity of the evaporation boundary is much lower than the gas velocity at the interface with vacuum, we can derive the following system of equations for λ_1 and λ_2:

$$\begin{aligned} \frac{(\varkappa-1)^2 \lambda_2^3}{2(\varkappa+1)^3} + \frac{(\varkappa-1)}{(\varkappa+1)^2}\lambda_2 &= \frac{\eta}{\beta} \exp\left[\frac{\varkappa(\varkappa+1)^2}{(\varkappa-1)^2 \lambda_2^2}\right] \\ \lambda_1 &= -\frac{\beta\lambda_2(\varkappa-1)}{(\varkappa+1)^2} \exp\left[-\frac{\varkappa(\varkappa+1)^2}{(\varkappa-1)^2\lambda_2^2}\right] \end{aligned} \tag{51}$$

where $\eta = q_0/\rho_0 L_b^{3/2}$ and $\beta = B/\rho_0$.

The system (51) has an approximate analytical solution. As shown by numerical analysis, λ_2 has a weak dependence on the parameter η/β, that is, on the incident power density in the range of 10^6-10^9 W/cm^2. Hence, we can find approximate expressions for the gas dynamic parameters at the interface with the condensed phase:

$$\rho_1 = \frac{(\varkappa+1)\rho_0|\lambda_1|}{(\varkappa-1)\lambda_2} \approx \frac{C(\varkappa+1)q_0[A-\ln(\eta/\beta)]^{1/2}}{(\varkappa-1)L_b^{1/2}\gamma^{1/2}}$$

$$v_1 = \frac{(\varkappa-1)L_b^{1/2}\lambda_2}{\varkappa+1} \approx \frac{(\varkappa-1)L_b^{1/2}}{(\varkappa+1)}\left(\frac{\gamma}{A-\ln(\eta/\beta)}\right)^{1/2} \qquad (52)$$

$$v_0 = L_b^{1/2}\lambda_1 \approx \frac{Cq_0}{\rho_0 L_b}$$

$$p_1 = \frac{(\varkappa-1)|\lambda_1|\lambda_2\rho_0 L_b}{\varkappa(\varkappa+1)} \approx \frac{C(\varkappa-1)q_0\gamma^{1/2}}{\varkappa(\varkappa+1)L_b^{1/2}[A-\ln(\eta/\beta)]^{1/2}}$$

The following notation is used in Eqs. (52):

$$A = \ln\left[\frac{(\varkappa-1)^2}{2(\varkappa+1)^3}\lambda_2^3 + \frac{(\varkappa-1)\lambda_2}{(\varkappa+1)^2}\right] \approx \text{const} \qquad (53)$$

$$C = \frac{1}{\left[1+\frac{(\varkappa-1)^2}{2(\varkappa+1)}\lambda_2^2\right]} \qquad (54)$$

$$\gamma = \frac{\varkappa(\varkappa+1)^2}{(\varkappa-1)^2} \qquad (55)$$

For the power density range from 10^6 to 10^9 W/cm^2 at $\varkappa = 5/3$ the following estimates have been made for iron [2]: $A = 1.9$, $C = 0.85$ and $\gamma = 26.7$.

The gas dynamic evaporation theory can be also applied to the physical processes at high power densities. The radiation with $q_0 \geqslant 10^9$ W/cm^2 is rather rarely used for machining the materials. The gas dynamic theory and the thermal model yield close estimates for temperatures and evaporation rates in the radiation power density range from 10^6 to 10^9 W/cm^2.

The theoretical predictions of both models are only in qualitative agreement with the experimental results.

(c) APPLICATION OF THE SIMILARITY THEORY

The thermal and gas dynamic models of disintegration can be used to estimate the depth of the craters in the material only for the one-dimensional heating problem:

$$r_f \gg h \qquad (56)$$

where r_f is the radius of the focussing spot and h is the depth of the crater. When $r_f \approx h$ a significant contribution to formation of the crater is made by melting and subsequent removal of the material from the walls of the crater by the vapour jet from the bottom of the crater.

The process of crater formation is essentially not one-dimensional, and its mathematical treatment is complicated. Application of the similarity theory and dimensional analysis [131] yields some general equations and recommendations which can sometimes prove to be useful. If we neglect the removal of the solid material the losses due to radiation to the external space, chemical reactions and ionization, and some other factors which are not significant for a number of materials in the process of crater formation we can see that the depth $l(t)$ and the diameter $d(t)$ of the crater and the mass $m(t)$ of the expelled material depend on the following basic parameters: the time t elapsed from the beginning of disintegration, the energy $E(t)=\int_0^t P_0(t)\,dt$ (here $P_0(t)$ is the radiation power), the specific energy of transformation of the material into liquid (L_m) and vapour (L_b), the material density ρ, the absorption coefficient α for the radiation and the focussing parameters (the focal ratio D_0/F of the objective and the focussing depth h, that is, the distance between the real focus and the surface).

If we select t, E and L_b as independent parameters we can derive some equations for dimensionless parameters according to the π theorem [148].

For the materials with the same ratio between the specific heats of melting and vaporization ($L_m/L_b=\text{const}$) these equations are

$$\begin{aligned} l(t) &= tL_b^{1/2} f_1\left(\frac{\rho t^3 L_b^{5/2}}{E(t)}\right) \\ d(t) &= tL_b^{1/2} f_2\left(\frac{\rho t^3 L_b^{5/2}}{E(t)}\right) \\ m(t) &= \frac{E(t)}{L_b} f_3\left(\frac{\rho t^3 L_b^{5/2}}{E(t)}\right) \end{aligned} \tag{57}$$

If one of the above parameters is a power function or can be approximated by a power function then the other parameters are also power functions or can be approximated by power functions. Then Eqs. (57) can be rewritten as

$$\begin{aligned} l &= \text{const}\left(\frac{\rho}{E}\right)^k t^{1+3k} L_b^{(5k+1)/2} \\ d &= \text{const}\left(\frac{\rho}{E}\right)^b t^{1+3b} L_b^{(5b+1)/2} \\ m &= \text{const}\,\rho^n E^{1-n} t^{3n} L_b^{5n/2-1} \end{aligned} \tag{58}$$

Analyzing the physical conditions we can find the following ranges of variation in the exponents at the parameters ρ, E, t and L_b in the above equations:

$$\begin{aligned} &0 \geqslant \frac{5k+1}{2} \geqslant -\frac{1}{3} \\ &0 \geqslant \frac{5n}{2} - 1 \geqslant -1 \end{aligned} \tag{59}$$

If we assume that as a crater develops it retains its shape so that $d = \text{const}\ l$ we can equate the limiting values of the exponents for $d(t)$ and $l(t)$. The limiting cases of material disintegration by powerful radiation yield the following values for the maximum exponents in Eqs. (59): $k = -1/5$ and $n = 2/5$. Then Eqs. (57) yield

$$\begin{aligned} l &= \text{const}\ (E/\rho)^{1/5}\, t^{2/5} \\ m &= \text{const}\ \rho^{2/5} E^{3/5} t^{6/5} \end{aligned} \tag{60}$$

These equations are the same as those derived by Sedov [131] for a powerful point explosion; they describe the development of a spherical cavity and the removal of the material from it. This situation is close to disintegration of a material by a single radiation pulse from a Q-switched laser. The process of mass ejection is explosive and practically does not involve evaporation of the ejected material.

The minimum exponents in Eqs. (59) correspond to $k = -1/3$ and $n = 0$, and Eqs. (58) yield

$$\begin{aligned} l &= \text{const}\,[E(t)/\rho L_b]^{1/3} \\ d &= \text{const}\,[E(t)/\rho L_b]^{1/3} \\ m &= \text{const}\, E(t)/L_b \end{aligned} \tag{61}$$

The system of equations (61) describes simple evaporation of the material with conservation of the initial shape ($d \approx l$). The disintegration process is quasistationary.

Generally, the actual process of material disintegration by laser radiation is more complicated than the above models as it involves simultaneously evaporation and melting (expelling) of the material.

(d) BULK EVAPORATION

The above thermal and gas dynamic models of material disintegration describe the radiation effects on materials which contain no impurities, gases and structural defects. In addition, the above models neglect the liquid phase present on the target surface and do not take into account the real structure of the radiation pulse.

Some experimental results [12] show that the depth and diameter of the hole produced by laser radiation depend on the purity of the material and the presence of the gases in it. Since small concentrations of impurities do not result in significant changes in such parameters as the specific heat of evaporation or density of the material the considerable effect of impurities on the size of the affected zone must be due to a process which differs from the surface evaporation and melting (washing-out) of the material. This part can be played by bulk evaporation.

As shown in [2], in the ideally absorbing mediums, that is in the mediums which do not contain impurities, gases and structural defects (microcracks, pores, etc.), the bulk evaporation can be significant and comparable to the surface evaporation only at the temperatures about $0.3\,\omega/k$, that is, about 10^4 K for typical metals. When the radiation power density is lower than 10^9 W/cm^2 the resulting temperatures are below this level so that the bulk evaporation in such materials is insignificant *.

Let us discuss a possible contribution of the bulk evaporation to the mechanism of disintegration of opaque materials under the laser radiation with $q_0 > 10^6$ W/cm^2.

While in the ideal liquids superheating is needed to produce a noticeable rate of nucleation for evaporation, under real conditions boiling of the liquid metal phase can be facilitated by the existence of artificial vaporization nuclei such as blisters, shrinkage cavities, impurity clusters, nonmetallic inclusions, or dissolved gases. Let us determine the conditions for the existence of the liquid phase on the surface of the irradiated material. When $q_0 > 10^6$ W/cm^2 evaporation dominates the energy balance of the process being more important than heat conduction. Depending on q_0 the thickness of the liquid phase layer where the bulk evaporation can occur varies from hundreds to a few micrometres. The liquid phase determines the dynamics of the bulk evaporation since its thickness is the upper limit of the critical nuclei size.

The critical size r^* of the bubble which is in the thermodynamical and mechanical equilibrium and capable of growing can be found from the stability conditions

$$
\begin{aligned}
&p = p_b + 2\sigma/r^* \\
&V_b \leqslant V_c \\
&p = p' \exp\left(-\frac{2\sigma V_l}{kTr^*}\right)
\end{aligned}
\qquad (62)
$$

where V_b is the bubble volume, V_c is the critical volume, p is the bubble pressure, p_e is the external pressure, p' is the liquid pres-

* Since $0.3\,\omega/k$ is typically higher than the critical temperature the contribution of the bulk evaporation to the disintegration process must be significant at the temperatures lower than $0.3\,\omega/k$.

sure, V_l is the liquid volume per molecule, and σ is the surface tension.

In a macroscopic examination the minimum bubble size must be about 0.1 μm. The upper limit of the sizes of inhomogeneities which can be the centres of bulk evaporation is determined by the thickness of the liquid phase layer at the irradiated surface which varies from hundreds to a few micrometres. At $q_0 = 10^6$-10^7 W/cm^2 the critical bubble size for copper is estimated to be 10^{-5}-10^{-6} cm [120]. In the uniformly heated liquid the bubble size increases as $\sqrt{t}$:

$$r(t) = A2c_p\gamma' \Delta T (\pi at)^{1/2}/\pi L_b\gamma'' \tag{63}$$

where γ' is the specific volume of the liquid, γ'' is the specific volume of the vapour at the given temperature, c_p is the specific heat, ΔT is the superheating temperature and A is a constant close to 1.7.

As the bubble grows the heat transfer to it increases with time owing to the surface evaporation (as the bubble travels towards the evaporation front in the coordinate system fixed to this front). If we neglect floating up of the bubbles the lifetime of a bubble depends on q. The lifetime of a bubble in copper as a function of the power density is given in Fig. 3. Figure 4 presents the calculated bubble radius as a function of the power density in the molten copper at the moment corresponding to the lifetime of a bubble. Calculations show that even with superheating by 1 K the size of a bubble at the moment of its emergence on the evaporation surface is close to the thickness of the liquid zone corresponding to the given incident power density.

It is possible to observe experimentally and to differentiate the drops of liquid expelled owing to the bulk evaporation processes and to melting and washing-out processes. This can be done because in the latter process drops are ejected mostly at an angle to the target surface while the drops due to bulk evaporation can be ejected normally to the surface [98].

With an increase in the radiation power density and a decrease in the pulse duration (that is, for Q-switching laser operation) the thickness of the molten zone at the surface decreases, the temperature in the narrow layer (tending to the absorption layer thickness) increases above the critical temperature, and the lifetime of the nucleation bubbles decreases. Under these conditions the artificial evaporation centres do not make a significant contribution compared to the spontaneous centres since the material in the absorption layer can be regarded as a high-density gas [2]. In this case the pulse duration must be decreased with the power density of $q < 10^8$-10^9 W/cm^2 (Fig. 3); when the power density is 10^8 W/cm^2 the lifetime of a bubble is 3×10^{-7} s; during this time the bubble can grow up to a size comparable to the thickness of the liquid phase zone (of the order of a few micrometres). If the mechanism of heat transfer to the bubble is assumed to remain unchanged under these conditions the contribution of the bulk evaporation remains the same though its effect on the removal of the material will decrease (the pulse duration is smaller than the lifetime of a bubble).

An increase in superheating the liquid phase stimulates the growth of bubbles and can result in ejection of the liquid phase from the heating zone [98]. Ejection of the liquid phase is enhanced not only by superheating the liquid phase but also by fluctuation of the reactive pressure owing to the spike structure of the radiation pulse [99].

Experiments [98] show ejection of liquid drops at the moments directly after a sharp drop in the power density in the free-generation regime.

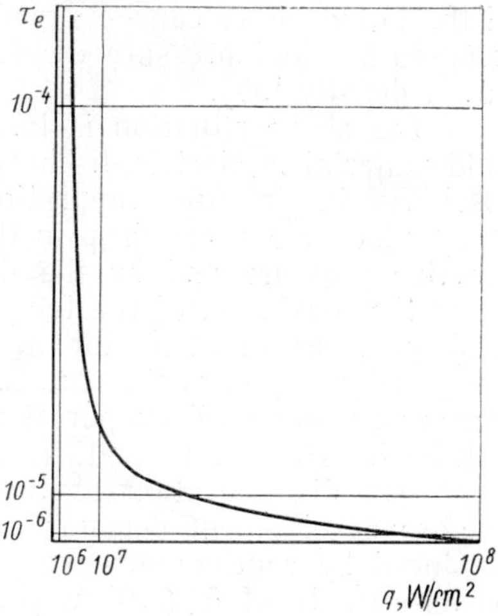

FIG. 3. Lifetime τ_e of a bubble in liquid phase (copper) as a function of incident power density q

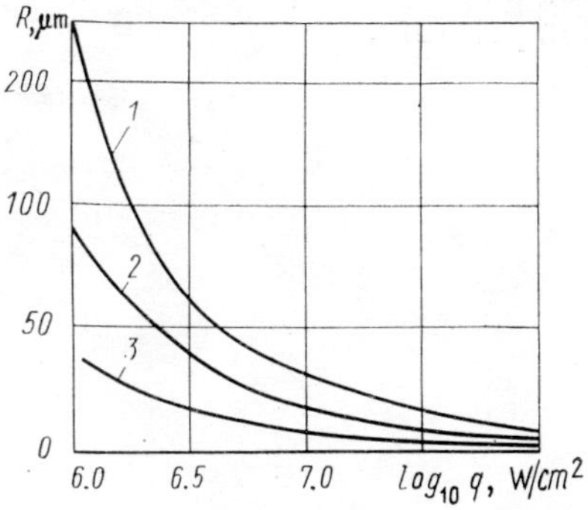

FIG. 4. Bubble radius R in liquid phase (copper) as a function of incident power density q

1 — $\Delta T = 1$ C; 2 — $\Delta T = 0.5$ C; 3 — $\Delta T = 0.2$ C

It would be interesting to compare the characteristic time of temperature variation at the interface with the time of temperature variation in the liquid phase directly below the absorption layer. As heat conduction is the highest-inertia process of temperature variation in the absorption layer and the only temperature-variation process in the heated surface layer the characteristic times of cooling of the absorption layer (τ_λ) and the molten zone (τ_h) are $\tau_\lambda \approx \lambda^2/a$ and $\tau_h \approx h^2/a$ (where h is the thickness of the molten zone). At the power density of $q_0 \approx 10^6$ W/cm^2 the thickness of the molten zone h is about 10^{-2}-10^{-3} cm so that even the most conservative estimates yield $\tau_h \gg \tau_\lambda$. Hence the inner layers of the liquid phase can be superheated relative to the equilibrium temperature and pressure owing to fluctuations of the radiation power density [99].

The above situation is largely similar to ejection of the liquid material in electric discharge [29] when after the end of the discharge the intense evaporation of the electrode material is accompanied by a sharp drop in the surface temperature and in the reactive pressure and by the development of bulk superheating when the maximum of the temperature is shifted inside the electrode creating conditions for the thermal explosion.

The concepts of the explosive character of the material ejection (with advanced evaporation) under the radiation fluxes resulting in critical temperatures have been developed in [85, 130]. Explosive disintegration of the metastable state of the liquids (whose development depends on the degree of superheating) is produced by continuous-wave radiation of a CO_2 laser with the power density of 6×10^3 W/cm^2 in ethanol and acetylene [85]. Samokhin [130] has found that an increase in the ejected mass and the recoil momentum owing to explosive evaporation can be observed in the narrow power density range, and the corresponding surface temperatures are expected to be close to the critical temperature.

2
TECHNIQUES FOR STUDYING LASER RADIATION EFFECTS ON OPAQUE MATERIALS

2.1. TECHNIQUES FOR STUDYING PROCESSES OF INTERACTION BETWEEN LASER RADIATION AND MATERIALS

Let us now discuss some experimental techniques for studying the laser radiation effects on materials and for measuring the radiation parameters. Using the optical system shown in Fig. 5 we can simultaneously use photography and oscillography for recording the structure of laser output. The near-field and far-field radiation zones are studied by means of photography using a photorecorder. The slit of the photorecorder (its width varies from 0.1 to 0.3 mm) cuts, along the diameter, a part of the light field pattern which is scanned on the photographic film with the rate from 3750 to 30,000 rpm. To eliminate the effect of the pumping light lamp a set of filters is placed before the objective of the photorecorder. Attenuation is selected to set the normal range of blackening on the characteristic curve of the film for $\lambda = 0.69$ μm or $\lambda = 1.06$ μm.

Photometry is carried out with microphotometers. Photometry of recording in the direction of scanning (along the time axis) yields the distribution of laser radiation intensity with a better time resolution than oscillography. If $s < v_1(\tau_1 - \tau)$ (where s is the size of the slit image on the film, v_1 is the rate of scanning and τ_1 is the time interval between successive light intensity fluctuations with the duration τ) then a microphotometer determines variations in the radiation pulse intensity without distorting its temporal structure.

Photometry of the laser radiation recording along the slit image yields the spatial distribution of the light field. Interferometers (for instance, an interferometer with the resolution about 0.2 Å for the 6-mm base) are used for analyzing the spectra of laser radiation. A highly sensitive coaxial photocell is used as a detector of the laser radiation. Another type of photocell is used as a detector of the pumping light.

The divergence of the laser beam is measured using a simple technique — the images of light spots are photographed on a semitransparent screen and then analyzed with a microphotometer. From the characteristic curve of the photographic film and its

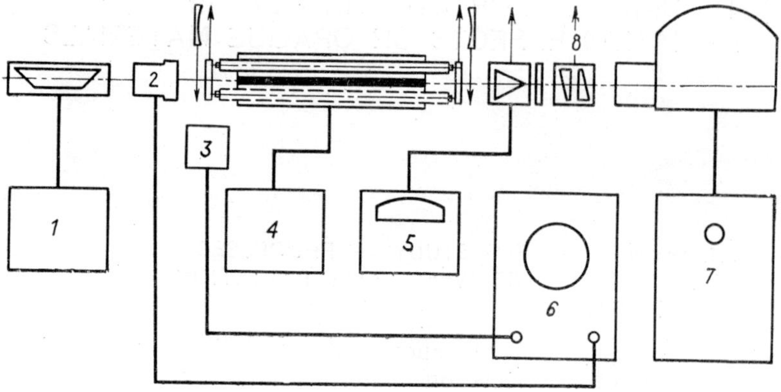

FIG. 5. Optical system for studying and recording energy parameters of laser radiation for free-generation operation

1—gas laser; *2*, *3*—photocells; *4*—laser power supply unit; *5*—calorimeter; *6*—oscillograph; *7*—high-speed photorecorder; *8*—interferometer

blackening the light intensities can be found which are used (in relative units) to plot the radiation pattern. The divergence angle of the beam can be determined from the half-widths of the patterns.

The system shown in Fig. 6 can be used for studying the effects of laser radiation on materials (including their disintegration) [40, 107]. This system makes it possible simultaneously to study the effects of laser radiation on materials and to make photorecordings of the character of radiation generation by the laser. The laser effects on solids are recorded with the photorecorder having special attachments so that it can operate both as a photorecorder (slit scanning) and as a high-speed camera. Such parameters of the radiation effects as the velocity of the gas jets and the ejection velocities of the solid disintegration products can be found from the recordings using the known scanning rate and measuring the slopes of the traces of disintegration products. The accuracy of such measurements is 10-15%.

A significant feature of the above system is its capability for simultaneous photographing of the ejection of the solid disintegration products and the emitting end of the laser active element on the same film. The beam is split by a parallel-sided glass plate *5* (see Fig. 6). The major portion of the beam power is focussed on the target, and a small part (about 4%) is directed into the objective of photorecorder *10*. The beam passing through plate *5* is deflected by a system of reflecting prisms vertically downwards, and lens *8* focusses the beam on target *7*. With this arrangement the gas jet from the target is directed upwards as required by the vertical position of the photorecorder slit; in doing so the target surface must lie in the central plane of the laser ruby rod.

This system makes it possible to photograph the traces of the disintegration products in the image plane of the photorecorder on one half of the film and to record emission of radiation at the output mirror or the active element end through plate *5* and mirror *9* on the other half of the film. The axonometric view of the optical system is given in Fig. 6 (right side). By varying the scanning rate we can record the entire process or its particular stages with a high time resolution. Filters are used to attenuate radiation.

To determine the pulse energy a fraction of the beam energy is deflected by a parallel-sided plate set after plate *5* to a calorimeter which can be also placed instead of target *7*. The shape of the radiation pulse and the optical pumping are recorded by photocells placed behind the end reflector of the laser or from the side of the pumping light lamp.

The system shown in Fig. 6 can be used to study the above processes separately; it has also an auxiliary objective and a set of deflecting mirrors for photographing the cross section of the hole from the back side of the target on the same film.

The development of a cavity can be studied using a composite target made up of a metal parallelepiped with a glass plate having polished surfaces glued on one of its sides [9, 34]. The light beam is focussed on the glass — metal interface, and high-

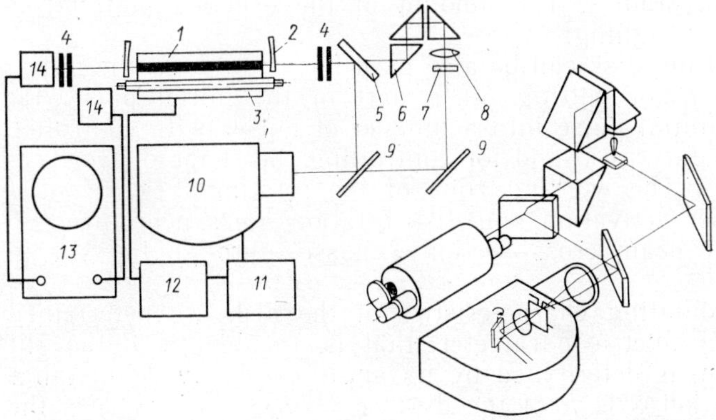

FIG. 6. Optical system for studying laser radiation and its interaction with materials

1—laser; *2*—flat or spherical resonator mirrors; *3*—pumping light lamp; *4*—filters; *5*—parallel-sided glass plate; *6*—system of reflecting prisms; *7*—target; *8*—focussing lens; *9*—mirror; *10*—high-speed photorecorder; *11*—photorecorder control unit; *12*—laser power supply unit; *13*—oscillograph; *14*—detectors of laser pulse and pumping light (photocells)

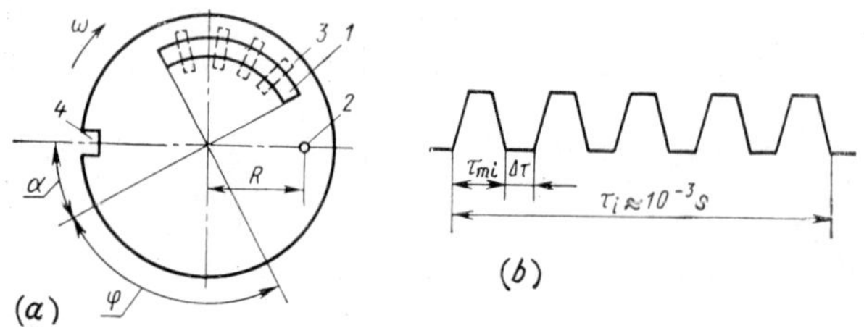

FIG. 7. Disk geometry for laser beam modulation (*a*); modulated laser pulse shape (*b*)
1—slit; *2*—laser beam cross section; *3*—plate; *4*—aperture for synchronization of disk rotation

speed filming of development of the hole under the effect of half of the beam is done from the side (the other half of the beam freely passes through the glass).

Scanning of the heated zone by the laser beam can be accomplished by using a disk of polished metal as a target; the disk is fitted on the shaft of a high-speed motor [39]. The laser pulse with the duration of 1-2 ms is focussed by a short-focal-length lens on the peripheral regions of the disk, and when the rotation speed is about 25,000 rpm and the disk diameter is about 100-150 mm the heated zone becomes an arc-like line about 100 mm long. This size of the heated zone facilitates investigation of the affected regions due to individual pulse spikes (in the form of individual microcavities). High-speed filming is used for the rotating-target studies. The amount of the expelled material is determined by weighing.

The rotating disk can be also used for modulating the initial laser radiation, for cutting off a part of the initial pulse, for dividing the initial pulse into a number of pulses with controlled duration and duty ratio and for controlling the shape of laser radiation pulses. The working zone of the rotating disk is slit *1* (Fig. 7). Laser activation and disk rotation are synchronized so that the laser beam cross section *2* passes through the slit in the disk.

The modulating characteristics of the disk are determined by its main geometric parameter, that is, the length of the slit which, in turn, is determined by its angle φ and by its distance from the rotation axis R of the disk: $\varphi = 2\pi\omega\tau_i$ (where ω is the angular rotation speed of the disk and τ_i is the duration of the laser pulse). For instance, for $\omega = 12{,}000$ rpm and $\tau_i = 1.5$ ms we have $\varphi = 108°$. Slit *1* in the disk must be displaced by a certain angle α' with respect to the beam spot *2* on the disk and the synchronization aperture *4*.

2.2. MEASUREMENT OF LASER BEAM PARAMETERS

The laser radiation parameters can be classified into the internal and the external ones [153]. We shall not discuss the main internal parameters such as the resonator mode spectrum, amplification, noise and modulation. We shall not also dwell upon such control techniques for the laser beams as the amplitude division, dissipation of the laser beam for controlling its parameters, induced fluorescence, transformations of harmonics, photochemical decomposition, etc.

The external parameters are: (1) radiation power (continuous-wave, averaged and instantaneous power): (a) power density in the near-field and far-field zones, (b) distribution of the radiation power density in the beam; (2) radiation energy: (a) energy density in the near-field and far-field zones, (b) distribution of the radiation energy in the beam; (3) angular divergence and diameter of the beam, variation in the angular divergence in the near-field and far-field zones; (4) wavelength of radiation, the discrete spectrum and its variation with time; (5) coherence: (a) spatial, (b) temporal, (c) total; (6) polarization.

Let us discuss some problems encountered in measuring the external laser parameters which are directly relevant to the technological laser applications.

(a) MEASUREMENT OF BEAM PARAMETERS. MEASUREMENT OF THE ANGULAR DIVERGENCE AND INTENSITY DISTRIBUTION OVER THE BEAM CROSS SECTION

The intensity distribution over the laser beam or the structure of the transverse modes is analyzed mostly with the following two techniques [153]: photographing of the beam and scanning of the cross section of the beam with a detector which determines the intensity of a small area of the cross section at a given moment.

The simplest technique for determining the mode structure, which yields a qualitative estimate of the intensity distribution at a given distance from the laser, is to direct the beam on an exposed photographic film or plate. The type of the photographic material and the time of exposure (which is important for a continuous-wave laser) are selected according to the output parameters of the laser (energy, power, duration, etc.). If necessary the beam is attenuated with appropriate filters. To study the mode structure or intensity distribution in the near-field and far-field zones a lens is used for exposure (or photographing) as shown in Fig. 8.

The radiation structure in the near-field zone (the Fresnel zone) can be recorded by placing the photographic film or plate at

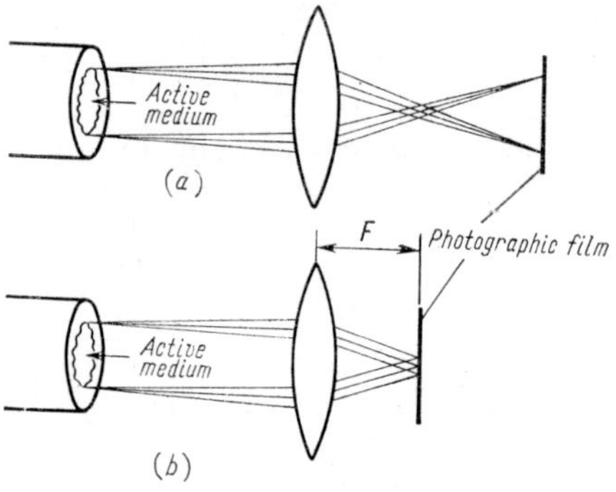

FIG. 8. Focussing conditions for photographing laser radiation patterns in near-field (*a*) and far-field (*b*) zones

a distance considerably smaller than D (see Chapter 1). In other words, the camera objective must be focussed on the emitting region of the active element end (Fig. 8*a*). To record the radiation structure in the far-field zone the photographic film or plate must be placed at a distance considerably larger than D in the focal plane of the lens or objective, that is, the objective must be set to infinity (Fig. 8*b*).

Electron-optical image converters (EOC) are used for converting the infrared or ultraviolet radiation of a laser into the visible light. The mode or intensity distribution in the infrared or ultraviolet laser beam can be visualized on the screen of the electron-optical image converter and photographed.

Since the available electron-optical image converters operate with radiation having the wavelength up to 1.2 μm the longer-wavelength radiation can be photographically analyzed using the evaporographs which can register radiation having the wavelength up to 18 μm [153].

The intensity distribution over the cross section of a laser beam can be determined quantitatively with a photometrical technique using the fact that the blackening of the photographic material is a linear function of exposure in a wide range of exposures. Using the blackening of the photographic material and the slope γ of the characteristic curve of the photomaterial we can determine the exposure and hence the radiation intensity.

To measure the radial intensity distribution we have to know only the relative radiation intensities. The ratio between the respective intensities (or exposures) is given by the following equation:

$$\log_{10}\frac{I_1}{I_2}=\frac{x_1-x_2}{\gamma} \tag{64}$$

where I_1 and I_2 are the intensities (or exposures) and x_1 and x_2 are the blackenings.

The accuracy of the photometric technique depends mainly on variation in the value of γ and the nonlinearity of the characteristic curve of the photographic material. The mean error in the typical range of blackenings (0.3-1.6) amounts to 0.5-2.0% [153].

The scanning techniques. The scanning techniques are most useful for determining the mode structure of the continuous-wave gas laser radiation. A photodetector is moved over the laser beam cross section; at some moment the detector measures the intensity in a small area, and the electric signal from it describes the radiation distribution over the beam. To obtain a high spatial resolution small-size detectors must be used.

The reference [153] describes the analysis of the intensity distribution in the beam of the cesium gas laser ($\lambda = 7.18$ μm) with a gold-alloyed germanium photodetector; for the beam of the helium-neon laser ($\lambda = 0.6328$ μm) the cross section is scanned with a silicon photodetector covered with a brass plate having an 0.038-mm aperture.

Divergence of the beam of the pulsed solid-state lasers. This can be measured by either one of the two photographic techniques. In the first case the photomaterial is placed at a distance which is not less than ten times the size of the emitting end or aperture of the laser. Then the divergence α of the laser beam can be found from the following equation:

$$\alpha \approx D/R \tag{65}$$

where D is the diameter of the effectively exposed spot on the photomaterial and R is the distance from the laser to the photomaterial.

The other technique makes possible a highly accurate determination of the beam divergence using a long-distance objective. When the objective is set to infinity the real image of the beam in the far-field zone is obtained. The divergence of the beam is here given by the following equation (decrease by a factor of e):

$$\alpha = l/F \tag{66}$$

where l is the diameter of the spot on the photomaterial (measured at the 1/e level of the intensity) and F is the focal distance of the objective.

The latter photographic technique can be used for analyzing the intensity distribution and measuring the divergence for the beam of the Q-switched laser. The laser beam intensity for such studies is attenuated by successive reflections in the dielectric-coated mirrors to the level at which we can use the normal filters, photomaterials, etc.

Experiments frequently involve measuring the diameter of the laser beam. The simplest way to do this is to direct the laser beam on a material which will be affected by the radiation. Such materials as the copying paper (the sensitive coating to the beam) and the exposed photographic film or paper (the emulsion coating to the beam) are most frequently used. The measurements should take into account the distance to the paper owing to the beam divergence. This technique is mostly used with ruby lasers and infrared (for instance, neodymium) lasers.

Another technique for measuring the laser beam diameter is most useful for continuous-wave gas lasers emitting visible light [153]. A short-focal-length objective (20-30 mm) with the aperture about 1 cm is used to focus the beam on a screen the distance to which is accurately measured. Then the beam diameter d_b is given by the following equation:

$$d_b = \frac{F d_s}{l - F} \tag{67}$$

where d_s is the diameter of the focussed spot and l is the distance from the entrance pupil of the objective to the screen.

(b) MEASUREMENT OF ENERGY AND POWER OF LASER RADIATION

The free generation of radiation by lasers can be classified into three groups by the energy and power levels: low power — 0.1 W and 1 J, medium power — up to 100 W and 100 J, and high power — over 100 W and 100 J [16].

The devices measuring the energy and power of the coherent radiation can be of the absorbing type and of the transit type. The absorbing detectors are calorimetric and photoelectric devices, and also measuring devices using the ponderomotive force. The device measuring the energy and power of the transit radiation makes use of the nonlinear crystal polarization under laser radiation. The energy and power of the transit radiation can be measured also by absorbing devices with the use of directional couplers.

The pulsed lasers are described by the energy radiated during the pulse, and the continuous-wave lasers are characterized by the radiation power.

The energy E of the radiation pulse is

$$E = \int_0^{\tau_i} P(\tau)\, d\tau \tag{68}$$

where $P(\tau)$ is the instantaneous radiation power at the moment τ and τ_i is the pulse duration.

The peak power of a single pulse is $P_m = P_{\max}(\tau)$. The mean power of the pulse is given by

$$P_{mean} = \frac{E}{\tau_i} \tag{69}$$

For the modulated pulse radiation the mean power is given by

$$P_{mean} = \frac{1}{T_r} \int_0^{T_r} P(\tau)\, d\tau \tag{70}$$

where T_r is the pulse repetition period.

Calorimetric techniques for measuring laser radiation power and energy. These techniques make use of the conversion of the laser energy into the thermal energy. The resulting increase in the temperature can be measured either directly or indirectly by determining the changes in volume, pressure or other parameters of the absorber (the variable-temperature calorimeters). In the constant-temperature (isothermal) calorimeters the liberation of the thermal energy in the absorber is accompanied by absorption of the same amount of the thermal energy (for instance, with phase transitions [16]).

The most widely used devices are the variable-temperature calorimeters in which radiation energy is converted into heat in solid-state absorbers of the black-body type; the absorbers have spherical, cone, and wedge-like shapes or combinations of them. The amount of the absorbed energy is determined by the temperature change.

For the constant power P_0 and the starting condition $T(0)=0$ the variation in the temperature in calorimeter is given by the following equation:

$$T(\tau) = \frac{\varepsilon P_0}{\alpha}\left[1 - \exp\left(-\frac{\tau}{\tau_a}\right)\right] \tag{71}$$

where ε is the absorption coefficient of the absorber, $\tau_a = c/\alpha$ is the time constant of the absorber and c is the specific heat of the absorber material.

A light pulse, for instance, of rectangular shape and duration τ_i (typically $\tau_i \ll \tau_a$), produces the maximum temperature of the absorber given by the following equation:

$$T_{\max} = \frac{\varepsilon P_i}{\alpha}\left[1 - \exp\left(-\frac{\tau_i}{\tau_a}\right)\right] \tag{72}$$

As $\tau_i \ll \tau_a$ Eq. (72) can be transformed into

$$T_{\max} = \frac{\varepsilon E}{c} \tag{73}$$

where $E = P_i \tau_i$ is the laser pulse energy.

To prevent disintegration of the absorber additional external attenuators (for instance, neutral filters) must be used when measuring the pulse energies exceeding 10 J.

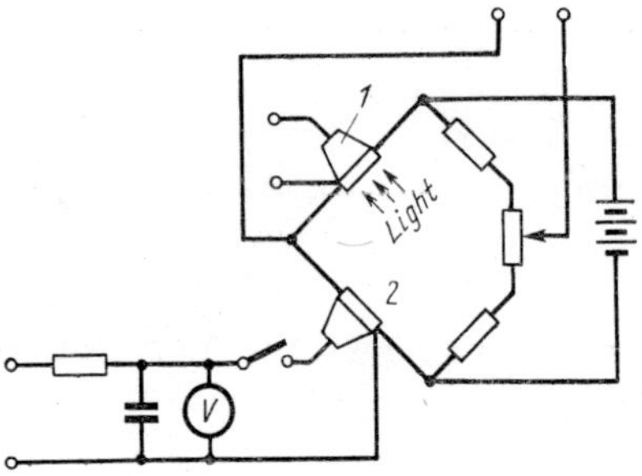

FIG. 9. Calorimeter with cone absorber
1—working cone; 2—absorbing cone

Let us discuss the most typical calorimeter designs. A device wherein laser radiation is directed on the internal surface of a carbon cone which is the absorber in the calorimeter is described in [38]. The increase in the cone temperature is measured with two thermistors which are in direct contact with the cone and connected to a measuring bridge circuit (Fig. 9). The output from the bridge circuit is fed via a dc amplifier to a recorder. The calorimeter is calibrated by discharging a condenser across the surface of the absorbing cone. As the resistance of the discharge circuit is known we can calculate the energy dissipated in the cone. This calorimeter model can be used for measuring the continuous-wave radiation power. Figure 10 presents the absorbing cone and its size, and Fig. 11 shows the calibration curve for this calorimeter. The curve is close to a straight line, its slope being about 45°, and some degree of nonlinearity is due to the nonlinearity of the thermistors used. The error in measuring energy is about 10%.

Liquid calorimeters can be used for measuring high radiation powers from heavy-duty lasers [153, 170].

A liquid calorimeter is a Dewar vessel with double walls (Fig. 12). Distilled water with a dye additive is used as the absorbing liquid; it provides for a satisfactory sensitivity of the calorimeter in the energy range up to a few tens of joules. For instance, green ink for the recorder is added to water to provide for the absorption capacity of about 80% per cm of path for the radiation with $\lambda = 0.694$ μm [170]. A 50-J rated calorimeter can be used to measure energies in the 1-100 J range.

High radiation energies can be measured with the use of liquid calorimeters by recording the liquid volume variations [153]. Since $\Delta V = V\beta\Delta T$ (where V is the liquid volume, ΔV and ΔT are the variations in the volume and temperature of the liquid, respectively, and β is the thermal expansion coefficient) the variation in the liquid volume yields the amount of absorbed energy. The heat ΔQ required to raise the temperature of the absorbing liquid with the mass m by ΔT is given by

$$\Delta Q = \frac{m}{V}\frac{c_p}{\beta}\Delta V = \rho\frac{c_p}{\beta}\Delta V \tag{74}$$

where ρ is the density of the liquid.

The efficiency of the calorimeter is determined by the ratio $\beta/\rho c_p$ since its sensitivity increases with increasing β and decreasing ρc_p. The calorimeter described in [153] has the copper nitrate solution in acetonitrile as the absorbing liquid. The transmission coefficient of a cell 75 mm long determined by the copper nitrate concentration is 10^{-4} for $\lambda = 0.69$ μm. With heating the liquid expands into an 0.1-mm capillary. Absorption of 2.6 J raises the liquid level in the capillary by 25 mm. A piston attachment is used to set the liquid level according to the ambient temperature.

Ponderomotive force technique. Very high energies can be measured with a technique using the pressure of the electromagnetic radiation. The radiation pressure is measured with transducers of various types: capacitive, piezoelectric and other; typically, transducers of the torsion (or pendulum) type are used.

A number of authors [10, 16, 53, 169, 204] have described the basic design and particular models of these devices. The diagram of the ponderomotive device for measuring the energy and power of radiation is given in Fig. 13. The continuous-wave power is gauged by the rotation angle θ of the moving system when the radiation falls perpendicular to the reflecting surface of the sensitive element. The energy of a radiation pulse is measured by the maximum angle of rotation of the moving system. In another model of ponderomotive device for measuring energy it is gauged by the displacement of the reflecting plate of the capacitive transducer, which can be measured in the range from 2×10^{-5} to 2×10^{-2} cm [10].

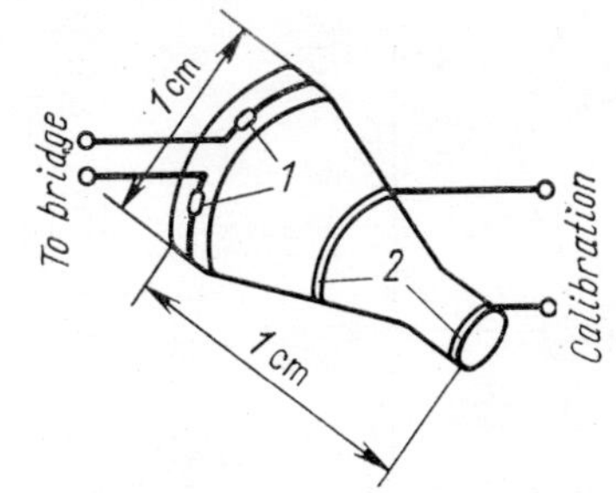

FIG. 10. Carbon cone absorber
1—thermistors; 2—contact rings for calibration

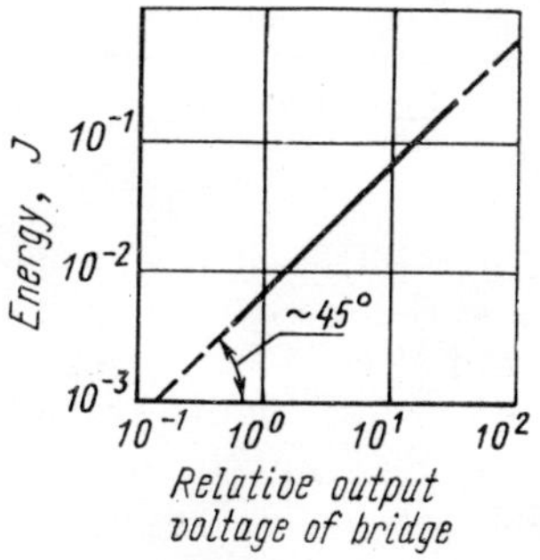

FIG. 11. Calibration curve for a calorimeter

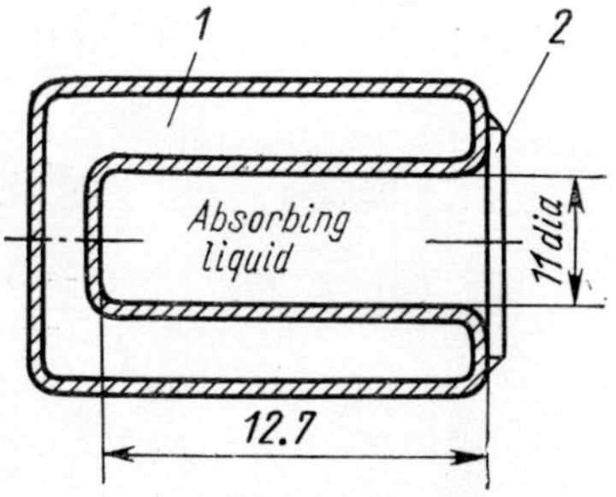

FIG. 12. Liquid calorimeter
1—vacuum; 2—glass window

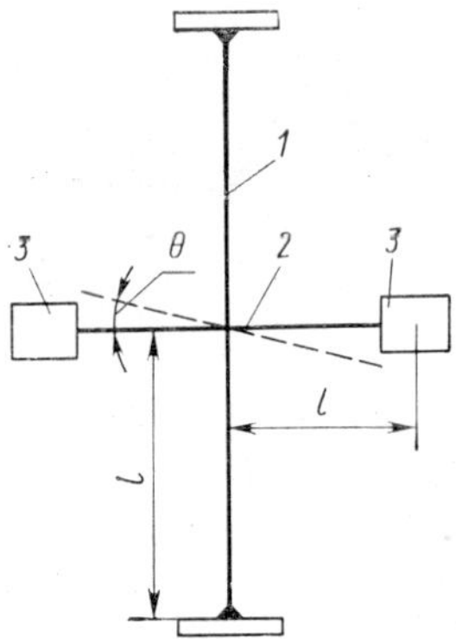

FIG. 13. Ponderomotive device for measuring radiation energy and power
1—quartz filament; 2—quartz rod; 3—reflecting blades

Photoelectric technique. Vacuum photocells and photomultipliers can be used as radiation detectors to measure the energy and power parameters of laser radiation [16, 38, 153]. The diagram of the measuring circuit with a photoelectric radiation detector is given in Fig. 14.

The capacitor C charges from the supply source to the voltage V. Just before the measurement the supply source is disconnected. The radiation gives rise to the current $i(\tau)$ in the photocell circuit resulting in discharge of the capacitor by the charge Q:

$$Q = \int_0^{\tau_i} i(\tau)\, d\tau \tag{75}$$

An electrostatic voltmeter is used for measuring Q; it determines the capacitor voltage difference before the radiation pulse and after it.

When the photocurrent i is proportional to the radiation intensity q the capacitor voltage difference ΔV is proportional to the energy E of the radiation pulse:

$$E = \frac{C}{R} \Delta V \tag{76}$$

where R is the sensitivity of the photocell.

Neutral filters and diffuse reflectors (for instance, plates coated with magnesium oxide or $BaSO_4$) [153] are used to measure high energies and the photodetector response within the linear region.

The energy parameters of visible light are measured with photodetectors using the extrinsic photoeffect. Such detectors have high sensitivity and low inertia (the response time can be as low as 10^{-9} s) but they are useless for measuring the power of continuous-wave radiation or high-frequency pulsed radiation since their sensitivity varies with prolonged radiation.

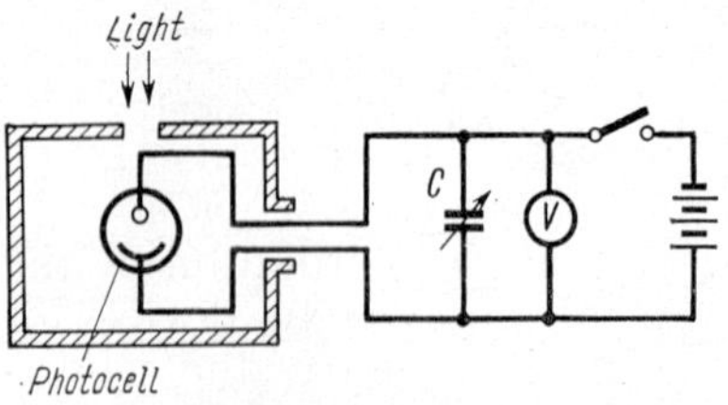

FIG. 14. Photoelectric radiation detector

The calorimetric, photoelectric and other types of measuring devices described above are the absorbing devices, and radiation of high intensity can destroy their absorbing elements. Attenuators cannot always be used to advantage for the electric fields of about 10^6 V/cm can give rise to nonlinear effects in dielectrics, which hamper calibration, etc. The same difficulties are encountered in measuring the energy parameters of giant pulses. Therefore, new techniques have to be developed for measuring high energies, which can make use of the pyroelectric, piezoelectric and other nonlinear elements in dielectrics transparent for laser radiation.

Calorimeters are most widely used out of the above measurement devices. However, a comparatively high inertia of calorimeters (10^{-4}-10^{-3} s) renders them useless for measuring the parameters of the giant nanosecond pulses or short high-frequency pulses.

As ponderomotive devices reflect, rather than absorb, light they are useful for measuring high pulsed energies (over 10 J) with short pulse durations (10^{-4}-10^{-8} s). These devices have a somewhat better accuracy than calorimeters which, however, show a better performance.

The photoelectric devices have poor accuracy (about 20-30%) but rather low inertia (10^{-6}-10^{-9} s). Photodetectors of all types must be previously calibrated for measuring the energy parameters. The pyroelectric devices have a number of advantages: low inertia, nonselectivity, high sensitivity, and a wide operational range. A number of measuring devices produced both at home and abroad are discussed in [38].

2.3. TECHNIQUES FOR STUDYING LASER-AFFECTED ZONES

The laser machining affects very small zones of the material; the typical size of the affected zone is a fraction of a millimetre. Since the radius of the focussed spot r_f is usually about 0.01 cm the small affected zone of the solid material exhibits a marked heterogeneity of structure, composition and properties owing to high rates of heating and cooling and to high temperature gradients and pulsed pressures. Let us discuss briefly the main techniques used for analyzing materials irradiated by the laser beam.

(a) METALLOGRAPHY

The metallographic samples are cut along the direction of the incident laser beam approximately in the centre of the hole produced by radiation of the material.

To prevent distortion of the edge in preparation of microsections the metallic samples for structural analysis and microhardness measurement are previously coated with iron in electrolyte of the following composition: ferrous sulphide — 400 g/l, sodium chloride — 200 g/l [115]. The electrolyte has pH 1. The anode is made of armco iron, and the cathode is the sample being studied. Iron plating is carried out at 80-90 C for 30-60 min with the current density of 2 A/dm^2. The samples with iron plating 20-50 μm thick are put into special holders with liner of soft material.

It is generally difficult to visualize the region of structural changes due to laser radiation of steel samples; therefore, multiple etching and polishing and sometimes electrolytic polishing are done. When the samples are to be machined by *Q*-switched laser radiation they are repeatedly etched and polished to remove the cold-hardened layers on the surface. The samples are photographed using the metallographic microscopes. The microhardness of materials in the affected zone is measured with a microhardness tester with the loading of 20 and 50 gf for the free-generation laser operation and 10 and 50 gf for the *Q*-switched laser operation.

Iron plating of samples prevents the distortion of the hole edges and makes it possible to measure the microhardness in the cross section of the affected zone starting from the depth of a few micrometres. For each sample microhardness is measured throughout the depth of the affected zone and at three points along its width. The spread of microhardness values for the steel samples is not more than 40-60 kgf/mm^2 and the measurement accuracy is $\pm$15-20 kgf/mm^2.

(b) ELECTRON-MICROSCOPIC ANALYSIS

Electron-microscopic analysis is done with microscopes of different types. The studies are made with one-step collodion replicas shadowed with chromium (sputtering angle 18°) and thin foils.

The dislocation structure of thin foils of a model material in the zone affected by the laser beam is studied by electron microscopy. Vacuum-melted carbonyl iron (about 0.02% C) is used as the model material. The 0.1-mm foil is previously annealled at 680 C for 2 h in vacuum (about 10^{-3} mm Hg) to reduce considerably the density of dislocations. The foil is irradiated with a laser beam, and 3.6-mm disks containing the affected zone are cut out from the foil. The disks are previously electrolytically polished (100 cm^3 of H_3PO_4 + 20 g of CrO_2) to produce a cavity in the centre of the zone, and finally they are chemically polished at 40-60 C in H_3PO_4 + 30% H_2O_2 to make a through hole in the centre of the sample. The thin edges of the hole about 1500 Å thick are studied in a transmission electron microscope at 100 kV to determine the dislocation structure.

(c) SPECTRAL ANALYSIS

Spectral analysis is used to analyze the profile of carbon and other alloying elements in the affected zone of the sample [72].

The samples with cavities produced by free-generation laser pulses are first polished to make the cavity depth of 0.1-0.2 mm. Spectral analysis is then done, and the sample is again polished. The layers are analyzed, polished off and again analyzed until the element concentrations reach the initial proportion.

Spectrographs with one-lens condensers are used for spectral analysis of the layers. The output voltage of a generator is 6 kV, the current in the primary circuit is 1.2-1.4 A, the slit width of the spectrograph is 0.01 mm, the duration of preparatory spark treatment is 50 s for the steel samples and 30 s for the armco iron samples, the exposure is 15 s; type II spectral photoplates are used with the sensitivity of 22 GOST (USSR State Standard) units. The calibration curves are plotted using the three-standard technique.

This spectral technique makes possible the determination of the carbon concentration C_c (in mass %) in iron alloys in the 0.06-0.2% range; the reproducibility of results is $\pm 20\%$ for $C_c < 0.2\%$.

Another technique for determination of profiles of the alloying elements in the heat-affected zones in some alloys is the spectral microanalysis using a "point" source of rectified high-frequency current [67, 115], which has the advantages of the constant-polarity spark discharge (localization of the disintegration sites on electrodes) and satisfactory reproducibility of results as given by the high-frequency spark discharge. The current in the spark circuit does not exceed 0.007 A. The specimen is clamped in the holder of a microhardness tester and acts as the cathode. The anode is a copper needle clamped in a special holder which is installed instead of the holder of the diamond pyramid in the microhardness tester.

The electrode-specimen gap is 0.1-0.3 mm. To determine the concentrations of various elements in steel specimens exposure of 30 s is required. A lens is used to focus the spark on the slit of a spectrograph, the slit width being 0.015 mm. The spark produces a pit in the material with the diameter up to 0.08 mm and the depth of a few micrometres. The analyzed volume is not more than 5×10^{-3} mm^3 and the mass of the evaporated material is about 3×10^{-7} g.

The sensitivity of this technique varies from 0.001 to 0.1% for various elements. For Mg, Ti and Nb the sensitivity is 0.007% while for Mn, Cr and Si it is only 0.1%.

Spectral microanalysis using this technique is done in the following way. A specimen with the laser-produced holes having the diameter about 1.5 mm and the depth about 2.7 mm (the laser pulse energy is approximately 150 J and the duration of the pulse focussed on the specimen surface is about 1.8 ms) is cut along the hole diameter, and the surface is polished. The spark pits are made on both sides of the hole at a distance of 0.05, 0.15 and 0.45 mm from the edge and at various depths.

In other steel specimens about 1 mm thick with the through 0.8-mm holes (produced by ruby laser radiation with the energy of 5 J and the 4×10^{-3}-s pulses focussed on the surface of the specimen) the surface of the specimen normal to the hole axis is polished. The centres of the spark pits are at the distances of 0.05, 0.1, 0.2 mm and more from the edge of the hole. For the pits which are at the same distance from the edge of the hole the concentration results are averaged. Then a metal layer about 3 μm thick (the pit depth) is polished off, and the measurement is repeated.

(d) X-RAY SPECTRAL MICROANALYSIS

X-ray apparatus are used to study the affected zone of laser-irradiated samples. The specimens with cavities produced by the laser beam are initially polished to remove an 0.1-0.2 mm layer, and analysis is done at about 10 points of the resulting surface (the focussed spot is about 2 μm in diameter). Then the same surface is subjected to local optical spectral analysis, in the latter technique the previously treated area is so large that the analytical accuracy is not affected by the previous X-ray analysis. After analysis the specimen is polished to remove a 10-200 μm layer, and then the analytical procedure is repeated.

Polishing and analyzing are repeated until the measured concentrations of the elements reach the initial levels.

The roughness of the internal surface of the holes drilled by the laser beam (free generation operation) and the depth of cavities produced by giant laser pulses are measured with profilographs and profilometers.

3
HEAT TREATMENT AND WELDING BY LASER RADIATION

3.1. HEATING OF OPAQUE MATERIALS WITHOUT DISINTEGRATION

The effect of the laser beam can be represented by the effect of an equivalent heat source which can be on the surface or in the bulk of the material and have the extended or the lumped character according to the model used, the problem formulation and the physical properties of the material. A special feature of laser heating is that the equivalent heat source is typically described by sharply varying temporal and spatial characteristics. Therefore, simplifications of the heat source properties, which are frequently used in mathematical models to reduce their complexity, can result in significant differences between predicted and experimental results. Therefore, careful attention should be paid to analyzing the laser parameters and the spatial and temporal characteristics of the equivalent heat sources when studying the heating of materials with laser beams.

Under most circumstances of laser heating the heat sources in opaque materials can be assumed to be distributed over the heating surface according to a certain function (the absorption coefficient α is about 10^4-10^5 cm^{-1}). The temporal structure of a laser pulse is determined by the laser type and generation regime. Generally, the specific power of a heat source on the material surface is a complicated function of coordinates and time:

$$q = q(x, y, \tau) \tag{77}$$

The function q in Eq. (77) can be approximated by a product of a function of time and a function of the surface coordinates:

$$q(x, y, \tau) = A\varphi(\tau)\, q^*(x, y) \tag{78}$$

where A is the absorptivity which, as a rule, depends on the state (finish) of the surface and its temperature, $\varphi(\tau)$ describes the temporal structure of the laser pulse and $q^*(x, y)$ is the spatial distribution of the radiation intensity.

If the inhomogeneities of the laser system are small the distribution $q^*(x, y)$ is given by the diffraction distribution [see

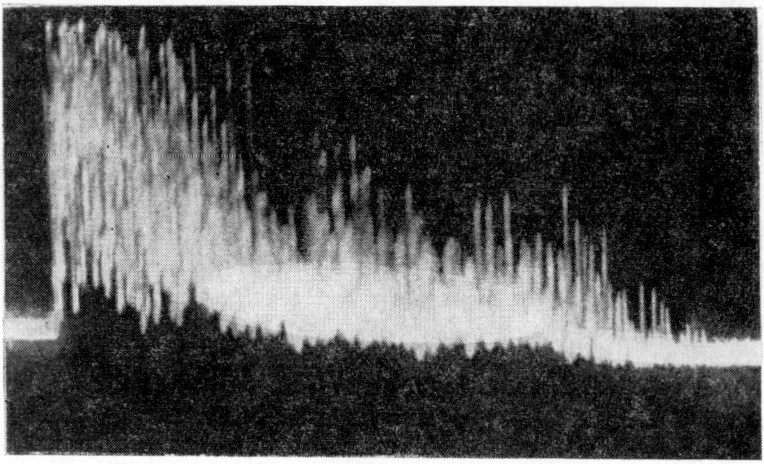

FIG. 15. Oscillogram of a spike pulse with chaotic generation

Eq. (9)]. In practice distribution of laser radiation intensity differs from Eq. (9) owing to nonuniform distribution of the phase and amplitude of radiation over the end of the active laser element, which is caused by a number of factors (generation of many modes of the optical resonator, asymmetry of excitation, optical imperfection of the crystal, etc.) [88]. Thus, it seems reasonable to use a simpler approximation of the spatial structure of a laser pulse in analysis of the laser heating processes (gaussian distribution, or uniform distribution of q over the focussing spot).

The spatial distribution of the laser radiation intensity (or power density) depends on the mode of operation of the laser. There are various experimental techniques for determination of the function $q^*(x, y)$, some of them being discussed in Chapter 2. It is not always that we can express the spatial and temporal structure of a laser pulse in the form (78). In these cases the structure of a laser pulse is analyzed with the techniques described in [7, 153].

The absorption capacity and the reflection coefficient depend on the wavelength and the intensity of radiation. However, even at relatively low radiation intensities (less than 10^6 W/cm^2) their values are known with a poor accuracy, especially, for the temperatures above the melting point of the material. The absorption coefficient can vary with increasing temperature. To simplify analysis, the absorption capacity is usually assumed to be independent of temperature, and the value used is the mean value in a certain temperature range.

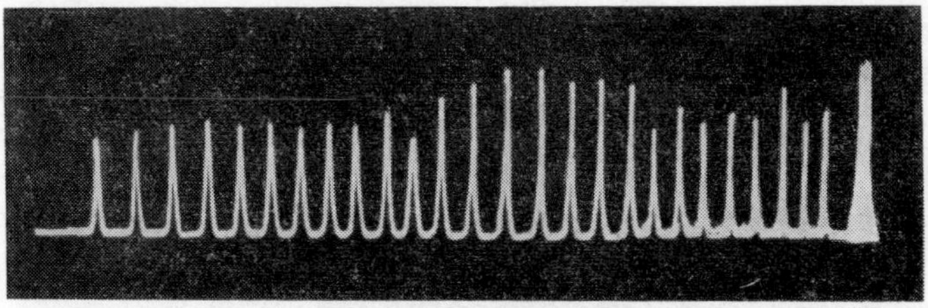

FIG. 16. Oscillogram of a spike pulse with ordered generation

Let us discuss in more detail the temporal structure of a laser pulse. The solid-state lasers have several typical regimes of radiation generation. The most widely used regimes are chaotic generation of the spike-structure radiation, ordered generation of the spike-structure radiation and quasi-continuous-wave generation. The duration τ_i of generation with these regimes is about 10^{-3} s if measures are not taken to reduce or increase it. Resonator Q-switching makes possible the generation of high-power pulses with the duration τ_i about 10^{-8} s. Finally, there exist special techniques for producing ultrashort laser pulses (10^{-10}-10^{-11} s) which so far have not been applied to laser machining.

A radiation pulse with duration about 10^{-3} s consists of a sequence of spikes with the duration about 1 μs and the duty ratio about 0.2 (Fig. 15). The amplitudes of various spikes are different. In a spike pulse with chaotic generation we can distinguish the front with a certain rise rate and some regularity in amplitude decrease towards the end of the pulse. Thus, the envelope of the spikes of a pulse can be a bell-like asymmetric curve. For instance, the envelope of spikes can be described by a product of an exponential function and a power function:

$$\varphi(\tau) = \tau^n \exp(-b\tau^m) \tag{79}$$

where n and m are integers or fractions.

A spike pulse with ordered generation (Fig. 16) is a sequence of individual spikes with the duty ratio about 0.2 and the total duration about 10^{-3} s similar to the chaotic generation. But the amplitude of the individual spikes and the time interval between successive spikes remain constant during a considerable part of the pulse. This laser radiation regime can be described by a periodic function, for instance,

$$\varphi(\tau) = 1 - \cos \omega\tau \tag{80}$$

where ω is the repetition rate of spikes in the pulse.

Heating of a semi-infinite body or plate by a heat source with the temporal structure [Eq. (80)] is analyzed in [127].

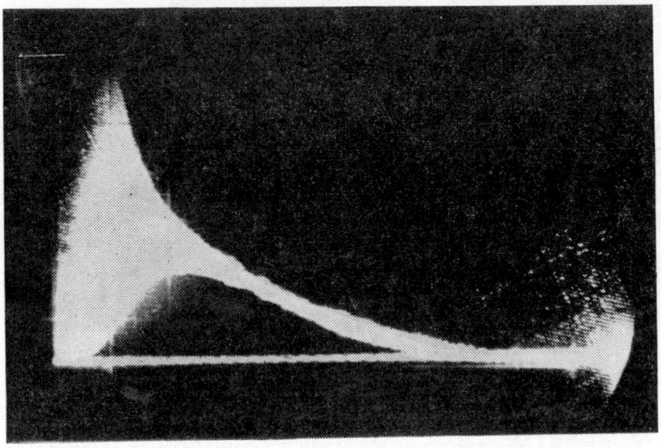

FIG. 17. Oscillogram of quasistationary laser radiation

Quasistationary radiation of pulsed lasers described in [47, 134] practically does not contain spikes with the exception of the initial period (Fig. 17). In the first approximation the temporal structure of the pulses of quasistationary radiation can be described by the step function

$$\varphi(\tau) = \begin{cases} 0 & \tau < 0 \\ 1 & \tau_i \geqslant \tau \geqslant 0 \\ 0 & \tau > \tau_i \end{cases} \tag{81}$$

where τ_i is the pulse duration.

Q-switching makes it possible to produce single pulses (Fig. 18) with the duration τ_i about 10^{-8} s whose temporal structure can be described by a sawtooth function in which the steepness of the front edge can differ from that of the trailing edge.

Let us discuss the mathematical formulation of the problems of laser heating of materials. Such problems are distinguished by the following features.

1. When the radius of the heated zone satisfies the condition

$$r_f \gg (a\tau_i)^{1/2} \tag{82}$$

the problem can be formulated in a one-dimensional form.

2. In most cases of practical interest the heated surface can be assumed to have zero heat conduction. For large heated zones it may become necessary to take into account heat transfer [83].

3. Typically, the thermal and optical parameters for metals have a weak dependence on temperature, which can be ignored in the first approximation when analyzing heating of metals.

4. When the spike structure of the chaotic generation pulse and the spatial nonuniformity of the laser pulse are not taken into account in the analysis the predicted results can differ significantly from the experimental data.

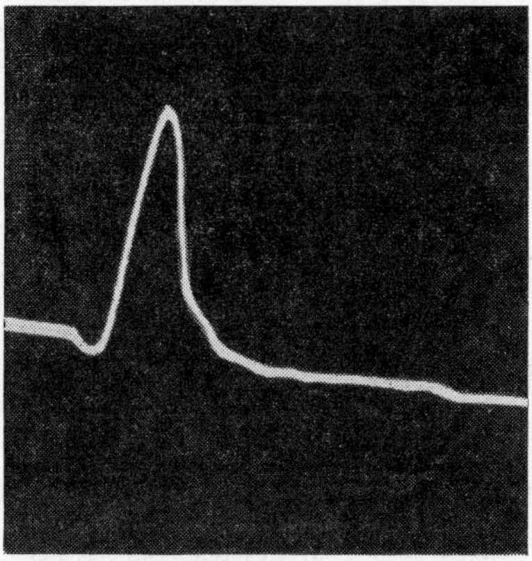

FIG. 18. Oscillogram of a Q-switched laser pulse

The main physical parameter of the heating process is the temperature field in the material, which should be determined experimentally or analytically. We can deduce from the temperature field such important parameters as the rates of cooling and heating, the temperature gradients in various directions, etc. and select correctly the parameters of lasers (energy, power, pulse duration, etc.) and the optimum operation modes.

Let us formulate the equation describing the temperature field of the semi-infinite body heated by a surface heat source in the following form:

$$\frac{1}{a}\frac{\partial T(x, y, z, \tau)}{\partial \tau} = \Delta T(x, y, z, \tau) \tag{83}$$

where $\tau > 0, \quad x, y, z \geqslant 0$

$$z = 0, \quad -\lambda \frac{\partial T}{\partial z} = A\varphi(\tau)\, q(r) \tag{84}$$

$$z = \infty, \quad \tau = 0, \quad T = 0 \tag{85}$$

(here $r = \sqrt{x^2 + y^2}$).

The heat source in this problem is assumed to be axially symmetrical.

The simplest approach to the analysis of heating the materials by laser beams is to make use of one-dimensional models with the condition (82) satisfied. Obviously, this condition cannot be satisfied in all the cases of laser heating but the significant fact is that the initial stages of heating both by continuous-wave and pulsed laser radiation can almost always be treated in the one-dimensional approximation.

In a one-dimensional model the solution of Eqs. (83)-(85) is

$$T(z,\tau)=\frac{q_0\sqrt{a}}{\lambda\sqrt{\pi}}\int_0^{\tau}\varphi(\tau-\xi)\,\xi^{-1/2}\exp\left(-z^2/4a\xi\right)d\xi \tag{86}$$

where q_0 is the maximum power density $q(r)$ multiplied by the absorption capacity A.

Let us discuss some variants of Eq. (86) corresponding to the initial stages of heating for different operation modes of pulsed lasers.

1. For ordered generation the radiation pulse is described by the function (80). Substituting Eq. (80) into Eq. (86) yields

$$T(z,\tau)=\frac{2q_0\sqrt{a\tau}}{\lambda}\,ierfc\left(\frac{z}{2\sqrt{a\tau}}\right)$$

$$-\frac{q_0\sqrt{a}}{\lambda\sqrt{\pi}}\int_0^{\tau}\cos\omega(\tau-\xi)\,\xi^{-1/2}\exp\left(-z^2/4a\xi\right)d\xi \tag{87}$$

The integral in Eq. (87) can be estimated for various ω, τ, z and a with the numerical techniques.

2. For the quasistationary pulse generation we can make use of Eq. (81) for $\tau < \tau_i$

$$T(z,\tau)=\frac{2q_0\sqrt{a\tau}}{\lambda}\,ierfc\left(\frac{z}{2\sqrt{a\tau}}\right) \tag{88}$$

3. For the Q-switched pulses at the initial stage $(\tau < \tau_i)$ we have the linear time dependence (for $\tau > 10^{-9}$ s). Then Eq. (86) can be transformed into

$$T(z,\tau)=\frac{q_0\sqrt{a}}{\lambda\tau_i}(4\tau)^{3/2}\,i^3erfc\,\frac{z}{2\sqrt{a\tau}} \tag{89}$$

The functions $ierfc(u)$ and $i^3erfc(u)$ which are the functions of the probability integral are given in the tabulated form in [82]. When $r_f \approx \sqrt{a\tau_i}$ we have to take into account the spatial distribution of the heat flux density. For the gaussian distribution the temperature field in a semi-infinite body heated by a heat source with the effective power P_0 at $\varphi(\tau)=1$ for $\tau > 0$ is given by [117]

$$T(r,z,\tau)=\frac{2P_0}{c\rho(4\pi a)^{3/2}}\int_0^{\tau}\frac{d\xi\,\exp\left[-\frac{1}{4a}\left(\frac{z^2}{\xi}+\frac{r^2}{\tau_0+\xi}\right)\right]}{\sqrt{\xi}\,(\tau_0+\xi)} \tag{90}$$

where $\tau_0=(4ak)^{-1}$ is the time constant.

If the intensity of the source does not exceed the critical value $q_c^{(1)}$ for melting the heating process can reach the limiting state when the temperature in the central point on the focussed spot of the semi-infinite body is given by

$$T(0, 0, \infty) = \frac{P_0}{2\lambda} \sqrt{\frac{k}{\pi}} \tag{91}$$

The relationship between the effective power P_0 and the intensity q_0 of a normal heat source is given by

$$P_0 = \frac{\pi q_0}{k}$$

Similar equations for the temperature field can be derived also for the uniform distribution of intensity over the heated spot [60].

It is very important to take into account the thickness of the heated material in analysis of the laser heating effects in a wide range of technological applications, such as welding, heat treatment, film preparation, etc. Heating of a thin sheet by a normal-distribution circular heat source with heat transfer on the back side of the sheet is described by Eq. (83), and Eq. (85) is transformed into the heat transfer condition on the surface $z = h$ where h is the thickness of the sheet, and the origin of coordinates is on the heating surface. This condition can be written in the following form:

$$z = h, \quad -\lambda \frac{\partial T}{\partial z} = \alpha_1 T \tag{92}$$

where α_1 is the heat transfer coefficient which determines the rate of heat loss.

Using the integral transformations in r and τ we obtain

$$T(r, z, \tau) = \frac{1}{\lambda} \int_0^\infty \frac{\Phi(s) J_0(sr)}{\lambda s \sinh(hs) + \alpha_1 \cosh(hs)} \times [\lambda s \cos h (h - z) s + \alpha_1 \sinh(h - z)]\, ds$$
$$- \frac{2}{h} \sum_{n=1}^{\infty} \int_0^\infty \frac{\Phi(s) J_0(sr) s}{(\mu_n^2/h^2 + s^2)[(\alpha_1 h + \lambda) \sin \mu_n + \lambda \mu_n \cos \mu_n]} \times [\cos \mu_n (1 - z/h) + \alpha_1 h \sin \mu_n (1 - z/h)]\, e^{-(\mu_n^2/h^2 + s^2) a\tau}\, ds \tag{93}$$

where μ_n are the roots of the transcendental equation

$$\mu \tan \mu = \frac{\alpha_1 h}{\lambda} \tag{94}$$

and

$$\Phi(s) = \frac{q_0}{2k} \exp(-s^2/4k) \tag{95}$$

Numerical solutions of Eqs. (93)-(95) have been found for a copper sheet [118]. If we neglect heat transfer from the back side of the sheet assuming the ideal thermal insulation the equations for the temperature field become much simpler. The temperature distribution in a sheet with thermal insulation from the back side has been found in [174] for uniform distribution of the power density over the focussed spot.

For welding applications it is interesting to analyze the heating of the materials consisting of two layers with different thermal parameters. If we take into consideration the spatial distribution of the heat flux the resulting equations for the temperature field in two-layer materials become very cumbersome [128, 129]. Therefore, first we shall analyze a one-dimensional model and determine the main features of the heating process.

The following equations can be written to describe the temperature field in the two-layer material with the ideal thermal contact between the layers heated by a source with the constant intensity q_0:

$$\begin{gathered} \frac{1}{a_1}\frac{\partial T_1}{\partial \tau} = \frac{\partial^2 T_1}{\partial z^2} \\ \tau > 0, \; h \geqslant z \geqslant 0 \\ \frac{1}{a_2}\frac{\partial T_2}{\partial \tau} = \frac{\partial^2 T_2}{\partial z^2} \\ \tau > 0, \; \infty > z \geqslant h \end{gathered} \tag{96}$$

The boundary conditions of the problem are

$$z = 0, \quad -\lambda \frac{\partial T_1}{\partial z} = q_0 \tag{97}$$

$$z = h \quad \begin{cases} T_1 = T_2 \\ \lambda_1 \dfrac{\partial T_1}{\partial z} = \lambda_2 \dfrac{\partial T_2}{\partial z} \end{cases} \tag{98}$$

$$\tau = 0, \quad T_1 = T_2 = 0 \tag{99}$$

Equations (98) describe the ideal thermal contact between the layers (the temperatures and the heat fluxes on both sides of the interface are identical). The time dependence of the radiation pulse is assumed to be $\varphi(\tau) = 1$ for $\tau > 0$.

Equations (96)-(99) have the following solution:

$$T_1(z,\ \tau)=\frac{2q_0}{\lambda_1}\sqrt{a_1\tau}\ ierfc\left(\frac{z}{2\sqrt{a_1\tau}}\right)$$

$$+\frac{2q_0\sqrt{a_1\tau}}{\lambda_1}\sum_{n=1}^{\infty}(-g)^n\left[ierfc\frac{2nh+z}{2\sqrt{a_1\tau}}+ierfc\frac{2nh-z}{2\sqrt{a_1\tau}}\right] \quad (100)$$

$$T_2(z,\ \tau)=\frac{2q_0\sqrt{a_1\tau}}{\lambda_1}(1-g)\sum_{n=1}^{\infty}(-g)^n\, ierfc$$

$$\times\left[\frac{(2n+1)h+(z-h)\sqrt{a_1/a_2}}{2\sqrt{a_1\tau}}\right] \quad (101)$$

where

$$g=\frac{\lambda_1/\sqrt{a_1}-\lambda_2/\sqrt{a_2}}{\lambda_1/\sqrt{a_2}+\lambda_2/\sqrt{a_2}} \quad (102)$$

The first term in Eq. (100) describes the surface heating of a homogeneous material, and the second term is the temperature change due to the effect of the second layer with different thermal properties and the finite width of the first layer.

The most important characteristics to be found are the temperatures on the heating surface and at the interface between the layers as functions of time. These functions are given below in the dimensionless form:

$$\frac{T_1(0,\ \tau)\lambda_1}{q_0h}=\theta(0,\ \tau)$$

$$=2\sqrt{\mathrm{Fo}}\left(\frac{1}{\pi}+\sum_{n=1}^{\infty}2(-g)^n\, ierfc\,\frac{n}{\sqrt{\mathrm{Fo}}}\right) \quad (103)$$

$$\frac{T_1(h,\ \tau)\lambda_1}{q_0h}=\theta(h,\ \tau)=$$

$$=2\sqrt{\mathrm{Fo}}\left\{ierfc\frac{1}{2\sqrt{\mathrm{Fo}}}+\sum_{n=1}^{\infty}(-g)^n\left[ierfc\frac{h(2n+1)}{2\sqrt{\mathrm{Fo}}}\right.\right.$$

$$\left.\left.+\,ierfc\,\frac{h(2n-1)}{2\sqrt{\mathrm{Fo}}}\right]\right\} \quad (104)$$

Here $\mathrm{Fo}=\frac{a_1\tau}{h^2}$ and $ierfc\,\frac{k}{2\sqrt{\tau}}=\frac{1}{\sqrt{\pi}}\exp(-k^2/4\tau)$ $-\frac{k}{2\sqrt{\tau}}\,erfc\,\frac{k}{2\sqrt{\tau}}$

The functions (103) and (104) for the parameter ranges which are of interest for laser heating are plotted in Fig. 19. They

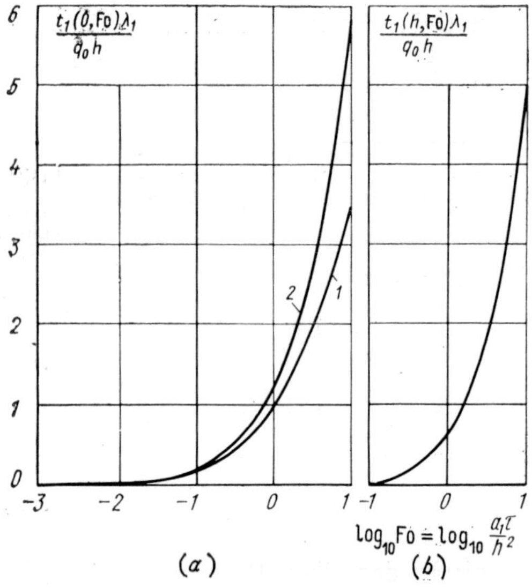

FIG. 19. Dimensionless temperature as a function of the log of Fourier criterion (*a*) *1*—on surface of semi-infinite Al body; *2*—on surface of semi-infinite Al-Si body ($h = 0.02$ cm); (*b*) at contact of two layers in Al-Si body

are compared with variation in the temperature of the surface of a semi-infinite body heated by a heat source of constant intensity q_0.

The functions (100) and (101) can be easily applied to the heat source varying arbitrarily in intensity with time by using Duhamel's integral

$$T_i^* = \int_0^\tau g(\tau - \xi)\, T_i(x, \xi)\, d\xi; \qquad i = 1, 2 \tag{105}$$

where T_i^* is the solution for the source with arbitrary time dependence of the intensity.

Welding can be done with continuous-wave lasers by moving the focussed spot along the material surface. When the focussed spot having the radius r_f moves along the surface at a constant velocity v the temperature will increase at the initial stage. After a certain time period which is determined by the properties of the heated material a quasistationary state will be established, and the heated zone of a constant size will move together with the heat source [117].

In analysis of the effects of moving heat sources the main cases of interest are the surface heating and welding by a normal-distribution circular source having the effective power P_0 moving at a constant velocity v along the surface of a plate with the thickness δ and melting of a semi-infinite body by the same heat source.

The temperature in the moving coordinate system fixed to the heat source is given by the following equation [117]:

$$T(x, y, t) = \frac{P_0}{2\pi\lambda\delta} \exp\left(-\frac{vx}{a} + bt_0\right)$$

$$\times K_0(\rho_2)\,[\psi_2(\rho, \tau + \tau_0) - \psi_2(\rho_2\tau_0)] \tag{106}$$

where

$$\tau_0 = \left(b + \frac{v^2}{4a}\right) t_0, \quad t_0 = 1/4ak$$

$$\tau = \left(b + \frac{v^2}{4a}\right) t, \quad \rho_2 = r\sqrt{\frac{v^2}{4a^2} + \frac{b}{a}}$$

$K_0(\rho_2)$ is the Bessel function of the imaginary argument and $\psi_2(\rho_2, \tau)$ is the heat saturation coefficient tabulated in [117].

The limiting temperature on the axis $OZ (t \to \infty)$ can be found from the equation

$$T(0, \infty) = \frac{P_0}{4\pi\lambda\delta} \exp(bt_0)\,[-E_i(-\tau_0)] \tag{107}$$

If the heat source moves with a high velocity the heat propagation equation is simplified because the heat will be propagated mostly normal to the path of the source:

$$T(y, t) = \frac{P_0}{v\delta\sqrt{4\pi\lambda c\gamma(t_0 + t)}} \exp\left[-\frac{y^2}{4a(t_0 + t)} - bt\right] \tag{108}$$

The temperature field in the moving coordinate system fixed to the heat source with normal distribution of intensity on the surface of a semi-infinite body is given by the following equation:

$$T(x, y, z, t) = \frac{2P_0 \exp(-vx/a)}{c\gamma(4\pi a)^{3/2}} \int_0^t \frac{d\xi}{\sqrt{\xi}\,(t_0 + \xi)}$$

$$\times \exp\left[-\frac{z^2}{4a\xi} - \frac{r^2}{4a(t_0 + \xi)} - \frac{v^2(t_0 + \xi)}{4a}\right] \tag{109}$$

To analyze the limiting stage of the heating process it is convenient to make use of the cylindrical coordinate system (r, φ, z) where φ is the angle between the radius-vector and the positive semiaxis OX along which the heat source moves.

The temperature at the limiting stage is given by

$$T(r, \varphi, z, \infty) = T_c \exp(-2\sqrt{np}\cos\varphi)\frac{2}{\pi}$$

$$\times \int_0^\infty \frac{d\omega}{1+\omega^2} \exp\left[-\frac{m}{\omega^2} - \frac{n}{1+\omega^2} - p(1+\omega^2)\right] \quad (110)$$

where $kz^2 = m$, $kr^2 = n$, $\frac{v^2 t_0}{4a} = p$, and

$$T_c = T(0, 0, 0, \infty) = \frac{P_0}{2\lambda}\sqrt{\frac{k}{\pi}} \quad (111)$$

If the velocity of the source is high the normal-distribution circular source on the surface of a semi-infinite body is transformed into the normal-distribution linear source, and the process of heat propagation is described by the equation

$$T(y, z, t) = \frac{2P_0}{vc\gamma} \frac{\exp\left(-\frac{z^2}{4at}\right)}{\sqrt{4\pi at}} \frac{\exp[-y^2/4a(t_0+t)]}{\sqrt{4\pi a(t_0+t)}} \quad (112)$$

where y and z are the coordinates of a point on the plane which goes through the centre of the source perpendicular to the direction of its motion, and the time t is reckoned from the moment the linear source crosses this plane.

The maximum temperature is found along the axis of the body:

$$T(0, 0, t) = \frac{P_0}{2\pi\lambda v\sqrt{t(t_0+t)}} \quad (113)$$

At the beginning of the process ($t_0 \gg t$) the temperature decreases as $1/\sqrt{t}$ and in the last stage ($t_0 \ll t$) as $1/t$.

3.2. CRITICAL LASER RADIATION POWER DENSITIES

If we know the temperature field produced by the laser radiation in the material we can determine the critical power density which is needed to reach a given temperature at a given point or in a given volume of the material in a given time interval τ_i. For instance, we can determine the power density resulting in disintegration of the material. By disintegration * we can mean the temperature increase on the surface of the material up to the melting point T_m or the boiling point T_b under normal pressure.

* We should emphasize that this definition of the term "disintegration" is arbitrary since some brittle materials disintegrate after cooling has ended.

To determine the radiation power density needed to obtain the temperature T_m on the surface we can use the one-dimensional model of heating of a semi-infinite body by a heat source with the intensity constant in time:

$$q_c^{(1)} = \frac{0.885 T_m \lambda}{(a\tau_i)^{1/2}} \tag{114}$$

where τ_i is the pulse duration.

Table 2 presents the estimated values of $q_c^{(1)}$ for a number of metals with different thermal properties.

TABLE 2. Critical Radiation Power Densities $q_c^{(1)}$ for a Number of Metals

Metal	λ, W/cm K	a, cm²/s	T_m, C	τ_i, s	$q_c^{(1)}$, W/cm²
Cu	3.89	1.12	1083	10^{-3}	1.1×10^4
				10^{-8}	3.5×10^7
Steel	0 51	0 15	1535	10^{-3}	3.5×10^3
				10^{-8}	1.8×10^5
Ni	0.67	0 18	1453	10^{-3}	6.5×10^3
				10^{-8}	2.0×10^5
Ti	0 15	0.06	1800	10^{-3}	3.0×10^3
				10^{-8}	10^5
W	1.69	0.65	3380	10^{-3}	2.0×10^4
				10^{-8}	6.2×10^5
Mo	1.41	0.55	2600	10^{-3}	1.3×10^4
				10^{-8}	4.4×10^5
Cr	0.70	0.22	1830	10^{-3}	7.7×10^3
				10^{-8}	2.7×10^5
Al	2.09	0.87	660	10^{-3}	4.2×10^3
				10^{-8}	$1 3 \times 10^5$

Equation (114) shows that the critical power density $q_c^{(1)}$ increases with increasing melting temperature, heat conductivity and specific heat of the material and decreases with increasing duration of the radiation pulse (see Table 2). Equation (114) can be used to estimate the critical power densities which must not be exceeded in the heat-treatment processes.

The one-dimensional model of heating of a semi-infinite body by a constant-intensity heat source can be used to find the time needed to obtain the temperature T_m on the body surface:

$$\tau_m = \frac{0.79 T_m^2 \lambda^2}{q_0^2 a} \tag{115}$$

The time required to reach a given temperature on the surface of the material increases with increasing melting temperature, heat conductivity and specific heat of the material and decreases with increasing radiation power density. For instance, in copper $\tau_m = 10^{-5}$ s at $q_0 = 10^6$ W/cm^2 and $\tau_m = 10^{-7}$ s at $q_0 = 10^7$ W/cm^2. In a similar way we can determine the radiation power density needed to obtain the temperature T_b on the surface of the material:

$$q_c^{(2)} = \frac{0.885 T_b \lambda}{(a\tau_i)^{1/2}} \tag{116}$$

When the temperature of the surface becomes T_b the material is intensely vaporized. For instance, for copper ($T_b = 2300$ C) we have $q_c^{(2)} = 2.1 \times 10^5$ W/cm^2.

Equation (116) can be used to estimate the critical radiation power density $q_c^{(2)}$ for laser welding of materials since in most cases the vaporization of the material from the melting zone is undesirable.

The following analysis yields an estimate of the radiation power density which results in considerable development of the vaporization processes during the pulse duration [34]. In the process of surface heating two waves are propagated into the bulk of the material — the thermal wave and the evaporation wave. If the power density is small the velocity of the thermal wave is much higher than the velocity of the evaporation wave. With increasing power density the evaporation rate v_b increases and becomes equal to the heating rate v_h at some power density $q_c^{(3)}$. Hence we can estimate $q_c^{(3)}$. We have $v_h \approx (a/\tau_i)^{1/2}$ and $v_b \approx q_c^{(3)}/L_b\,\rho$ where L_b is the evaporation heat. As v_h is equal to v_b we obtain

$$q_c^{(3)} = L_b \rho\,(a/\tau_i)^{1/2} \tag{117}$$

The estimated values of $q_c^{(3)}$ for various metals are given in Table 3.

The critical power density $q_c^{(3}$ increases with increasing evaporation heat and temperature conductivity of the material and decreasing radiation pulse duration. For most materials $q_c^{(1)} < q_c^{(2)} < q_c^{(3)}$.

Let us estimate the rates of heating and cooling of the material in the framework of the simple one-dimensional model of heating of a semi-infinite body by a constant-intensity heat source. The rate of heating is found by differentiating the function (88) with respect to τ

$$\frac{\partial T(z,\tau)}{\partial \tau} = \frac{q_0}{\lambda}\left(\sqrt{\frac{a}{\tau}}\ ierfc\,\frac{z}{2\sqrt{a\tau}} + \frac{z}{2\tau}\,erfc\,\frac{z}{2\sqrt{a\tau}}\right) \tag{118}$$

TABLE 3. Critical Power Densities $q_c^{(3)}$ for a Number of Metals

Metal	ρL_b, J/cm³	a, cm²/s	τ_i, s	$q_c^{(3)}$, W/cm²
Cu	42.88	1.12	10^{-3}	1.4×10^6
			10^{-8}	4.6×10^8
Steel	54.76	0.15	10^{-3}	6.7×10^5
			10^{-8}	2.1×10^8
Ni	55.3	0.18	10^{-3}	7.5×10^5
			10^{-8}	2.4×10^8
Ti	44.27	0.06	10^{-3}	3.4×10^5
			10^{-8}	1.1×10^8
W	95.43	0.65	10^{-3}	2.4×10^6
			10^{-8}	7.7×10^8
Mo	69.05	0.55	10^{-3}	1.6×10^6
			10^{-8}	5.1×10^8
Cr	54.17	0.22	10^{-3}	8.4×10^5
			10^{-8}	2.5×10^8
Al	28.09	0.87	10^{-3}	8.6×10^5
			10^{-8}	2.7×10^8

Here

$$erfc\,\frac{z}{2\sqrt{a\tau}} = 1 - erf\,\frac{z}{2\sqrt{a\tau}} = 1 - \frac{2}{\sqrt{\pi}} \int_0^{z/2\sqrt{a\tau}} \exp(-\xi^2)\,d\xi$$

For instance, the rate of heating v_h on the surface of the material ($z = 0$) is

$$v_h = \frac{q_0}{\lambda} \sqrt{\frac{a}{\pi\tau}} \tag{119}$$

Equation (119) indicates that the heating rate is proportional to the power density and decreases with increasing heat conductivity λ, specific heat of a unit volume $c\rho$ and the time of existence of the heat source.

Figure 20 presents the heating rate as a function of the time of existence of the heat source for various power densities.

To find the cooling rate we shall introduce the concept of a heat sink [117]. The heat sink is a heat source with the negative intensity equal in magnitude to the intensity of our source which starts operating later than our source by the time τ_i. Then the one-dimensional temperature field due to the effect of a constant-intensity heat source in a semi-infinite body can be written for the time $\tau \geqslant \tau_i$ as

$$T(z, \tau) = \frac{2q_0}{\lambda} \left(\sqrt{a\tau}\; ierfc\,\frac{z}{2\sqrt{a\tau}} - \sqrt{a(\tau - \tau_i)}\; ierfc\,\frac{z}{2\sqrt{a(\tau - \tau_i)}} \right) \tag{120}$$

Hence we obtain an expression for the rate of cooling of the surface $z = 0$ after the end of the radiation pulse:

$$v_c = \frac{\partial T(0, \tau)}{\partial \tau}\bigg|_{\tau > \tau_i} = \frac{q_0}{\sqrt{\pi \lambda c \rho}} \left(\frac{1}{\sqrt{\tau}} - \frac{1}{\sqrt{\tau - \tau_i}} \right) \tag{121}$$

The predicted rates of cooling of the body surface for $\tau > \tau_i$ are plotted as functions of time in Fig. 21 for various q_0 values. The cooling rate similar to the heating rate is proportional to the power density q_0.

When a semi-infinite body is heated by a normal-distribution circular heat source on its surface we can find the expression for the heating rate v_{hn} from the temperature field (90):

$$v_{hn} = \frac{2q_0 \exp\left[-\frac{z^2}{4a\tau} - \frac{r^2}{4a\tau + k^{-1}} \right]}{\lambda \sqrt{\pi}\,(1 + 4\pi a k)} \sqrt{\frac{a}{\tau}} \tag{122}$$

Using the concept of a heat sink we can derive the equation for cooling of a semi-infinite body by a normal-distribution cir-

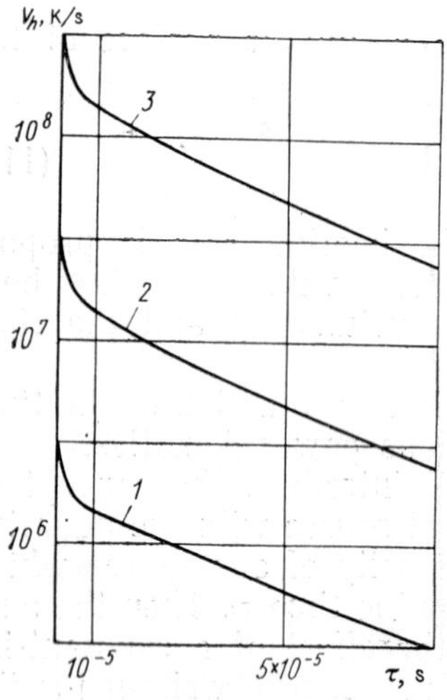

FIG. 20. Heating rate v_h in steel ШХ15 as a function of radiation time τ (from 10^{-6} to 10^{-4} s) at various laser radiation power densities

1—10^4 W/cm²; 2—10^5 W/cm²; *3*—10^6 W/cm²

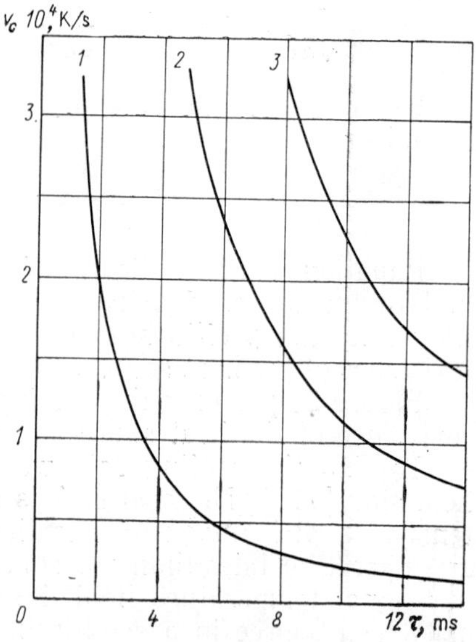

FIG. 21. Predicted cooling rate v_c in steel ШХ15 as a function of radiation time τ at various laser radiation power densities ($\tau_i = 1$ ms)

1—10^4 W/cm²; *2*—5×10^4 W/cm²; *3*—10^5 W/cm²

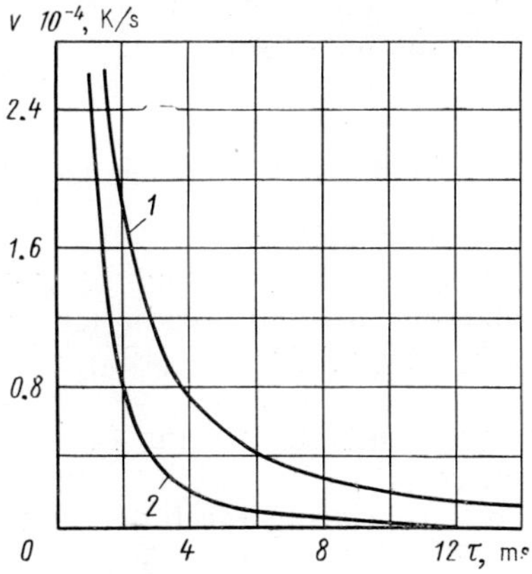

FIG. 22. Heating and cooling rates v in steel ШХ15 heated by a normal-distribution circular heat source on surface as functions of time ($q_0 = 10^4$ W/cm², $\tau_i = 1$ ms)

1—heating rate; 2—cooling rate

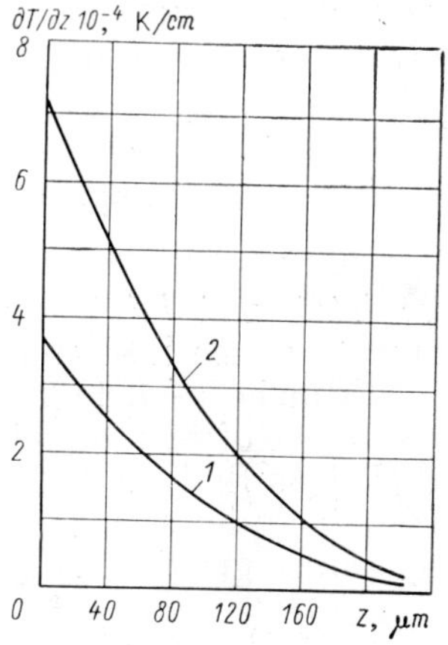

FIG. 23. Distribution of temperature gradient in semi-infinite body (steel ШХ15, $\tau_i = 1$ ms)

1—10^4 W/cm²; 2—2×10^4 W/cm²

cular heat source with the finite lifetime τ_i. Equation (90) yields for $\tau \geqslant \tau_i$

$$T(r, z, \tau) = \frac{2q_0}{c\rho(4\pi a)^{3/2}} \left\{ \int_0^{\tau} \frac{d\xi \exp\left[-\frac{1}{4a}\left(\frac{z^2}{\xi} + \frac{r^2}{\tau_0+\xi}\right)\right]}{\sqrt{\xi}\,(\tau_0+\xi)} - \int_0^{\tau-\tau_i} \frac{d\xi \exp\left[-\frac{1}{4a}\left(\frac{z^2}{\xi} + \frac{r^2}{\tau_0+\xi}\right)\right]}{\sqrt{\xi}\,(\tau_0+\xi)} \right\} \tag{123}$$

Differentiation of Eq. (123) with respect to τ yields the rate of cooling

$$v_{cn} = \frac{q_0\sqrt{a}}{\lambda\sqrt{\pi}} \left\{ \frac{\exp[-z^2/4a\tau - r^2/(4a\tau + k^{-1})]}{\sqrt{\tau}\,(1+4ak\tau)} - \frac{\exp\left[-z^2/4a(\tau-\tau_i) - r^2/4a(\tau-\tau_i) + k^{-1}\right]}{\sqrt{\tau-\tau_i}\,[1+4ak(\tau-\tau_i)]} \right. \tag{124}$$

The predicted rates of cooling and heating in the semi-infinite body heated by a normal-distribution circular heat source on its surface are plotted in Fig. 22.

The same procedure can be used for deriving the expressions for heating and cooling rates when the intensity of the heat source is uniform over the heated spot with the radius r_f. Given below is the respective expression for the heating rate:

$$v_{hr} = \frac{q_0 \sqrt{a} \exp\left(-\frac{z^2}{4a\tau}\right) r_f}{2\lambda \sqrt{\pi\tau}} \int_0^\infty I_1(sr_f) I_0(sr) \exp(-s^2 a\tau)\, ds \quad (125)$$

Here $I_1(u)$ and $I_0(u)$ are the Bessel functions of the imaginary argument of the first and zero order, respectively. Numerical techniques can be used to compute the integral in Eq. (125) for various values of r_f, r, a and τ.

Let us find the temperature gradients for heating of a semi-infinite body by a surface heat source of a constant intensity. The temperature gradient is found by differentiating the function (88):

$$\frac{\partial T(z, \tau)}{\partial z} = -\frac{q_0}{\lambda}\, erfc\, \frac{z}{2\sqrt{a\tau}} \quad (126)$$

The temperature gradient with $\tau > 0$ remains constant on the body surface where $z = 0$:

$$\frac{\partial T(0, \tau)}{\partial z} = -\frac{q_0}{\lambda} \quad (127)$$

The temperature gradient as a function of the body depth z is plotted in Fig. 23 for two values of q_0. Estimates show that the temperature gradient decreases sharply enough with the depth. The temperature gradient on the body surface can be very large. For instance, at $q_0 = 10^6$ W/cm² in copper $\frac{\partial T(0, \tau)}{\partial z} = 3 \times 10^5$ C/cm and at $q_0 = 10^7$ W/cm² it is 3×10^6 C/cm. Using the temperature fields and the temperature gradients we can estimate the size of the heat-affected zone in the material. In qualitative analysis we can use the simple estimate $d \approx \sqrt{a\tau_i}$ where d is the size of the heat-affected zone.

For many materials the heat source cannot be assumed to have the surface character. Later we shall derive a criterion for defining a heat source as a surface one or a bulk one. Absorption of the radiation decreases exponentially with depth (Bouguer's law) so that variation in the heat source power $P(z)$ with depth is given by

$$P(z) = P_0 \exp(-\alpha z) \quad (128)$$

where α is the absorption coefficient.

The solution of the one-dimensional equation for the semi-infinite body heated by the heat source with the power of Eq. (128)

can be found by using the Laplace transformation in time:

$$T(z, \tau) = \frac{\alpha q_0}{c\rho}\left[\frac{\exp(-\alpha z + a\alpha^2\tau)}{a\alpha^2}\, erfc\left(\frac{2a\alpha\tau - z}{2\sqrt{a\tau}}\right)\right.$$

$$\left. + \frac{\exp(\alpha z + a\alpha^2\tau)}{a\alpha^2}\, erfc\left(\frac{2a\alpha\tau + z}{2\sqrt{a\tau}}\right) + \frac{4\sqrt{\tau}}{\alpha\sqrt{a}}\, ierfc\,\frac{z}{2\sqrt{a\tau}}\right] \qquad (129)$$

In the case of a pulsed bulk heat source with the lifetime τ_i the temperature field in the body can be found by introducing a heat sink which starts operating later than the heat source by the period τ_i:

$$T(z, \tau) = \frac{\alpha q_0}{c\rho}\left\{\frac{e^{-\alpha z}}{a\alpha^2}\left[e^{a\alpha^2\tau}\, erfc\left(\frac{2a\alpha\tau - z}{2\sqrt{a\tau}}\right)\right.\right.$$

$$\left. - e^{a\alpha^2(\tau-\tau_i)}\, erfc\left(\frac{2a(\tau-\tau_i) - z}{2\sqrt{a(\tau-\tau_i)}}\right)\right] + \frac{e^{\alpha z}}{a\alpha^2}$$

$$\times\left[e^{a\alpha^2\tau}\, erfc\left(\frac{2a\alpha\tau + z}{2\sqrt{a\tau}}\right) - e^{a\alpha^2(\tau-\tau_i)}\, erfc\left(\frac{2a(\tau-\tau_i) + z}{2\sqrt{a(\tau-\tau_i)}}\right)\right]$$

$$\left. + \frac{4}{\alpha\sqrt{a}}\left[\sqrt{\tau}\, ierfc\,\frac{z}{2\sqrt{a\tau}} - \sqrt{\tau-\tau_i}\, ierfc\,\frac{z}{2\sqrt{a(\tau-\tau_i)}}\right]\right\} \qquad (130)$$

Equations (129) and (130) describe the temperature distribution in the body and can be used to determine the rates of heating and cooling and the temperature gradients in the material.

Let us analyze Eqs. (129) and (130) in special cases. If the penetration depth $\delta = \alpha^{-1}$ of radiation is much larger than the thickness of the heated zone, about $\sqrt{a\tau}$, that is, if $\alpha\sqrt{a\tau} \ll 1$ then we have

$$T(z, \tau) = \frac{\alpha q_0 \tau}{c\rho}\exp(-\alpha z) \qquad (131)$$

and for the surface $z = 0$

$$T(0, \tau) = \frac{\alpha q_0 \tau}{c\rho} \qquad (132)$$

that is, the temperature increases with time. For such materials as ferrite ($\alpha = 2 \times 10^3$ cm^{-1}, $a = 5 \times 10^{-3}$ cm^2/s) Eqs. (131) and (132) are valid at $\tau < 10^{-4}$ s. If $\alpha\sqrt{a\tau} \gg 1$ the heat source can be assumed to have the surface character, and the temperature field for it is described by Eq. (88). For such materials as ferrite this condition is satisfied when $\tau \gg 10^{-4}$ s. For metals the condition $\alpha\sqrt{a\tau} \gg 1$ is typically satisfied.

For the bulk heat source with the intensity varying according to Bouguer's law [Eq. (128)] the critical power density for initiation of disintegration of the material can be found from Eq. (129):

$$q_c^v = \frac{\lambda a T_b}{\exp(a\alpha^2\tau)\, erfc(\alpha\sqrt{a\tau}) - 1 + 2\pi^{-1/2}\alpha\sqrt{a\tau}} \tag{133}$$

For instance, for ferrite (the reflectivity is 0.23) Eq. (133) yields the critical power density of 3×10^4 W/cm². To take into account the spatial distribution of the radiation power density for bulk absorption of radiation we can formulate the temperature-distribution problem in terms of dimensionless variables [122]:

$$\frac{\partial\Theta}{\partial \mathrm{Fo}} = \frac{\partial^2\Theta}{\partial\rho^2} + \frac{1}{\rho}\frac{\partial\Theta}{\partial\rho} + \frac{\partial^2\Theta}{\partial\xi^2} + A_1\exp(-\beta\rho^2 - \gamma\xi) \tag{134}$$

$$\mathrm{Fo} > 0,\ \infty > \rho \geqslant 0,\ \infty > \xi \geqslant 0$$

$$\frac{\partial\Theta}{\partial\xi} = 0 \quad \text{for} \quad \xi = 0$$

$$\Theta = 0 \text{ for } \xi,\ \rho \to \infty,\ \mathrm{Fo} = 0$$

where

$$\mathrm{Fo} = a\tau/r_f^2,\quad \rho = \frac{r}{r_f},\quad \xi = \frac{z}{r_f}$$

$$\Theta = \frac{T}{T_l},\quad \beta = kr_f^2,\quad \gamma = r_f\alpha$$

and $A_1 = r_f^2 q_0 \alpha/\lambda T_l$ are the dimensionless variables.

The temperature T_l is, for instance, the limiting temperature in the centre of the axially symmetrical source on the surface of an opaque material for $\mathrm{Fo} \to \infty$, which is given by Eq. (107).

The solution of Eq. (134) can be found by using the Laplace and Hankel integral transformations in Fo and ρ, respectively:

$$\Theta(\rho, \xi, \mathrm{Fo}) = \frac{A_1}{2\beta}\exp(-\gamma\xi)$$

$$\times \int_0^\infty \frac{sJ_0(s\rho)\exp(-s^2/4\beta)\left(1 - e^{-(s^2-\gamma^2)\mathrm{Fo}}\right) ds}{(s^2-\gamma^2)}$$

$$+ \frac{\gamma A_1}{2\sqrt{\pi}} \int_0^\infty \frac{sJ_0(s\rho)\exp(-s^2/4\beta)}{s^2-\gamma^2}$$

$$\times \left[\int_0^{\mathrm{Fo}} \frac{dy\exp(-\xi^2/4y)}{4y}\left(e^{-(s^2-\gamma^2)\mathrm{Fo}-\gamma^2 y} - e^{-s^2 y}\right)\right] ds \tag{135}$$

Equation (135) is rather involved but it can be simplified for small values of Fo corresponding to large r_f or small τ. For Fo $\ll 1$ Eq. (135) can be transformed into

$$\Theta(\rho, \xi, \mathrm{Fo}) = A_1 \exp(-\beta\rho^2)\left\{\gamma^{-2}e^{-\gamma\xi}(e^{\mathrm{Fo}\gamma^2} - 1) + \frac{1}{\gamma\sqrt{\pi}}\right.$$

$$\left.\times\left[\int_0^{Fo} \frac{\exp(-\xi^2/4x)}{\sqrt{x}}(1 - e^{\mathrm{Fo}-x}\gamma^2)\,dx\right]\right\} \tag{136}$$

Equation (136) shows that for small Fo values the temperature field in the material is given by the product of the function which describes the intensity distribution of the heat source over the heated spot and the appropriate solution of a one-dimensional problem. For the centre of the heated spot Eq. (136) yields [36, 122]

$$\Theta = \frac{2\sqrt{\mathrm{Fo}}}{\sqrt{\pi}} A_1\left[1 - \frac{\sqrt{\pi}}{2\varepsilon}(1 - e^{\varepsilon^2} erfc\,\varepsilon)\right] \tag{137}$$

where $\varepsilon = \gamma\sqrt{\mathrm{Fo}}$.

At the initial heating stages in the bodies with surface or bulk heat sources the isotherms of the heated zone are congruent with the spatial distribution of the heat source intensity. Mathematically this means that at the initial heating stages the temperature field can be described by a product of two functions one of which is the intensity distribution. This feature can be used for determining the radial distribution of the heat source intensity in the laser-irradiated material which undergoes phase transitions in the solid state and also for evaluating the integral mean of the absorption coefficient during the radiation pulse [11].

Let us analyze the limiting case of Eq. (135) for long time periods (Fo $\to \infty$). If the heat source operates continuously the maximum temperature at the point $(0, 0, \infty)$ is

$$\Theta_m = \frac{A_1}{2\beta}\int_0^{\infty} \frac{\exp(-u^2)\,du}{u + \gamma/2\beta^{1/2}} \tag{138}$$

Equation (138) indicates that the maximum temperature is determined by two dimensionless parameters A_1 and γ (β is a constant). This equation suggests that the extent of the surface character of a heat source is determined not by the absorption coefficient itself but rather by its product by the radius of the focussed spot. Similarly, the extent of approximation of a heat source by the point source depends not only on the radius of the

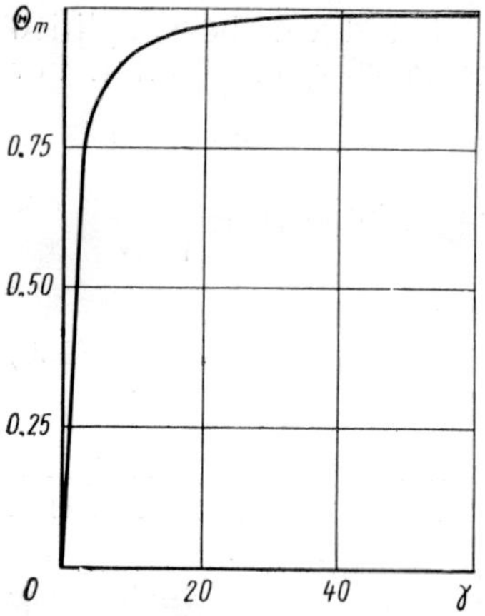

FIG. 24. Parameter Θ_m as a function of γ

focussed spot but also on the absorption coefficient of the material.

This can be readily seen from the plot of Θ_m as a function of γ (Fig. 24). For $\gamma \gtrsim 15$ the heat source can be assumed to be a practically surface one. In metals with $\alpha \approx \approx 10^4$ cm^{-1} we find $r_{fs} \approx 15$ μm where r_{fs} is the limiting radius of the focussed spot in metals — for larger radiuses we have the surface heat sources. For short radiation pulses the respective value of γ depends on Fo, that is, on the pulse duration.

In semiconductors and dielectrics the heat source with various radiuses r_f can be assumed to be the surface or the bulk one according to the value of γ. For instance, for the wavelength about 1.2 μm we have in silicon $\alpha = 1$ cm^{-1}, and the respective focussed spot radius r_f is too large for the heat source to be regarded as the surface one and to satisfy the condition of local heating ($r_f \lesssim 0.1$ cm). This condition is valid only for relatively low temperatures which do not produce a significant increase in α.

The above estimates show that even in metals where absorption is high ($\alpha \approx 10^4$-10^5 cm^{-1}) the heat source due to the laser radiation effect cannot be treated as a point source, at least in analysis of the melting zone.

3.3. NONLINEAR PROBLEMS IN LASER HEATING

Let us consider some nonlinear problems encountered in analysis of laser heating of materials. Most problems related to the effects of highly concentrated sources ($q_0 > 10^4$ W/cm^2) must be formulated in the nonlinear form. In the nonlinear heat-transfer problems some of the following parameters depend on temperature: (a) the heat conductivity; (b) the specific heat; (c) the heat-transfer coefficient; (d) the heat flux on the surface; (e) the intensity of the internal heat sources (sinks); and (f) the position of the body boundaries [65].

For the sake of convenience the following classification of the problems can be introduced: the problems in which thermal parameters depend on temperature are the problems with the I order nonlinearities; the problems with nonlinear boundary conditions are the problems with the nonlinearities of order II, and the problems with the temperature-dependent internal heat sources are the problems with the nonlinearities of order III.

In the analysis of laser heating of materials we encounter problems of orders I, II and III. If the radiation power density is below $q_c^{(1)}$ (heat treatment) the main problems contain nonlinearities of orders I and II when we have to take into account the temperature dependence of the thermal parameters (order I) and the temperature dependence of the optical parameters of the material and, hence, the intensities of the heat sources (order II). If the radiation power density exceeds $q_c^{(1)}$ then after a period τ_c we have to consider a problem with the order III nonlinearities, or the problems involving determination of the interface positions known as the Stefan problems.

The most general form of the problem contains at once all three types of nonlinearities. Since even one type of nonlinearity makes a problem difficult to solve we shall not analyze the problems which contain all types of nonlinearities.

Let us discuss briefly the techniques used for solving nonlinear problems.

Nonlinearities of order I. Various linearization techniques are widely used for solving nonlinear problems since the methods for solving the linear heat-transfer problems are developed much better. The most typical technique is to make use of various substitutions to simplify equations with nonlinear terms.

Substitution of the type

$$\lambda = \lambda_0^{-1}(T_0) \int_{T_0}^{T} \lambda(T)\, dT \tag{139}$$

is done in the problems with temperature-dependent heat conductivity $\lambda = \lambda(T)$. This substitution is known as Kirchhoff's transformation or sometimes as Varshavsky's substitution or transformation. This technique linearizes the nonlinearity in the stationary heat-conduction equation and introduces it into the boundary condition if it is of order III. No change occurs in the boundary conditions of order II which are the most significant for the laser heating processes. Substitution of Eq. (139) into nonstationary heat-conduction equations transforms them by introducing nonlinear terms.

Goodman's substitution

$$v = \int_{0}^{T} c(T)\, dT \tag{140}$$

is generally used to take into account the temperature dependence of the specific heat $c = c(T)$.

Boltzmann's substitution or transformation is sometimes used for infinite or semi-infinite bodies:

$$\xi = \frac{x}{2\sqrt{\tau}} \tag{141}$$

This technique makes it possible to transform a partial differential equation into a total differentiation equation.

Analogue computers are widely used for solving the nonlinear problems of order I. The following procedures are typically used: (a) the basic equation is simplified or linearized by means of Kirchhoff's or Goodman's substitution; (b) the resistances, capacitances and other electric parameters of the analogue computer are varied as $\lambda(T)$, $c(T)$ and $\rho(T)$; (c) additional currents are introduced into the electric model for simulating the effect of nonlinearities.

Among other techniques used for solving nonlinear problems of order I are the integral technique, iterative technique, etc. which we shall not discuss here.

Nonlinearities of order II. An arbitrary temperature dependence $q(T)$ makes the problem so complicated that it can be solved practically only by analogue or net techniques. For the thermally thin bodies ($Bi < 0.25$ where Bi is Biot's parameter) the problem can be described by an ordinary differential equation. If the body cannot be assumed thermally thin, in some cases the temperature field in a finite body can be represented by the temperature distributions along the coordinate axes or over the body surface. The type of coordination is determined by the conditions of the process. The temperature field can be determined also with the zone technique in which the temperature is found by successive "steps in time", that is, the known temperature distribution at the $(m-1)$th moment is used to find the distribution at the mth moment. Another technique suggested by Ivanov and Vidin [65] is to use nonlinear integral substitutions which transform the initial linear differential equation into a nonlinear equation wherein the nonlinear terms can be omitted if certain conditions are satisfied; the resulting equation yields an approximate solution of the problem with a known accuracy. This technique is used for analyzing heating of a body whose radiation absorption capacity depends on temperature.

Other techniques for solving the nonlinear problems of order II are the variational technique, the integral technique [175], the iterative technique, the finite differences technique, etc.

Nonlinearities of order III. In the nonlinear problems of order III the intensities of the sources inside the body and on its surface depend on temperature. This class of problems includes what is called Stefan's problems in which the position of the moving phase interface and the intensity of the internal heat sources depend on the temperature (or temperature range) of the phase transition.

Various techniques have been used by numerous authors [30, 116] for solving Stefan's problem, such as the variational technique, the integral technique, the finite differences technique, the substitution technique, the straight lines technique, the dimensional analysis, etc.

Let us analyze the effect of temperature dependence of the optical parameters of a metal on the process of laser heating. As the absorption capacity A of the metal depends on temperature this is a nonlinear problem of order II. In the first approximation $A \approx T$ where T is the temperature of the absorbing surface. Taking into account the anomalous skin effect [80] we obtain

$$A = a_0 + bT \tag{142}$$

where $a_0 = 0.75\, v_0/c$, $b = \Omega/2\pi\sigma_0$, v_0 is the velocity of the conduction electron at the Fermi surface, c is the velocity of light, Ω is the plasma frequency and σ_0 is the static electrical conductivity of the metal *.

Typically, $a_0 \ll bT$ and the first term in Eq. (142) can be omitted. The following one-dimensional equation can be written to describe heating of a metal with a variable absorption capacity [80]:

$$\frac{\partial T}{\partial \tau} = a\frac{\partial^2 T}{\partial z^2} + Aq_0\alpha \exp(-\alpha z) \tag{143}$$

$$z = 0; \quad \frac{\partial T}{\partial z} = 0 \tag{144}$$

$$\tau = 0, \quad z = \infty, \quad T = T_i$$

where α is the absorption coefficient and T_i is the initial temperature of the metal.

Note that α in this approximation is independent of temperature. Equation (143) with the boundary condition (144) has two limiting solutions (for the large and small parameters $S = 4bq_0/\alpha\lambda$):

$$T - T_i = \frac{A_0}{b}\left[\exp\left(\frac{bq_0\alpha\tau}{c_p}\right) - 1\right] \tag{145}$$

$$T - T_i = \frac{A_0 q_0}{c_p\sqrt{a}}\left\{\frac{c_p\sqrt{a}}{bq_0}\left[e^{\left(\frac{bq_0}{c_p a}\right)^2\tau} erfc\left(-\frac{bq_0\sqrt{\tau}}{c_p a}\right) - 1\right]\right.$$

$$\left. + \frac{1}{\alpha\sqrt{a}}\left[1 + \exp(\alpha^2 a\tau)\, erfc\left(\alpha\sqrt{a\tau}\right)\right]\right\} \tag{146}$$

where $A^0 = A(T_i)$ is the initial absorptivity.

* Since A has a linear temperature dependence for many metals the problem is formulated in the linear form.

The surface temperature and its variation with time for the time-independent absorption capacity can be found from Eq. (146) at $b \to 0$ [or from the dimensionless Eq. (137)]:

$$T - T_i = \frac{A_0 q_0}{c_p a}\left\{2\sqrt{\frac{\tau}{\pi}} + \frac{1}{\alpha\sqrt{a}}\right.$$
$$\left.\times\left[\exp(\alpha^2 a\tau)\, erfc\left(\alpha\sqrt{a\tau}\right) - 1\right]\right\} \qquad (147)$$

The temperature dependence of the absorption capacity of a metal gives rise to the following features of the heating process. When $S \gg 1$ the resulting heating process differs essentially from the heating processes with the constant $A = A_0$. When $S \ll 1$ (the typical case) two types of heating can occur. If $\frac{bq_0\tau}{c_p\sqrt{a}} \ll 1$ the surface temperature increases with time according to Eq. (147), that is, the time dependence of $A(\tau)$ for the time intervals $\tau \ll \frac{c_p\sqrt{a}}{bq_0}$ can be neglected. For $\tau \gg \frac{c_p\sqrt{a}}{bq_0}$ Eq. (146) yields

$$T \approx 2\frac{A_0}{b}\exp\left(\frac{bq_0}{c_p\sqrt{a}}\right)^2\tau + T_i \qquad (148)$$

A comparison of Eqs. (147) and (148) shows that the difference between the process with $A = A_0$ and the process with $A = A(\tau)$ becomes significant starting from the moment $\tau \gg \frac{c_p\sqrt{a}}{bq_0}$. In the first case T increases with $\sqrt{\tau}$ and in the second case

$$T \approx \exp\left[\left(\frac{bq_0}{c_p\sqrt{a}}\right)^2\tau\right]$$

The temperature variation becomes practically exponential according to Eq. (148) starting from the moment $\tau \approx 2\Big/\left(\frac{bq_0}{c_p\sqrt{a}}\right)^2$.

The temperature dependence of A changes the power density required for obtaining a given temperature of the surface by the end of the radiation pulse, for instance T_m. It has been estimated in [80] that for temperature-dependent A in silver we have $q_c^{(1)} = 9 \times 10^5$ W/cm^2 $\left(\tau_i = 10^{-3}\text{ s}\right)$ while for temperature-independent A we have $q_c^{(1)} = 3 \times 10^6$ W/cm^2.

The problem of laser heating of materials with the temperature-dependent $A(T)$ can be formulated in the three-dimensional form. For many materials Eq. (142) seems to be a reasonable approximation of the temperature dependence of the absorption capacity; this can be substantiated by the experimental results if we allow for a considerable spread of the experimental data and the significant effects produced by the state of the irradiated surface and other factors which are difficult to control on the absorption capacity.

Table 4 presents the factors a_0 and b * for a number of materials. Absolute temperatures are used, and the temperature range is below the melting point.

TABLE 4. Parameters of Temperature Dependence of A for Some Materials

Material	a_0	$b \times 10^4$	Material	a_0	$b \times 10^4$
Tungsten	0.024	1.03	Copper	0.028	0.14
Molybdenum	0	0.99	Graphite	0.890	—0.40
Tantalum	0.048	0.92	Alumina	0.770	—4.28
Platinum	0.099	0.64	Germanium	0.747	—0.92

In the nonmetallic materials the absorption capacity can drop with increasing temperature.

Let us consider the axially symmetrical radiation flux with the gaussian distribution of intensity. Absorption of radiation is assumed to occur on the surface. If we neglect phase transitions, heat losses from the heated surface and the temperature variation in thermal parameters we can describe the heating problem by the following equation and boundary conditions:

$$\frac{1}{a}\frac{\partial T}{\partial t}=\frac{\partial^2 T}{\partial r^2}+\frac{1}{r}\frac{\partial T}{\partial r}+\frac{\partial^2 T}{\partial z^2}$$

$$z=0, \quad \lambda\frac{\partial T}{\partial z}+(a_0+bT)\,q_0\exp(-kr^2)=0 \tag{149}$$

$$T(r,\ z,\ 0)=T(\infty,\ z,\ t)=T(r,\ \infty,\ t)=T_0$$

Let us introduce the dimensionless variables

$$T=\frac{q_0A_0}{\lambda\sqrt{k}}u+T_i, \quad t=\frac{\tau}{ak}, \quad r=\frac{\rho}{\sqrt{k}} \tag{150}$$

$$z=\frac{\xi}{\sqrt{k}}, \quad \beta=\frac{q_0b}{\lambda\sqrt{k}}$$

* The factors a_0 and b are derived from the extrapolation of the experimental plot $A(T)$ to the temperature range below T_m.

and transform Eqs. (149) into

$$\frac{\partial u}{\partial \tau}=\frac{1}{\rho}\frac{\partial}{\partial \rho}\left(\rho\frac{\partial u}{\partial \rho}\right)+\frac{\partial^2 u}{\partial \xi^2}$$

$$\xi=0,\quad \frac{\partial u}{\partial \xi}+\beta u\exp(-\rho^2)=\exp(-\rho^2) \tag{151}$$

$$u(\rho,\ \xi,\ 0)=u(\infty,\ \xi,\ \tau)=u(\rho,\ \infty,\ \tau)=0$$

Equations (151) indicate that in the dimensionless form the system depends only on one parameter β which is proportional to the factor b describing the variation in the absorption capacity with time.

Let us introduce a new unknown function w:

$$u=\frac{1}{\beta}(e^{\beta w}-1) \tag{152}$$

and transform the system (151) into the following system which is nonlinear in w:

$$\frac{\partial w}{\partial \tau}=\frac{1}{\rho}\frac{\partial}{\partial \rho}\left(\rho\frac{\partial w}{\partial \rho}\right)+\frac{\partial^2 w}{\partial \xi^2}+\beta\left[\left(\frac{\partial w}{\partial \rho}\right)^2+\left(\frac{\partial w}{\partial \xi}\right)^2\right]$$

$$\xi=0,\quad \frac{\partial w}{\partial \xi}=-\exp(-\rho^2) \tag{153}$$

$$w(\rho,\ \xi,\ 0)=w(\infty,\ \xi,\ \tau)=w(\rho,\ \infty,\ \tau)=0$$

This substitution yields a convenient approximation of Eqs. (151) for real values of β and τ. The system (153) can be solved by using successive approximations technique:

$$\frac{\partial w_{n+1}}{\partial \tau}=\frac{1}{\rho}\frac{\partial}{\partial \rho}\left(\rho\frac{\partial w_{n+1}}{\partial \rho}\right)+\frac{\partial^2 w_{n+1}}{\partial \xi^2}$$

$$+\beta\left[\left(\frac{\partial w_n}{\partial \rho}\right)^2+\left(\frac{\partial w_n}{\partial \xi}\right)^2\right] \tag{154}$$

$$\xi=0,\quad \frac{\partial w_{n+1}}{\partial \xi}=-\exp(-\rho^2) \tag{155}$$

$$w_{n+1}(\rho,\ \xi,\ 0)=w_{n+1}(\infty,\ \xi,\ \tau)=w_{n+1}(\rho,\ \infty,\ \tau)=0 \tag{156}$$

The first approximation is found by solving Eq. (154) with the boundary conditions (155) and (156) when the nonlinear terms in Eq. (154) are omitted. When the value of the parameter β is less than 10 the estimates show that the difference between w_2 and w_1 is not more than 15% and decreases with decreasing β and τ. This is explained by the fact that the terms $\partial w/\partial \xi$ and $\partial w/\partial \rho$ are less than 1 everywhere, and their squares are, of course, even smaller.

In the first approximation the solution of Eqs. (154) and (155) is identical to the solution (80) for a normal-distribution circular source; it has the following form in the variables ρ, ξ and τ:

$$w_1(\rho, \xi, \tau) = \frac{2}{\sqrt{\pi}} \int_0^{\sqrt{\tau}} \exp\left[-\left(\frac{\rho^2}{1+4x^2} + \frac{\xi^2}{4x^2}\right)\right] \frac{dx}{1+4x^2} \qquad (157)$$

For the centre of the heated spot we obtain

$$w_1(0, 0, \tau) = \frac{1}{\sqrt{\pi}} \arctan 2\sqrt{\tau} \qquad (158)$$

The maximum temperature is found from

$$u_{\max} = \frac{1}{\beta}\left(e^{\frac{\beta}{\sqrt{\pi}} \arctan 2\sqrt{\tau}} - 1\right) \qquad (159)$$

To express the power density required for obtaining the temperature T_1 in time t in the centre of the focussed spot we use the dimensional variables of Eqs. (150) and find

$$q_0 = \frac{\lambda\sqrt{\pi k}\,(T_1 - T_i)(\ln A_1 - \ln A_0)}{(A_1 - A_0)\arctan 2\sqrt{akt}} \qquad (160)$$

where A_1 is the absorptivity at the temperature T_1.

Posing A_1 to A_0 in Eq. (160) we can find the equation for the power density q_0^* required to obtain the temperature T_1 in time t with the constant absorptivity A^*:

$$q_0^* = \frac{\lambda\sqrt{\pi k}\,(T_1 - T_i)}{A^* \arctan 2\sqrt{akt}} \qquad (161)$$

From Eqs. (160) and (161) we can find the constant absorptivity $\bar{A}$ which is equivalent to the variable absorptivity in producing the same maximum temperature in the same time period with the same power density. The identity $q_0 = q_0^*$ yields the averaged value of $\bar{A}$:

$$\bar{A} = \frac{A_1 - A_0}{\ln A_1 - \ln A_0} \qquad (162)$$

For instance, for molybdenum in the temperature range from 20 to 2600 C and at $A_0 = 0.029$ and $A_1 = 0.284$ we have $\bar{A} = 0.112$. Expanding the term $\exp(\beta w_1)$ in Eq. (152) in β series and retaining four terms of this series for small β we get

$$u \approx w_1 \left(1 + \frac{\beta}{2} w_1 + \frac{\beta^2}{6} w_1^2\right) \tag{163}$$

It can be seen that the temperature dependence of A makes a contribution of less than 10% for $|\beta| < 1/5$. Hence we can neglect the temperature variation of A when $|\beta| < 1/5$.

Let us compare the q_0 value calculated according to Eq. (160) for silver and the respective value reported in [80]. For $\lambda = 4.2$ W cm^{-1} K, $a = 1.74$ cm^2 s^{-1}, $b = 10^{-4}$ K^{-1}, $A_0 = 0.03$, $T = 300$ K, $T_1 = T_m = 1233$ K, $t = 10^{-3}$ s and $k = 10^4$ cm^{-2} we obtain $q_0 = 7.2 \times 10^6$ W/cm^2 ($\beta = 1.71$) while the value reported in [80] is 9×10^5 W/cm^2.

The plots of u_{max} as a function of $\sqrt{\tau}$ lead to the following conclusions. First of all, a qualitative difference can be seen in temperature variation with time when compared with a one-dimensional model. In the latter model for $b \geqslant 0$ the temperature tends to infinity with time [see Eqs. (147) and (148)] and for $b < 0$ the temperature tends to a steady-state value while in a three-dimensional model u_{max} always tends to a steady-state value when $t \to \infty$:

$$u(0,\ 0,\ \infty) = \frac{1}{\beta}\left(e^{\frac{\beta\sqrt{\pi}}{2}} - 1\right)$$

Note also that for $\beta < 0$ the temperature tends to its limiting value faster than when $\beta > 0$. For instance, by the moment $\tau = 1$ for $\beta = 2$ we have $u_{max} = 0.51\ u_{lim}$ and for $\beta = -2$ $u_{max} = 0.86\ u_{lim}$.

When the radiation power density is higher than $q_c^{(1}$ the body surface is heated to the temperature over T_m after the time period τ_m. The melting boundary which is the interface between the solid and liquid phases starts moving inwards.

In the case of ablation when the liquid phase either is not produced or immediately removed from the surface of the material the analysis is simplified as the interface is at the same time the radiation-absorbing surface. Stefan's problem consists in finding the temperature distribution in the solid and liquid phases and the velocity of the interface (the rate of vaporization or melting).

Let us formulate the one-dimensional problem of heating and subsequent melting of a plate by a heat source on its surface assuming that the liquid phase is removed as soon as it is produced. If the thermal parameters of the material are assumed to be independent of temperature [184] we can write

$$\frac{1}{a}\frac{\partial T}{\partial \tau} = \frac{\partial^2 T}{\partial z^2} \tag{164}$$

and for $s(\tau) < z < l$

$$T(x, 0) = T_0(x) \leqslant T_m \text{ for } \tau = 0, \ 0 \leqslant z \leqslant l$$

$$\frac{\partial T}{\partial x} = 0 \quad \text{for } z = l, \ \tau > 0 \tag{165}$$

$$q_0(\tau) = -\lambda \frac{\partial T}{\partial x} + \rho L_m \frac{ds}{d\tau} \quad \text{for } x = s(\tau), \ \tau > 0 \tag{166}$$

where $s(\tau)$ is the position of the phase transition boundary, L_m is the latent heat of melting and l is the thickness of the plate.

Equation (166) is known as Stefan's condition. Actually, it is the condition of energy conservation at the interface. This condition is satisfied both with melting and in the absence of melting when the following conditions are satisfied on the heated surface:

$$\begin{aligned} &\frac{ds}{d\tau} \geqslant 0 \quad \text{for } T[s(\tau), \ \tau] = T_m \\ &\frac{ds}{d\tau} = 0 \quad \text{for } T[s(\tau), \ \tau] < T_m \end{aligned} \tag{167}$$

The time interval τ_m between the beginning of the pulse effect and the moment when the surface temperature increases up to T_m is

$$\tau_m = \left(\frac{\rho L_m}{q_0} \sqrt{a}\, m \right)^2 \tag{168}$$

where $m = \frac{\sqrt{\pi}}{2} \frac{c(T_m - T_0)}{L_m}$ and the radiation power density q_0 is assumed to be constant.

If we introduce new variables and eliminate the moving boundary using the transformation

$$\xi = \frac{l - z}{l - s(\tau)} \tag{169}$$

we can reduce Eqs. (164)-(166) to the form which depends only on the parameter m. The general solution of this equation has been found in [184] by means of numerical integration.

Let us analyze some special cases of the above problem. In the steady state the melting rate is

$$v_m = \frac{ds}{d\tau} = \frac{q_0}{\rho [L_m + c(T_m - T_0)]} \tag{170}$$

Taking Eq. (170) as the first approximation (neglecting the time of transition to the constant melting rate) we can find the thickness of the molten layer after a period τ:

$$s(\tau) \approx v_m (\tau - \tau_m) \tag{171}$$

and the steady-state temperature distribution

$$T \approx T_0 + (T_m - T_0)\,e^{-\frac{v_m}{a}[z - v(\tau - \tau_m)]} \tag{172}$$

Let us now analyze the solutions of the problems for the limiting values of m.

Case $m = 0$. Since m is proportional to the enthalpy of the material heated from T_0 to T_m divided by the melting heat L_m the case with $m = 0$ is not realized physically. However, this is the limiting case for the situations when the latent heat of melting or vaporization is much higher than the enthalpy of the material.

Using the dimensionless variables we can write the temperature distribution $v_0(x, y)$ and the melting rate in the following form:

$$\begin{aligned} v_0(x, y) &= \pi^{1/2}\, erfc\,(x/2y) - (x/\pi)\arctan(y^{-1/2}) \\ &+ [(y+1)/\pi^{1/2}]\exp[-x^2/4(y+1)]\, erfz/2[y(y+1)]^{1/2} \\ &+ 2xW\,(x/[2(y+1)]^{1/2},\, y^{1/2}) \end{aligned} \tag{173}$$

$$\mu_0(y) = (2/\pi)\ \arctan\,(y^{1/2})$$

where $x = \dfrac{z - s(\tau)}{\sqrt{a\tau_m}}$; $y = \dfrac{\tau}{\tau_m} - 1$; $v_0(x, y) = \dfrac{T - T_0}{\sqrt{\pi}\,(T_m - T_0)}$

$$W(\alpha, \beta) = (2\pi)^{-1}\int_0^{\alpha}\int_0^{\beta x}\exp[-(x^2 + y^2)/2]\,dy\,dx \tag{174}$$

At the initial stage of melting when $x^{1/2}/y$ is high we have

$$v_0(x, y) \approx (1 + y)^{1/2}\, ierfc\left[z/2(1 + y)^{1/2}\right] \tag{175}$$

A comparison of Eqs. (175) and (88) shows that at the beginning of melting or vaporization (y is small compared to 1) the temperature inside the body is close to the temperature which would be found without melting. This result found in the infinite melting heat model explains a good agreement between the predicted temperatures and the melting boundary positions derived from the linear equations for short-duration pulses (and finite melting heat) and the nonlinear predictions.

Case $m = \infty$. Let us consider another limiting case when $L_m = 0$. In the quasistationary (equilibrium) state we have

$$v_e = \pi^{1/2}\exp(-\pi^{1/2}z/2) \tag{176}$$

The position of the melting boundary (in the dimensionless variables) in the steady state is

$$\omega_e = \nu_e\left(1 + y - \frac{\pi}{4}\right) \tag{177}$$

and the melting rate is

$$\nu_e = \sqrt{\pi}/2 \tag{178}$$

where

$$\omega_e = m\sigma, \quad \sigma = \rho\frac{L_m}{q_0}\frac{s}{\tau_m}$$

$$\nu_e = m\mu_e, \quad \mu_e = \frac{\rho L_m}{q_0}\frac{ds}{d\tau}$$

When $m \neq 0$ the solution is found by numerical integration. The resulting plots are given in [184]. The kinetics of formation of the liquid phase under the effect of a point heat source (the phase transition heat being taken into account) has been analyzed in [81]. It can be said with some reservations related mainly to the small liquid-phase volumes that the process of formation of the liquid phase under the effect of laser heating is described by equations given in [81].

The temperature distributions $T_1(r, \tau)$ in the liquid phase and $T_2(r, \tau)$ in the solid phase are described by the system of equations

$$\Delta T_1 + \frac{F(r)}{\lambda_1}\delta(r) = \frac{1}{a_1}\frac{\partial T_1}{\partial \tau} \quad (0 \leqslant r \leqslant \rho(\tau))$$

$$\Delta T_2 = \frac{1}{a_2}\frac{\partial T_2}{\partial \tau} \quad (\rho(\tau) \leqslant r < \infty) \tag{179}$$

$$F(\tau) = \begin{cases} F & \tau \geqslant 0 \\ 0 & \tau < 0 \end{cases}$$

with the initial and boundary conditions

$$T_1[\rho(\tau), \tau] = T_2[\rho(\tau), \tau] = T_m$$

$$-\lambda_1\frac{\partial T_1}{\partial r} = -\lambda_2\frac{\partial T_2}{\partial r} + L_m\gamma\frac{d\rho}{d\tau} \tag{180}$$

for $r = \rho(\tau)$

$$T_2(\infty, \tau) = T_2(r, 0) = T_m$$

$$\rho(0) = 0$$

Here the subscript 1 denotes the liquid phase, the subscript 2 denotes the solid phase and γ is the material density. The system (179) with the boundary conditions (180) is a spherically symmetrical Stefan's problem. For the small time periods (the small radiuses of the molten zone) we have

$$L_m\gamma \frac{d\rho}{d\tau} \ll -\lambda_2 \frac{\partial T_2}{\partial r} \quad \text{for } r=\rho(\tau) \tag{181}$$

The energy balance equation for the molten zone yields

$$\rho(\tau) = \sqrt{\frac{3F_\tau}{2\pi L_m \gamma}} \tag{182}$$

It has been estimated [81] that at small times and low kinematic viscosity ν of the melt, for instance, in liquid Ni and Fe the viscosity is about 7×10^{-3} cm²/s [21], the Reynolds number $\mathrm{Re} \approx lv\nu^{-1} \gg 1$ (v is the characteristic velocity of the liquid phase). This means that in the liquid phase there is turbulent motion which results in rapid mixing and temperature equalization over the bulk of the melt. Thus we can assume that there is no superheating in the liquid phase, and the heat flux from the liquid phase can be expressed by the effect of the equivalent surface heat source with the intensity $F/2\pi\rho^2(\tau)$ distributed over the spherical interface; in this way Stefan's problem for two phases can be reduced to a one-phase problem with the boundary condition

$$\frac{F}{2\pi\rho^2(t)} = -\lambda_2 \frac{\partial T_2}{\partial r} + L_m\gamma \frac{d\rho(\tau)}{d\tau} \tag{183}$$

for $r = \rho(\tau)$

The problem with the boundary condition (183) has the following solution

$$T_2(r,\tau) = T_m - \frac{1}{r} \sum_{n=0}^{\infty} \frac{2}{a_2^n (2n+1)!}$$

$$\times \frac{\partial^n}{\partial \tau^n} \left\{ [r-\rho(\tau)]^{2n+1} \left(-\frac{L_m\gamma}{\lambda^2} \rho(\tau) \frac{d\rho(\tau)}{d\tau} + \frac{F}{2\pi\lambda_2\rho(\tau)} \right) \right\} \tag{184}$$

The function (184) can be used to derive equation for $\rho(\tau)$. If in Eq. (184) we leave only the zeroth term of the derivative series and tend r to infinity we find

$$\frac{d\rho}{d\tau} = \frac{\alpha}{\rho^2(t)} - \frac{\beta}{\rho(\tau)} \tag{185}$$

where $\alpha = F/2\pi L_m\gamma$, $\beta = \lambda_2(T_m - T_0)/L_m\gamma$.

Equation (185) shows that at $\rho = \alpha/\beta$ the process is at the stable steady-state stage when $d\rho/d\tau = 0$, and all the heat flux from the source is transported into the solid phase. Hence we obtain

$$\rho_c = \frac{\alpha}{\beta} = \frac{F}{2\pi\lambda_2(T_m - T_0)} \tag{186}$$

The substitution of the dimensionless variables $\xi = \beta^3\alpha^{-2}\tau$ and $X(\xi) = (\beta/\alpha)\rho(\tau)$ yields the following solution of Eq. (185):

$$X(\xi) = (3\xi)^{1/3} - \frac{1}{4}(3\xi)^{2/3} \tag{187}$$

Equation (187) without the second term is equivalent to Eq. (182); the second term is the contribution of the heat diffusion into the solid phase. The time of transition to the steady state is

$$\tau_c = 4\frac{\alpha^2}{\beta^3} = \frac{F^2 L_m \gamma}{\pi^2 \lambda_2^3 (T_m - T_0)^3} \tag{188}$$

The time τ_c depends on the melting heat $L_m\gamma$ apart from the power density and the thermal parameters. Equations (186) and (188) give the size of the molten zone and the time of its formation. Table 5 presents the estimates of ρ_c and τ_c for a number of metals [81]. As the heat source is assumed to be a point one the estimates are only qualitative.

TABLE 5. Estimates of Maximum Radius ρ_c of Molten Zone and Time of Transition τ_c to Steady State

Metal	Predicted			Experimental	
	$\beta=\lambda(T_m-T_0)/L_m\gamma$, cm²/s	$\tau_c=4a^2\beta^{-3}$, ms	$\rho_c=\alpha\beta^{-1}$, μm	ρ_c, μm	$\tau_c \approx l^2/2a_2$, ms
Al	1.28	10^{-2}	21	28	2×10^{-4}
Cu	2.21	7×10^{-4}	6	4.3 7.5 8.5	7×10^{-5}
Mo	1.38	7×10^{-3}	15	8	2×10^{-3}
W	0.88	7×10^{-3}	14	—	2×10^{-2}
Ni	0.43	0.1	42	16 42	4×10^{-2}
Steel	0.14	40	410	250	10

Note. Source power is 100 W, wavelength 0.69 μm, reflectivities: Al—0.8, Cu—0.85, W, Mo and steel—0.6, Ni—0.7 (Experimental results have been reduced to the source power of 100 W).

3.4. HEAT TREATMENT OF MATERIALS WITH LASER BEAMS

The surface hardening of some materials was found at the early stages of laser applications. For instance, steels have been hardened by radiation of the free-generation laser [28, 93, 138]. In the laser treatment experiments the microhardness of the laser-treated surface of steels У8, 45, ХВГ, etc. increases from 300 up to 1000 and sometimes 2200 kgf/mm². This explains the interest in laser application for heat treatment of steels and alloys, especially tool steels.

Both free-generation operation (with various temporal structures of the radiation pulses) and Q-switched operation of lasers can be used for local hardening of the material surface. Free-generation laser radiation with an intensity higher than the disintegration threshold $q_c^{(3)}$ [see Eq. (117)] produces cavities or holes whose walls are hardened by the effect of a short thermal pulse.

As a free-generation laser pulse consists of individual spikes with the duration of a few microseconds (both for the chaotic and ordered generation) the spike intensity can be sufficient for vaporization of the surface layer of the material; this layer can be melted or partially vaporized even if the mean pulse intensity is lower than the melting threshold.

The thermal analysis results can be used for estimating the sizes of the heat-affected zones and predicting the rates of heating and cooling.

Let us estimate the depth of the heated zone z_0 which can be defined as the depth at which the temperature is 5% of the surface value T_0:

$$T(z_0) = 0.05T(0) \tag{189}$$

where $T(0)$ is the temperature on the material surface.

If $T(0) = T_b$ the temperature $0.05T(0)$ is low enough to assume that no significant changes in structure and properties of the material occur at the depth z_0. Equations (88), (131) and (132) for the one-dimensional model of heating of a semi-infinite body by a heat source with q_0 = const yield

$$z_0 = 3/\alpha \quad \text{for } \alpha\sqrt{a\tau} \ll 1 \tag{190}$$

$$z_0 = 2.4\sqrt{a\tau} \quad \text{for } \alpha\sqrt{a\tau} \gg 1 \tag{191}$$

where α is the absorption coefficient.

Hence, for small-duration pulses the depth of the heated zone is determined by direct penetration of radiation, and when $\tau > > 10^{-5}$ s the depth increases owing to heat conduction which becomes the predominant factor at $\tau > 10^{-3}$ s.

When the material is heated by a laser pulse the heating of the internal layers proceeds after the end of the pulse owing to propagation of the thermal wave. In a one-dimensional model the maximum depth of the layer heated to a given temperature (for instance, the temperature of thermal hardening T_h) can be estimated from the following equation [19]:

$$z_{\max} = \sqrt{\frac{4at_0(t_0 - \tau)\ln\sqrt{t_0(t_0 - \tau)}}{\tau}} \tag{192}$$

Here $\delta = T_h/T(0, \tau) < 0.3$, t_0 is the time during which the isotherm reaches the maximum depth and $T(0, \tau)$ is the temperature of the surface.

The following expression is a good approximation of Eq. (192)

$$\frac{z_{\max}}{\sqrt{a\tau}} = \frac{1}{\sqrt{2e\delta}} \tag{193}$$

A giant laser pulse with the intensity lower than the threshold for disintegration will produce negligible thermal hardening as the maximum depth of heating is small ($z_{\max} < 1$ μm). If the laser pulse duration τ_i is about r_f^2/a the spatial distribution of the pulse intensity on the surface of the material has a significant effect on the development of the heated zone.

When the radiation intensity is higher than the disintegration threshold $q_c^{(3)}$ it produces a cavity in the material with hardened walls. The zone of hardening is adjacent to the molten zone.

If the depth of the cavity is considerably larger than its diameter the heat-affected zone can be determined from the model of a semi-infinite cylinder with the radius r_0 in the body with the radius r_c and with a constant-intensity heat source on its internal surface.

This problem can be readily solved in cylindrical coordinates by means of the Laplace transformation in τ [101]:

$$T(r, \tau) = \frac{q_0 r_0}{\lambda}\sqrt{\frac{a\tau}{r_0 r}} \times \sum_{n=0}^{\infty}\left\{2\,\mathrm{ierfc}\left[\frac{(2n+1)(r_c/r_0-1)-(r_c/r_0-r/r_0)}{2\sqrt{a\tau/r_0^2}}\right] + 2\,\mathrm{ierfc}\left[\frac{(2n+1)(r_c/r_0-1)+(r_c/r_0-r/r_0)}{2\sqrt{a\tau/r_0^2}}\right]\right\} \tag{194}$$

where r_0 is the radius of the cavity in the body.

When $r_c/r_0 \gg 1$ as is typical for most applications Eq. (194) can be approximated by

$$T(r, \tau) = \frac{2q_0 r_0}{\pi\lambda}\sqrt{\frac{a\tau}{r_0 r}} \tag{195}$$

For $r \approx r_0$, that is near the surface of the cavity, we have

$$T(r_0, \tau) = \frac{2q_0\sqrt{a\tau}}{\lambda\pi} \tag{196}$$

The estimates (195) and (196) yield the shape of the heat-affected zone and the size of the zone of hardening. For instance,

Eq. (195) yields for the laser pulse with the duration τ_i

$$r_h = \frac{4q_0^2 r_0 a \tau_i}{\pi^2 \lambda^2 T_h^2} \tag{197}$$

where T_h is the temperature of hardening, which depends on a number of factors.

Estimates (197) for steel agree with experimental data.

The main difficulty in making accurate estimates is due to the fact that the power density of the source is known with a poor accuracy. The heat source on the walls of the cavity is produced both by the laser beam (direct and reflected from the walls) and the vapour jet (vapour condensation, and convective and radiation heat transfer to the walls [34]). The intensity of the source on the walls of the cavity is definitely not higher than the intensity of the initial source which produces the cavity and is sufficiently high to melt the walls of the cavity.

Let us analyze the surface hardening of steels by high-intensity Q-switched laser pulses. A short laser pulse (about 10^{-8} s) with the intensity below the threshold $q_c^{(3)}$ will produce negligible hardening as the hardened layer will be very thin. Short pulses (about 10^{-9} s) with the intensity much higher than $q_c^{(3)}$ can alter the material structure mostly by the effect of the shock wave with the pressure at its front about 10^2-10^3 kbars or more rather than by the thermal effect.

For pulses about 10^{-7} s [24] a significant contribution is made by the thermal effects due to the heating of the surface by a plasma cloud [68]. The zone of hardening in steels has a depth of about 10 μm and the area about 0.03 cm² which is much larger than the area of the focussed spot. The depth of the heat diffusion $(a\tau_i)^{1/2}$ is about 1 μm.

The structural changes in the thin surface layer of steel about 10 μm thick are due only to the thermal effects of the laser pulse. A new iron phase has been found to be produced by the shock wave with the pressure over 130 kbars (the temperature being not very high) generated by explosion on the metal surface. This is described in [35] wherein the surface hardening of metals by explosions is studied.

However, a laser pulse of about 10^{-7} s generates a shock wave with the front pressure not much higher than 100 kbars. Therefore, the structural changes in the 10-μm layer cannot be determined by the shock wave pressure.

It has been estimated that the electron and radiative heat conduction cannot provide for significant heating of the bulk of the solid phase during the pulse. It can be assumed therefore that hardening of the 10-μm layer takes place mainly after the pulse has ended by propagation of the heated zone from the surface into the bulk of the material. The intensity of the surface heat source which produces this effect must be much lower than the average intensity of the laser pulse (about 10^{11} W/cm²) but then such a source must operate for a longer period. Hence, the main thermal effect can be caused not directly by the laser pulse but by the plasma cloud produced by expulsion of a thin surface layer from the condensed phase. The depth of the hardened zone for various z and τ can be found from the graphical solution of the temperature equation

$$T_h(0, z, \tau) = \frac{2q^*}{\lambda\sqrt{\pi}}\{\sqrt{a\tau}\exp(-z^2/4a\tau) - \sqrt{a(\tau-\tau_i)}\exp[-z^2/4a(\tau-\tau_i)] - z[erfc(z/2\sqrt{a\tau}) - erfc(z/2\sqrt{a(\tau-\tau_i)})]\} \quad (198)$$

where q^* is the power density on the surface of the condensed phase.

Equation (198) is in fact applicable in 10-15 ns after the beginning of the pulse when a plasma cloud is formed in the affected zone. This provides a qualitative explanation of the fact that the affected zone due to a single laser pulse is much larger than the focussed spot (the focussed spot radius r_f is about 80 μm and the radius of the affected zone is about 500 μm).

This experimentally observed fact has important implications for precision machining of materials with high-power laser beams. Laser beams with $q_0 > 10^9$ W/cm² generally give rise to secondary thermal effects of the plasma cloud, which result in extension of the heat-affected zone and, typically, a loss in machining precision.

The power density q^* can be found from the experimentally measured maximum depths of hardening. As estimated in Eq. (198) a laser pulse with τ_i about 10^{-7} s must have $q^* = 5 \times 10^6$ W/cm² to produce a 10-μm hardened layer in steel.

Differentiation of Eq. (198) yields the rate of temperature variation in the condensed phase

$$v = \frac{\partial T(0, z, \tau)}{\partial \tau} = \frac{q^*\sqrt{a}}{\lambda\sqrt{\pi}}[\tau^{-1/2}(1 - 0.064z^2/a\tau)\exp \times (-z^2/4a\tau) - (\tau-\tau_i)^{-1/2}\exp[-z^2/4a(\tau-\tau_i)] \times (1 - 0.064z^2/a(\tau-\tau_i))] \quad (199)$$

Shown in Fig. 25 is the temperature variation rate v plotted as a function of the time for $z = 10$ μm and the pulse duration $\tau_i = 10^{-7}$ s. At the depth of 10 μm the heating rate reaches its

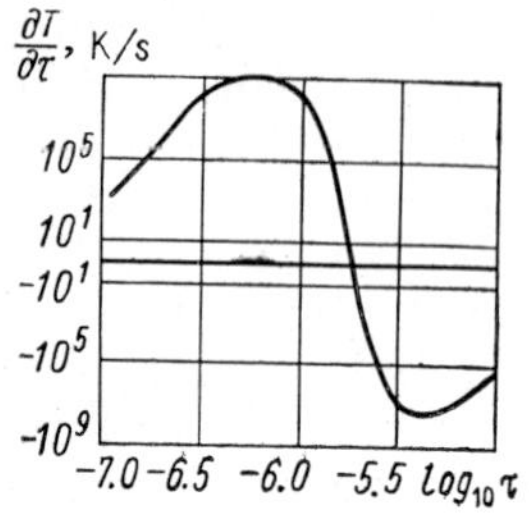

FIG. 25. Temperature variation rate as a function of time at depth of 10 μm in steel irradiated by a *Q*-switched laser pulse

maximum which is over 10^8 K/s in about 6×10^{-7} s. The maximum cooling rate about 10^7 K/s is obtained in 6×10^{-6} s after the beginning of the pulse. Such high cooling rate can result in the "freezing" of the structure produced by compression of the material with rapid heating since the contribution of diffusion is not high owing to the shortness of heating.

Let us estimate the power of the plasma radiation. The power of plasma radiation due to the free — free and free — bound transitions is [171]

$$P = (1 - \chi/T)\, 1.7 \times 10^{-32} n_e^2 Z^3 \sqrt{T/V} \tag{200}$$

where χ is the ionization potential, T is the mean plasma temperature, n_e is the electron density, Z is the multiplicity of ionization and V is the plasma volume.

The following equation [46] can be used for estimating the multiplicity of ionization in the plasma cloud with the temperature $T \approx 10^5$ K and the ion density 10^{19} cm^{-3}:

$$\chi\left(Z + \frac{1}{2}\right) = kT \ln\left(AT^{3/2}/Zn_e\right) \tag{201}$$

Here $A = 6 \times 10^{21}$ cm^{-3} eV$^{-3/2}$. Equation (201) is used in combination with the plot of the ionization potential as a function of Z for iron.

At $n_e = 10^{17}$-10^{19} cm^{-3} Eq. (201) yields $Z \approx 4$ for iron. Then Eq. (200) gives $P \approx 10^5$ W. If we assume that the plasma radiation intensity is uniformly distributed over the heated spot with the diameter about 0.5 mm we find that the power density of the heat source does not exceed 10^7 W/cm^2.

When the radiation power density is higher than 10^{11} W/cm^2 the energy absorbed in the thin surface layer is so high that the pressure at the front of the shock wave can exceed 130 kbars, and the thickness of the phase transition zone in iron can increase owing only to the shock compression of the iron.

The metallographic data agree with the theoretical predictions if we take into account the specific features of laser heating. For instance, hardening experiments have been carried out to determine the temperature profile in the laser-affected zone in steel [19] and to determine the spatial distribution of the heat source equivalent to the laser effect and the radiation absorption coefficient of steel averaged over the pulse duration [11].

These experimental techniques are based on the determination of the temperature in the bulk of the material produced by the laser pulse [11, 19] which is found from structure transformations. It is assumed that the temperature of the boundaries of phase transition and structure transformation are T_m and T_h, respectively, if the rates of heating and cooling are low. Higher rates of heating and cooling shift the critical points A_c and A_r in the iron alloys and other materials resulting in the change of the material volume undergoing phase transformations [61].

The shift of the critical points A_c and A_r depends apart from the rates of heating and cooling, on the following factors: chemical composition of the alloy, particle size in the material, microinhomogeneity in the distribution of some alloying elements (carbon, chromium, manganese, silicon), the starting dislocation structure, etc. For instance, practically no shift of the critical point A_{c3} is found in [49] for steel ШХ15 with the thin lamellar pearlite structure at the heating rates about 10^3 C/s. Pearlite spheroidizing in the steel of this grade results in the shift of this critical point. A similar trend of increasing shift of the critical point with "roughening" of the structure is found also for other materials, for instance, titanium alloys [32].

Let us estimate the shift of critical points in steels irradiated with laser beams taking into account nonuniform distribution of the radiation intensity over the surface, which results in differences between the heating rates at different points of the material surface. To do this we have to find the temperature field in the material during the pulse and variation in the heating rates at various points.

The heating rate in the bulk of the material is given by

$$v = \frac{q(r)}{\lambda\sqrt{\pi}}\left[\frac{1.06}{\sqrt{\tau}}\left(\sqrt{a} + \frac{z^2}{\tau\sqrt{a}}\right)\right]\exp(-z^2/4a\tau) \qquad (202)$$

Numerical calculations according to Eq. (202) have been made for steel ШХ15. The radiation power density $q(r)$ is assumed to have the gaussian distribution and the values of λ and a averaged over the temperature range from the normal level to the melting point are $\lambda = 0.28$ W/cm K and $a = 0.06$ cm^2/s. Equation (202) yields

$$v = 0.522q(r)\left(1 + \frac{z^2}{0.06\tau}\right)\exp(-z^2/0.24\tau) \qquad (203)$$

Figure 26 presents the results for various values of z and r at $k = 80$ cm^{-2}. The peak heating rate for a given r is at a certain distance from the heated surface which is given by

$$z_{max} = 0.424\sqrt{\tau} \qquad (204)$$

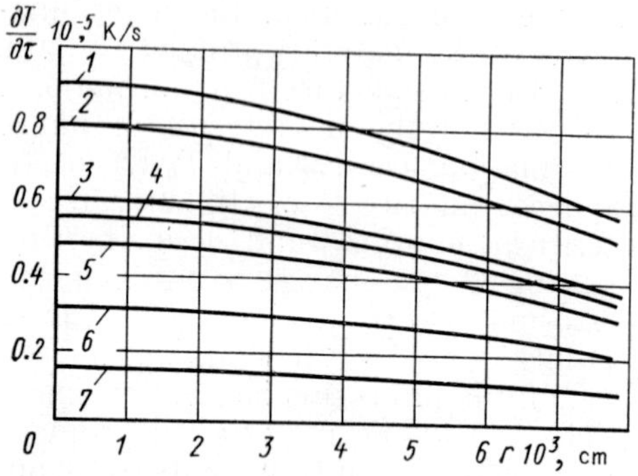

FIG. 26. Heating rate as a function of r at various values of z
1—3×10^{-3} cm; 2—2×10^{-3} cm; 3—10^{-3} cm; 4—5×10^{-3} cm; 5—0; 6—6×10^{-3} cm; 7—7×10^{-3} cm

To find z_{max} Eq. (203) is differentiated with respect to z, and the result is equated to zero. The point of the maximum heating rate shifts deeper into the bulk of the material with time according to Eq. (204). For $q_0 = 0.9 \times 10^5$ W/cm² the heating rate is close to 10^5 C/s.

Of the most interest is the variation of the heating rate in the affected zone and in the regions where the temperature can exceed A_{c1}. The plots show that in this zone the heating rate is comparatively low; it varies from 0.9×10^5 to 0.18×10^5 C/s with z increasing from 4×10^{-3} cm ($r=0$) to 7×10^{-3} cm ($r = 0.12$ cm).

To estimate the shifts of the critical points in steel we shall use the technique [87] based on the equations derived by Kolmogorov [69] for the kinetics of phase transformations, which express the relative fraction of the transformed material as a function of time (for the isothermal conditions). Hence the shift of the critical point is given by

$$T_x = \left(\frac{K^2x^2}{4\bar{D}}\right)^{1/3} v^{1/3} \tag{205}$$

where T_x is the temperature at which the diffusion front of austenite advances a distance x, K is the slope of the plot of relative concentration of Fe_3C as a function of $(x/2\sqrt{a\tau})$ for steel found from Fe—Fe_3C phase diagrams, v is the heating rate in the critical temperature range, $\bar{D}$ is the effective coefficient of carbon in austenite. The starting temperature for steels corresponds to the critical point $A_{c1} = 727$ C.

We have to find how well the results predicted in Eq. (205) agree with the actual shift of the critical points in the material heated by a laser beam when heating is not uniform (nonisothermal conditions), and the heating rate v varies with time and in the bulk of the material. It would be very difficult to take into account all the above factors, and Eq. (205) yields, generally, only qualitative estimates.

The parameter x^2 in Eq. (205) can be estimated [87] as

$$x^2 \approx 10/\rho_d \tag{206}$$

where ρ_d is the density of dislocations, cm^{-2}.

Equation (205) indicates that the temperature shift of the critical point depends significantly on the starting composition of steel. The higher the dislocation density, the smaller the shift of the critical point. Equation (205) yields $T_x = 200$ C if we substitute into it the following parameter values: $\bar{D} = 10^{-7}$ cm²/s, $v = 0.5 \times 10^5$ C/s, and $\rho_d = 10^{10}$ cm^{-2} (in the hardened steel [87]). According to the above discussion it becomes clear that it is not always reasonable to study the temperature fields and heat sources under conditions of rapid heating in steels of complex composition. Since the temperature of the critical point of phase transformation depends on the heating rate the calculations become more complicated and even inaccurate owing to the approximate character of Eq. (205). It seems more reasonable to carry out such studies in the materials in which the phase transition with heating or cooling is polymorphic.

The main applications of laser heating with the intensities below $q_c^{(1)}$ ($< 10^4$ W/cm²) are based on the fact that a temperature increase in the bulk of the material gives rise to the phase transition in the solid state so that the surface properties of the material are changed. If the laser radiation gives rise to a chemical reaction on the surface this produces new physical properties of the material.

If the material is heated to a temperature T in the range $T_f < T < T_m$, where T_f is the temperature of the solid-state phase transition, the required parameters of the laser pulse can be found from the above equations. The laser radiation with the above parameters is used mainly for surface hardening of the materials, stabilizing the parameters of films and controlling their properties, and recrystallization of semiconductor films.

Surface hardening. It is often needed to produce a zone with high mechanical properties on the surface of a machine component, for instance, on the cutting edge of a tool. This surface hardening can be done directly by laser heating or by alloying a thin molten layer on the surface of the material. Both techniques of local hardening have been tested with steels of various grades [56]. The tests are done with the quasistationary radiation of a ruby laser (the pulse energy being 20-22 J, the pulse duration about 10^{-3} s, the laser power density 10^5-10^6 W/cm²). Steel 45 used in the test is in the normalized state with the fine-grained ferrite-pearlite structure and hardness $H_{50} = 175$ kgf/mm².

The laser radiation pulse acting on a polished metal surface gives rise to a thin molten layer; the microhardness in this zone is different from the microhardness of the bulk metal. On the surface of this zone the microhardness is $H_{50} = 350$ kgf/mm^2 and it increases with depth up to 450 kgf/mm^2. In the heat-affected zone the microhardness increases up to 450-500 kgf/mm^2 and decreases in the bulk of the material to 158 kgf/mm^2 and then increases back to the initial value of 175 kgf/mm^2.

Laser alloying of steel 45 is done by applying to the steel surface a thin layer of the alloying fine powder mixture and water glass for cementing. The alloying mixture contains C, Mn, Co, Cr, Nb, Ni and Mo.

This process changes the chemical composition of the steel surface with the resulting change in the microhardness which is shown to be higher than in steel 45 heat-treated with the laser beam. The depth of the changed structure is estimated as a few tens of micrometres.

Stabilization of film parameters. Laser radiation can produce ageing of the film by changing its structure. As a result of laser radiation the electric conductivity, density and temperature resistance coefficient of the film approach the values for the bulk specimen. The experiments [142] have been carried out with 300-Å chromium films sputtered on glass substrates. The stability of the parameters of laser-treated films is two-three times that of the films conventionally treated by heating in vacuum ovens.

Recrystallization of films. Recrystallization of thin semiconductor and metal films on substrates of a foreign material can be used in film technology for transforming amorphous structure of thin films into mono- or polycrystalline structure. Thin germanium films sputtered on unheated structureless substrates are typical amorphous semiconductors. These films have lower densities than crystals, and the grain size in them is less than 100 Å. Such films typically have the hole conductivity with low carrier mobilities (100-200 cm^2/V s).

Recrystallization of germanium films produces a sharp increase in their conductivity. When the crystal size increases up to about 100 μm the semiconductor has the electron conductivity in case the starting germanium has the *n*-type conductivity [45].

All conventional techniques for treatment of semiconductor films consist essentially of heating and development of a certain temperature gradient giving rise to the controlled growth of single crystals.

The heat treatment is a prerequisite for transformation of the film microstructure; an important feature of this process is the heat exchange between the film and the substrate. In most techniques it is essential to maintain the film temperature near the melting point; it should be higher than the substrate temperature. It is practically unfeasible to obtain the controlled growth of single crystals over large areas of the film on a massive substrate by applying pulsed heating. The pulsed heating of "unsupported" films (a thin semiconductor film supported by a net) can produce large whiskers in it.

Laser radiation can give rise to crystal growth in the amorphous germanium film [45]. The germanium films are sputtered on cold and heated glass substrates in vacuum about 5×10^{-6} mm Hg. Then the films are washed off with water and put upon 50×50-μm support nets. A free-generation ruby laser is used with the pulse energy about 1 J.

The laser radiation focussed on the surface of thin germanium films gives rise to the growth of single crystals within the regions of the individual meshes of the support net. The length of the crystals is 15-20 μm.

The laser beam focussed on the carbon film [45] produces its annealling, which results in the growth of hexagonal graphite layers oriented mainly in the plane of the film and in elimination of microstresses.

3.5. THEORY OF LASER WELDING PROCESSES

Laser welding is especially useful when it is essential to limit the size of the heat-affected zone (for instance, for metal-semiconductor welding), to reduce the roughness of the welded surface and to eliminate mechanical effects.

Lasers are generally used for welding multilayer materials in which there are discontinuities in thermal properties at the interfaces where the layers come into contact.

The one-dimensional model of heating of a two-layer material has been discussed above [Eqs. (100) and (101)]. Estimates show that the temperature in the bulk of the material is highly dependent on the relationship between the thermal properties of the layers. Pulsed laser heating can be approximated by a one-dimensional model if the thickness of the upper layer which absorbs the radiation is small. The laser welding processes with pulses about 10^{-3} s and the heated spot radius r_f about 10^{-2} cm must be analyzed in the framework of the three-dimensional model since then $r_f \approx \sqrt{a\tau_i}$ (for $a \approx 0.1$-1 cm^2/s).

The following model is used for analysis [129]. The spatial and temporal structure of the heat source produced by the laser beam is assumed to have the time-independent intensity with the gaussian distribution over the surface of the material during the pulse duration. The thermal parameters of each of the materials in contact are assumed to be independent of temperature; the heat transfer from the heated surface outside is neglected. The contact between the layers can be ideal (the temperatures and heat fluxes at all the points are identical) and nonideal (there is a temperature discontinuity at the interface and the heat fluxes are identical). With these assumptions heating of a two-layer material with the ideal contact between the layers is described the following system of differential equations:

$$\frac{1}{a_1}\frac{\partial T_1}{\partial \tau}=\frac{\partial^2 T_1}{\partial r^2}+\frac{1}{r}\frac{\partial T_1}{\partial r}+\frac{\partial^2 T_1}{\partial z^2} \tag{207}$$

for $\tau > 0,\ r_0 \geqslant r \geqslant 0,\ h \geqslant z \geqslant 0$

$$\frac{1}{a_2}\frac{\partial T_2}{\partial \tau}=\frac{\partial^2 T_2}{\partial r^2}+\frac{1}{r}\frac{\partial T_2}{\partial r}+\frac{\partial^2 T_2}{\partial z^2}$$

for $\tau > 0,\ r_0 \geqslant r \geqslant 0,\ l \geqslant z \geqslant h$

The problem has the following boundary conditions:

$$z=0, \quad -\lambda_1\frac{\partial T_1}{\partial z}=q_0\exp(-kr^2)$$

$$z=h\begin{cases} T_1=T_2 \\ \lambda_1\dfrac{\partial T_1}{\partial z}=\lambda_2\dfrac{\partial T_2}{\partial z}\end{cases} \tag{208}$$

$$\begin{matrix} z=l \\ r=r_0,\ \tau=0 \end{matrix} \quad \begin{cases} T_2=0 \\ T_1=T_2=0 \end{cases}$$

where l and h are the thicknesses of the layers.

As the heat source produced by the laser beam is highly concentrated the boundary conditions at $r=r_0$ and $z=l$ for large specimens have a small effect on the temperature field in the zone of interest. For practical cases of laser welding the last condition in Eq. (208) must be changed into adiabatic condition.

Equation (207) with the boundary conditions (208) has the following solution [129]:

$$T_1(r,z,\tau)=\sum_{k=1}^{\infty}\left(C_{1k}e^{-s_k z}+C_{2k}e^{s_k z}\right)$$

$$\times J_0(s_k r)+\sum_{k=1}^{\infty}\sum_{n=1}^{\infty}D_{nk}J_0(s_k r)Z_{1nk}(z)\exp(-\mu_{nk}^2\tau) \tag{209}$$

$$T_2(r,z,\tau)=\sum_{k=1}^{\infty}\left(C_{3k}e^{-s_k z}+C_{4k}e^{s_k z}\right)J_0(s_k r)$$

$$+\sum_{k=1}^{\infty}\sum_{n=1}^{\infty}D_{nk}J_0(s_k r)Z_{2nk}(z)\exp(-\mu_{nk}^2\tau) \tag{210}$$

$$C_{1k}=R_k\left[(\lambda_1-\lambda_2)e^{(2h-l)s_k}-(\lambda_1+\lambda_2)e^{s_k l}\right]$$

$$C_{2k}=R_k\left[(\lambda_1+\lambda_2)e^{-s_k l}-(\lambda_1-\lambda_2)e^{(l-2h)s_k}\right] \tag{211}$$

$$C_{3k} = -2\lambda_1 R_k \exp(s_k l)$$

$$C_{4k} = 2\lambda_1 R_k \exp(-s_k l) \qquad (211\ cont.)$$

$$R_k = \frac{\Phi(s_k)}{\lambda_1 s_k \left\{\left[(\lambda_1 - \lambda_2)\,e^{2s_k h} - (\lambda_1 + \lambda_2)\right] e^{-s_k l} + \left[(\lambda_1 - \lambda_2)\,e^{-2s_k h} - (\lambda_1 + \lambda_2)\,e^{s_k l}\right]\right\}}$$

$$\Phi(s_k) = q_0 \frac{\exp\left(-s_k^2/4k\right)}{2k}$$

μ_{nk} are the roots of the transcendental equation:

$$-\lambda_1 \delta_{nk} \tan h\delta_{nk} = \lambda_2 \varepsilon_{nk} \cot(h - l)\,\varepsilon_{nk}$$

$$Z_{1nk} = \frac{\sin \varepsilon_{nk}(h - l)\cos \delta_{nk} Z}{\cos \varepsilon_{nk} h \cos \varepsilon_{nk} l}$$

$$Z_{2nk} = \frac{\sin \varepsilon_{nk}(z - l)}{\cos \varepsilon_{nk} l}$$

The parameter D_{nk} in Eqs. (209) and (210) is given by

$$D_{nk} = -\frac{\sin[\varepsilon_{nk}(h - l)]\,\Phi(s_k)}{\cos \delta_{nk} h \cos \varepsilon_{nk} l N_{nk} \mu_{nk}^2} \qquad (212)$$

where

$$\varepsilon_{nk} = \sqrt{\mu_{nk}^2/a_2 - s_n^2}, \quad \delta_{nk} = \sqrt{\mu_{nk}^2/a_1 - s_n^2}$$

and

$$N_{nk} = \frac{\sin^2 \varepsilon_{nk}(h - l)}{2\cos^2 \delta_{nk} h \cos^2 \varepsilon_{nk} l}\left[\frac{\lambda_1 h}{a_1} - \frac{\lambda_2 (h - l)}{a_2}\,\frac{\cos^2 \delta_{nk} h}{\sin^2 \varepsilon_{nk}(h - l)}\right]$$

The temperature field in the material is found for the moment corresponding to the end of the pulse $\tau_i = 8 \times 10^{-3}$ s. By this moment practically a steady-state temperature distribution exists in the affected zone.

Figure 27 presents the predicted temperature profile for the Al-Si system. A comparison is made with the predicted temperature profiles for a one-layer Al plate. The temperature in the Al plate decreases with depth faster than in the two-layer plate at the same q_0 (see isotherms *1* and *2* in Fig. 27).

Shown in Fig. 28 are the radial temperature profiles for one- and two-layer plates. The depth of the molten pool decreases in a one-layer plate and changes slightly in a two-layer plate with increasing size of the focussed spot (decreasing k) while the radiation power density remains constant.

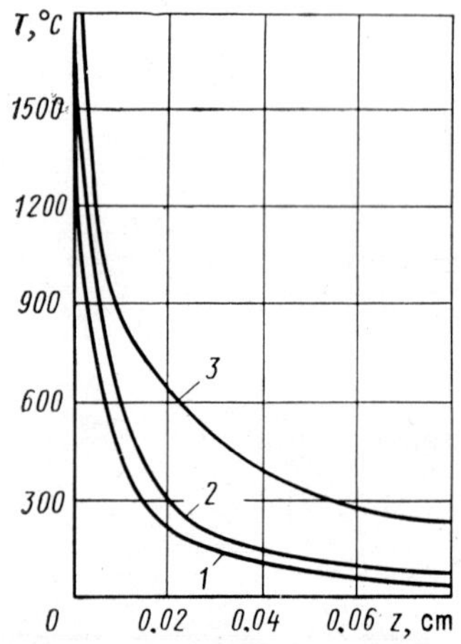

FIG. 27. Temperature distribution along axis ($r = 0$) in an aluminium plate and a two-layer Al-Si plate ($k = 3 \times 10^{-4}$ cm^{-2}, $\tau_i = 8 \times 10^{-3}$ s) 1—Al plate, q_1; 2—Al-Si plate, $q_1 = q_2$; 3—Al plate, $q_1 > q_2$

Let us now analyze heating of two-layer materials with short-duration pulses. Under such circumstances the temperature distribution in the two-layer plate is the product of the solution of one-dimensional heat conduction equation with the constant-intensity heat source and the spatial intensity distribution function as discussed above for the bulk heat sources.

For the normal-distribution circular heat sources the small-times approximation can be effectively used when

$$\tau < (20ak)^{-1} \tag{213}$$

For instance, for aluminium condition (213) is satisfied with pulse duration less than 10^{-5} s when $k = 10^4$ cm^{-2} ($r_f \approx 170$ μm).

Let us now analyze some features of laser heating of two-layer materials with nonideal contact. This problem differs from the ideal-contact problem by the boundary condition which, instead of Eq. (208), is

$$\begin{aligned} &\lambda_1 \frac{\partial T_1}{\partial z} = \lambda_2 \frac{\partial T_2}{\partial z} \\ &z = h \\ &-\lambda_1 \frac{\partial T_1}{\partial z} = \alpha_1 (T_1 - T_2) \end{aligned} \tag{214}$$

where α_1 is the reciprocal of the thermal resistance of the contact. When $\alpha_1 \to \infty$ the boundary condition (214) tends to the ideal-contact condition (208).

The solution of this problem is very complicated and we shall limit ourselves to analyzing the steady-state stage of the heating process. If the origin of the coordinates is at the interface the solution $T_1(r, z)$ can be written as [128]

$$T_1(r, z) = \frac{1}{\lambda_1} \int_0^\infty \frac{\bar{q}_s J_0(sr)\,[\lambda_1 (1 + \lambda_2 s/\alpha_1) \cosh(sz) - \lambda_2 \sinh(sz)]\, ds}{\lambda_1 \sinh(hs)(1 + \lambda_2 s/\alpha_1) + \lambda_2 \cosh(hs)}$$

where $-h \leqslant z \leqslant 0$ and

$$\bar{q}(s) = q_0 \frac{\exp(-s^2/4k)}{2k} \tag{215}$$

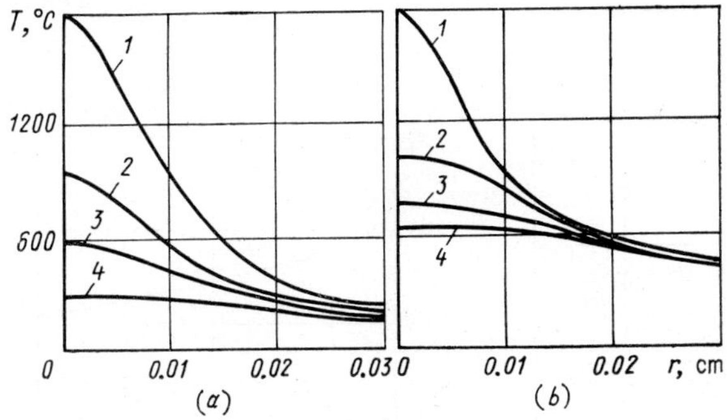

FIG. 28. Temperature distribution along axis z in an aluminium plate (a) and in a two-layer Al-Si plate (b)
$1 - z = 0$; $2 - z = 5 \times 10^{-3}$ cm; $3 - z = 10^{-2}$ cm; $4 - z = 2 \times 10^{-2}$ cm

To make estimates using Eq. (215) we have to know the parameter α_1 which is fairly difficult to find accurately in the case of laser welding. Attempts have been made [155] to predict the value of α_1. If the layers are not pressed together or the pressing force is not sufficient for plastic deformation of the microprojections on the surface we have

$$\alpha_1 \approx 2\lambda_m / h_{max} \tag{216}$$

where λ_m is the heat conductivity of the ambient medium, for instance, the air and h_{max} is the maximum height of the microprojections on the contacting surfaces.

If $\lambda_m = 0.58 \times 10^{-4}$ cal/cm² s C (the air) and $h_{max} = 2 \times \times 10^{-4}$ cm we find $\alpha_1 \approx \approx 1.2$ cal/cm² s C.

The predicted temperature variations with time for the two-layer Al-Si system (Al plate thickness is 100 μm and the heated spot diameter is 50 μm) are shown in Fig. 29 for the surface of an Al plate and in Fig. 30 for the contact of the layers with the pulse duration $\tau_i < 3 \times 10^{-3}$ s. Calculations are made with the above small-duration model. The value of the parameter α_1 has a marked effect on the heating process —

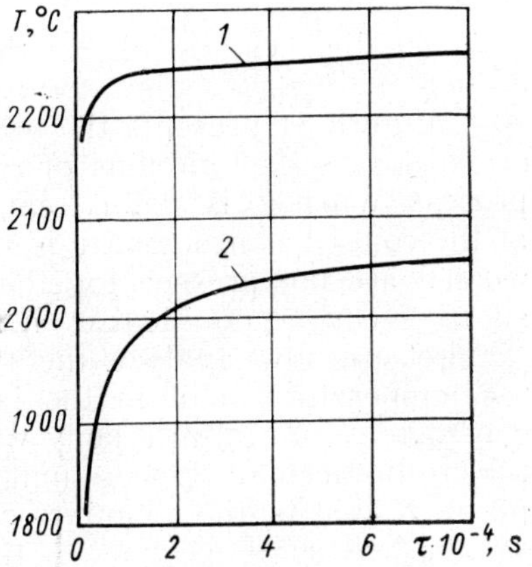

FIG. 29. Temperature on upper plate (Al) surface as a function of radiation time at $r = 0$
$1 - \alpha_1 = 1$ cal/cm² s C; $2 - \alpha_1 = 10$ cal/cm² s C

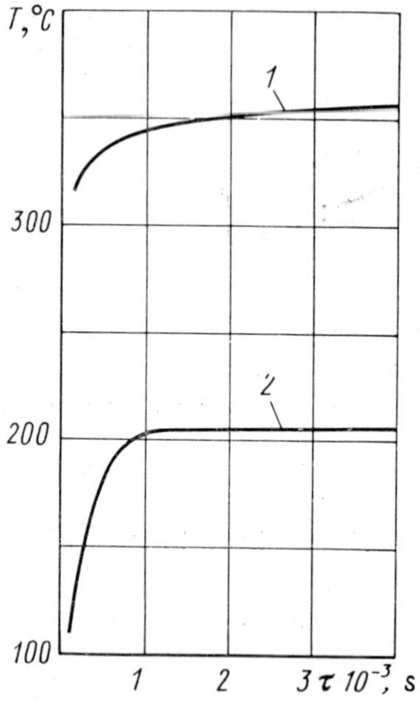

FIG. 30. Temperature at the contact in upper plate (Al) as a function of radiation time at $r = 0$

$1-\alpha_1 = 1$ cal/cm² s C; $2-\alpha_1 = 10$ cal/cm² s C

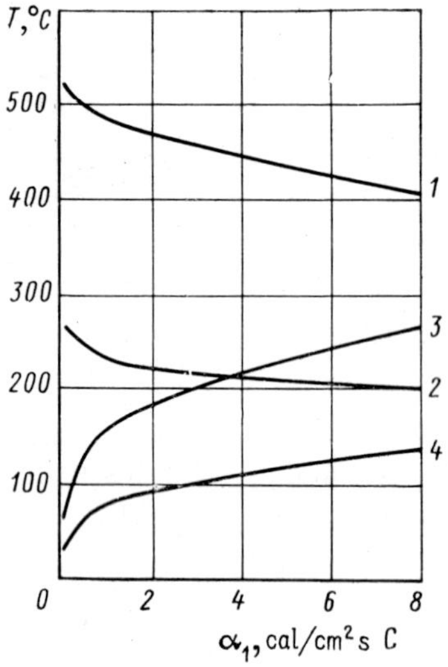

FIG. 31. Temperature at the contact in upper (Al) plate (T_1) and in lower (Si) plate (T_2) as a function of α_1

$1, 2-T_1$; $3, 4-T_2$; $1, 3-q_1 = 10^6$ W/cm²; $2, 4-q = 0.5 \times 10^6$ W/cm²

with smaller values of α_1 the steady-state stage is obtained earlier.

Figure 31 presents the predicted temperatures at the contact of the layers as functions of α_1 for the steady-state stage of the process. When α_1 is about 1 cal/cm² s C the temperature difference at the contact can be as high as 400 C, that is, if the plates being welded are not pressed together the thermal contact between the surfaces can be nonideal even for highly polished surfaces.

Let us now analyze the theoretical models for determining the temperature field in the laser-welded plates [83]. We shall consider a semi-infinite body whose surface heat transfer must be taken into account. Let us limit ourselves to considering the linear problem, that is, the boundary condition on the surface of the body will be written as

$$z = 0, \quad -\lambda \frac{\partial T_1}{\partial z} + \alpha_1 T_1 = q_0 \exp(-kr^2)\chi(\tau) \tag{217}$$

where α_1 is the surface heat-transfer coefficient.

The temperature distribution in the semi-infinite body with the zero initial temperature and the boundary condition (217) on the heated surface can be written as

$$T_1(r, z, \tau) = \frac{1}{4\lambda k \sqrt{a\pi}} \int_0^{\tau} \varphi(r, z, \tau, \xi)[1 - \psi(r, z, \xi)]\, d\xi \qquad (218)$$

where

$$\varphi(r, z, \tau, \xi) = \chi(\tau - \xi) \frac{\exp[-r^2/4a(\tau_0 + \xi) - z^2/4a\xi]}{\sqrt{\xi}\,(\tau_0 + \xi)} \qquad (219)$$

$$\psi(r, z, \xi) = \exp(z^2/4a\xi)\sqrt{\pi a\xi}\,\frac{\alpha_1}{\lambda}\exp\left(\frac{\alpha_1 z}{\lambda} + \frac{\alpha_1^2 a\xi}{\lambda^2}\right)$$
$$\times\, erfc\left(z/2\sqrt{a\xi} + \alpha_1\sqrt{a\xi}/\lambda\right) \qquad (220)$$

When $\psi(r, z, \xi) = 0$ Eq. (218) describes heating of a semi-infinite body without the heat transfer on the surface. In the centre of the focussed spot we have

$$\psi(0, 0, w) = w\sqrt{\pi}\exp(w^2)\, erfc(w) \qquad (221)$$

where $w = \alpha_1\sqrt{a\xi}/\lambda$.

Let us find when $\psi(0, 0, w)$ is different from zero, for instance, when

$$w\sqrt{\pi}\exp(w^2)\, erfc(w) = 0.1 \qquad (222)$$

The solution of the transcendental equation (222) is $w \approx \approx 0.061$. At the limiting stage of the process when $\tau_c \approx 10\, r_f^2/a$ we obtain [83]

$$\alpha_1 \approx \frac{0.02\lambda}{r_f} \qquad (223)$$

Equation (223) gives the estimate of the radius of the focussed spot whose heat transfer becomes significant. Assuming that the heat transfer at high temperatures about T_b is mainly due to the radiative heat transfer (heating in vacuum) we obtain

$$r_f = \frac{0.02\lambda}{\varepsilon\sigma_0 T_b^3} \qquad (224)$$

where ε is the degree of blackness of the body and σ_0 is the Stefan — Boltzmann constant.

In the materials with high heat conduction and comparatively low boiling point at normal pressure (copper, aluminium and other metals) the heat transfer from the laser-heated surface practically does not affect the temperature of the specimen while for such metals as W, Mo, Ti and steel heat transfer can have a certain effect on the specimen temperature if the laser pulse is sufficiently long. For titanium the critical radius r_f is 140 μm, for tungsten 200 μm, for molybdenum 400 μm, and for steel 500 μm in case of vacuum heating.

The following models are used for predicting the temperature field and estimating the required laser energy parameters for welding plates with the laser beam [83]: (1) semi-infinite body; (2) one-layer plate with various boundary conditions (adiabaticity, third-order condition from the back surface, isothermality); (3) two-layer plate with ideal or nonideal contact between the layers and one of the boundary conditions of Item 2 from the back surface. Other theoretical models can also be used for analyzing the laser welding processes.

The simplest approach to calculation of the temperature field is the model of a semi-infinite body with the instantaneous or continuous point heat source on the body surface [193]. In terms of this model it is impossible to estimate the temperature in the centre of the focussed spot and the minimum plate thickness for which this approximation is valid; hence, we shall not discuss this model.

The most universal approach is to make use of the two-layer plate model with the following boundary conditions: upper surface

$$z = 0, \; -\lambda_1 \frac{\partial T_1}{\partial z} + \alpha_1 T_1 = q(r, \tau) \tag{225}$$

interface

$$\lambda_1 \frac{\partial T_1}{\partial z} = \lambda_2 \frac{\partial T_2}{\partial z}$$
$$z = h_1 \tag{226}$$
$$-\lambda_1 \frac{\partial T_1}{\partial z} = \alpha_2 (T_1 - T_2)$$

back surface

$$z = h, \qquad -\lambda_2 \frac{\partial T_2}{\partial z} = \alpha_3 T_2 \tag{227}$$

where $h = h_1 + h_2$ is the total thickness of the plate and h_1 and h_2 are the thicknesses of the upper and lower layers, respectively.

Since the general solution of the system of differential equations for the temperature fields T_1 and T_2 with the zero initial condition and the boundary conditions (225-227) is too complicated we shall analyze simpler special cases of our model with the boundary conditions (225) and (226).

1. A semi-infinite body with the condition (225) and $\alpha_1 = 0$, that is

$$z = 0, \; -\lambda_1 \frac{\partial T_1}{\partial z} = q(r, \tau) \tag{228}$$

2. A one-layer plate with the condition (228) on the upper surface and one of the following conditions on the back surface ($z = h$):

$$T = 0 \tag{229}$$

$$\frac{\partial T}{\partial z} = 0 \tag{230}$$

$$-\lambda \frac{\partial T}{\partial z} = \alpha_3 T \tag{231}$$

3. A two-layer plate with the condition (228) on the upper surface and the nonideal contact (226) or ideal contact at the interface.

Let us define three characteristic times of the laser heating process: τ_1 is the time when the distribution of temperature on the surface is close to the heat source intensity distribution; τ_2 is the time required for obtaining the temperature T_b in the centre of the focussed spot; τ_3 is the time when the heating process becomes steady-state at the points in the body whose distance from the centre of the heated zone does not exceed the diameter of the focussed spot. The times τ_1, τ_2 and τ_3 are determined by the thermal properties of the material and the diameter of the focussed spot, and τ_2 additionally depends on the absorptivity of the material. The time τ_1 is found from the following condition [83]

$$\tau_1 \leqslant 1/20ak \approx 10^{-2} r_f^2/a \tag{232}$$

This estimate is valid both for a semi-infinite body and for a one-layer plate. When the pulse duration is less than τ_1 or equal to it the temperature field function is a product of the function of r and the function of z and τ found for a one-dimensional model with the uniform-intensity heat source. For a semi-infinite body the time τ_2 can be found from the following condition:

$$\tau_2 = \tau_0 \tan\left(\frac{\lambda \sqrt{\pi k}\, T_b}{q_0}\right) \tag{233}$$

Then the condition

$$\frac{\lambda \sqrt{\pi k}\, T_b}{q_0} \leqslant \frac{\pi}{2} \tag{234}$$

gives the minimum intensity $q_{0\,\min}$ providing for the temperature T_b on the heated surface

$$q_{0\,\min} \geqslant \frac{2.44 T_b \lambda}{r_f} \tag{235}$$

For instance, for copper with $r_f = 10^{-2}$ cm we find $q_{0\,\min} = 4.5 \times 10^5$ W/cm^2.

The time τ_3 gives the optimum laser pulse duration:

$$\tau_3 = \frac{10r_f^2}{a} \tag{236}$$

For instance, for copper with $r_f = 10^{-2}$ cm we find $\tau_3 = 1.2 \times 10^{-3}$ s. The estimate (236) can be used in other theoretical models.

The laser pulse duration τ_i is a contributing factor in selecting a model for analyzing laser heating. When $\tau \leqslant \tau_1$ the semi-infinite body model is applicable if the plate thickness h satisfies the condition $h \geqslant h_\infty$ where h_∞ is found from the following equation:

$$\frac{T(0, h_\infty, \tau_1)}{T(0, 0, \tau_1)} = \exp(-u^2) - \sqrt{\pi}\, erfc(u) = 0.05 \tag{237}$$

where $u = 9.6\, h/d_f$. Hence we find

$$h_\infty = 0.12 d_f \tag{238}$$

At $\tau \geqslant \tau_3$ the semi-infinite body model can be used if the plate thickness $h \geqslant H_\infty$ where H_∞ is found from

$$\frac{T(0, H_\infty, \infty)}{T(0, 0, \infty)} = \exp(v)\; erfc(\sqrt{v}) = 0.05 \tag{239}$$

where $v = H_\infty^2\, k$. Hence

$$H_\infty = 2.6 d_f \tag{240}$$

If the upper layer has the thickness less than H_∞ the heat transfer to the second layer must be taken into account. If the contact between the layers is almost ideal and the thickness of the upper layer is less than H_∞ then the model of the ideal contact must be used. If the total plate thickness $h = h_1 + h_2$ is larger than H_∞ we can assume $h = H_\infty$ with a sufficient accuracy. If $\tau > \tau_1$ we should define the time τ_4 when the thermal properties of the second layer do not affect the temperature field in the first layer, that is, the temperature field in the first layer is determined primarily by its thermal properties and the coefficient α_2.

The time τ_4 can be estimated in the following way [83]:

$$\tau_4 \leqslant \frac{\lambda_1^2 \times 10^{-2}}{\alpha_2^2 a_1} \tag{241}$$

For instance, for aluminium we find $\tau_4 \approx 9 \times 10^{-4}$ s. The estimate (241) shows that for small α_2 the time τ_4 is larger than τ_3 corresponding to the beginning of the steady-state stage of the heating process. This means that when the contact between the layers is poor the parameters of the second layer practically do not affect the temperature field in the first layer.

In laser applications for spot welding we should select the appropriate laser type and radiation parameters (pulse energy, pulse duration and the position of the focussed spot) and prepare the surfaces to be welded. Depending on the technological application the heat-affected zone can be subject to the following restrictions: (1) limited size of the heat-affected zone (for instance, in welding of metal foil and semiconductor, welding in the vicinity of a heat-sensitive component, etc.); (2) restriction of the material removal from the affected zone (this restriction is typically needed to obtain strong welds); (3) limited temperature gradient; (4) limited time of contact between solid and liquid phases to prevent the formation of intermetallic layers and embrittlement of the heat-affected zone; (5) sufficient depth of the molten pool without the removal of the material to obtain strong welds; (6) prevention of surface oxidation of the materials by using protective atmosphere or vacuum.

It is rather difficult to satisfy simultaneously all (or most) conditions discussed above. If the material is a solid we have to find the optimum pulse duration, the radiation power density and the spatial distribution of the power density. This problem can be considered as a special case of the problem of regulating a system with distributed parameters [15] but the general solution of this problem is difficult to find since not all the restrictions are compatible. For instance, an increase in power density can increase the mass of the removed material, apart from the depth of the molten pool.

The following conditions for the laser pulse parameters can be derived from the analysis of laser heating of materials.

Pulse duration. The optimum pulse duration range can be determined for each material and the depth of the molten zone, and within this range a weld is obtained without excessive removal of the material from the heated zone. The optimum pulse duration for a number of materials has been determined for a semi-infinite body model [94]. The optimum pulse duration ranges differ for various materials. For the higher pulse energies the optimum pulse durations are longer. For a given pulse energy there is a pulse duration at which the depth of the molten pool is the largest (Fig. 32). For shorter pulse durations there is not enough time for melting and for longer duration the pulse power density is not sufficient. Similarly, the optimum pulse duration can be found for the "pure" vaporization. To prevent the removal of the material the pulse duration must not be longer than τ_3.

Pulse energy. This determines essentially the welding process influencing the total amount of heat transferred to the affected

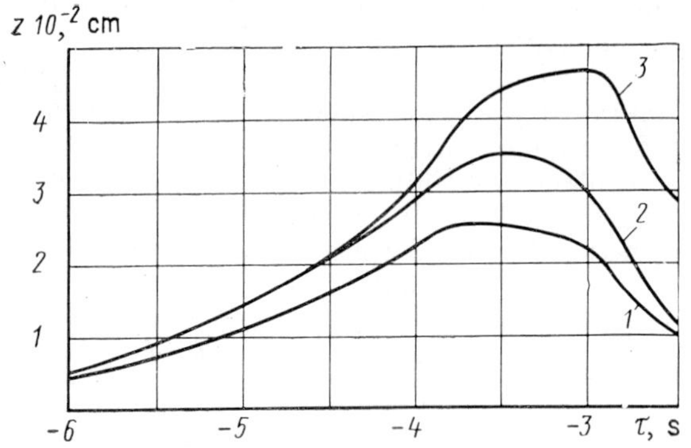

FIG. 32. Maximum melting depth as a function of the log of pulse duration for constant pulse energy
E—1.5 J; 1—W; 2—Cu; 3—Al

zone and some parameters of radiation (pulse duration, the structure and divergence of the pulse). Therefore the pulse energy is, as a rule, varied by means of filters or diaphragms.

Radiation power density. The required radiation power density can be calculated if we know the absorption coefficient for the given wavelength, the angle of incidence and the state of the absorbing surface. To select the variation ranges of the pulse energy and pulse duration, of the laser apparatus first of all, we have to estimate $q_c^{(1)}$. To do this we have to know the thermal and optical properties of the material and the radius r_f of the focussed spot. The requirements for the properties of the weld determine the choice of the radius r_f and the pulse duration τ_i. The selected τ_i value is compared to one of the characteristic times to choose a model for the thermal process. Then the optimum pulse duration and energy can be found. The predicted parameters are tested in experiments.

Given below is an estimate of the depth of the molten pool h along the axis of a semi-infinite body at the moment τ_b when the temperature at the surface becomes T_b [τ_b is given by Eq. (233)]:

$$h_{(cm)} \approx 3.54/T_m$$

More accurately, h can be found from the transcendental equation

$$hT_m = 3.54 \; erfc\left(h/2\sqrt{a\tau_b}\right) \tag{242}$$

Let us analyze the relationship between the depth of the molten pool and the duration of the laser pulse τ_i. We shall make use of the model of a semi-infinite body with the surface heat source with gaussian intensity distribution. The temperature distribution along the axis OZ can be found from Eq. (90):

$$T(0, z, \tau) = \frac{P_0}{4\pi\lambda\sqrt{\pi a}} \int_0^{\tau} \frac{\exp(-z^2/4a\theta)\, d\theta}{\sqrt{\theta}\,(\tau_0 + \theta)} \tag{243}$$

where P_0 is the effective radiation density assumed to be constant.

The following procedure can be used for making estimates from Eq. (243). The temperature is assumed to be T_m and the integral is calculated for various z and τ. Figures 33 and 34 present estimates for copper, aluminium, tungsten and steel at $k = 0.74 \times 10^6$ cm^{-2} ($r_f = 25$ μm) and with various pulse energies. The maximum pulse duration for which there is no melting

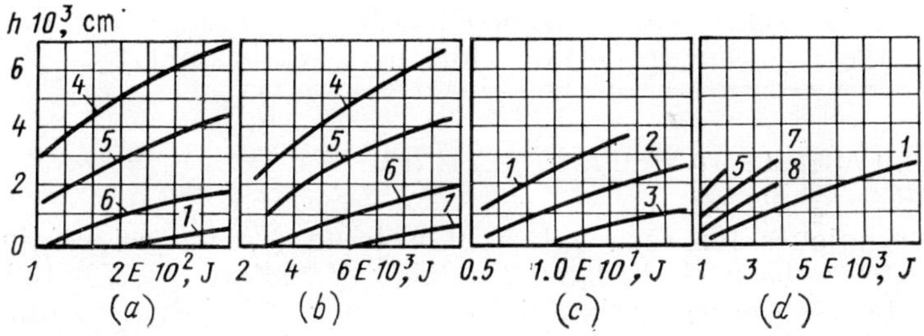

FIG. 33. Temperature in the centre of focussed spot on surface of irradiated material and depth of molten pool as functions of pulse duration at various pulse energies

(a) copper; 1—10^{-2} J; 2—2×10^{-2} J; 3—3×10^{-2} J; (b) aluminium; 1—2.5×10^{-3} J; 2—5×10^{-3} J; 3—10^{-2} J; (c) tungsten; 1—5×10^{-2} J; 2—10^{-1} J; 3—0.3 J; (d) stainless steel; 1—10^{-3} J; 2—2×10^{-3} J; 3—4×10^{-3} J; 4—10^{-2} J

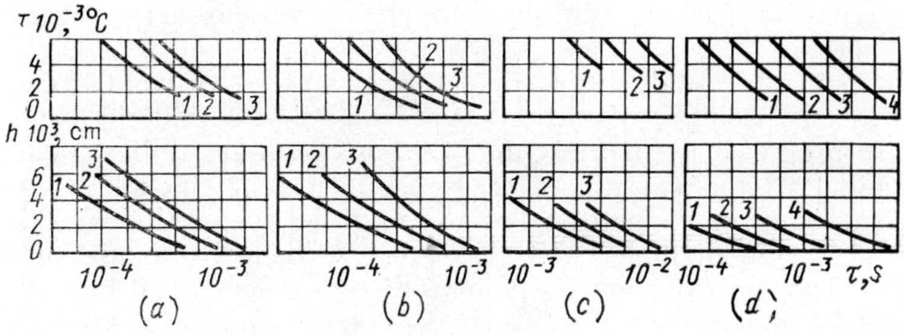

FIG. 34. Molten pool depth as a function of pulse energy of various pulse durations

(a) copper; (b) aluminium; (c) tungsten; (d) stainless steel; 1—10^{-3} s; 2—$2\times10^{-}$ s 3—4×10^{-3} s; 4—10^{-4} s; 5—2×10^{-4} s; 6—5×10^{-4} s; 7—4×10^{-4} s; 8—6.5×10^{-4} s increases with increasing pulse energy. If we attempt to increase the depth of the molten pool by increasing the power density through shortening the pulse duration we increase the vaporization of the material. Therefore for a given pulse energy the pulse duration should lie in a certain range. The optimum pulse duration range is $10^{-4} < \tau_i < 5 \times 10^{-4}$ s for copper, $5 \times 10^{-4} < \tau_i < 2 \times 10^{-3}$ s for aluminium and $5 \times 10^{-3} < \tau_i < 8 \times 10^{-3}$ s for steel, that is, each material has its own optimum pulse duration range.

Let us discuss briefly the ejection of the metal during laser welding. The metal ejection in laser welding is typically attributed to two factors, namely, the removal of the metal from the welding pool by the reactive pressure of the vapour owing to nonuniform distribution of radiation power density over the heated surface and ejection owing to the presence of impurities, dissolved gases, etc. in the bulk of the material. The removal of the liquid phase can be controlled by decreasing the power density and by varying the focussing conditions for the laser beam [76].

3.6. LASER SPOT WELDING TECHNIQUES

Laser techniques for such a typical welding process as welding of wires have been studied by a number of authors [5, 70, 139]. Let us discuss several pulsed laser techniques for wire welding.

Laser welding of wires is done using four configurations: butt welding, overlapping welding, crosswise welding and T-configuration welding. For butt welding the ends of two wires are contacted, and the laser beam is focussed on the contact producing melting and welding. For overlapping welding two wires are placed together, and the laser beam is focussed on their contact spot. For crosswise welding the wires are placed one upon another, and for T-configuration welding one wire is perpendicular to another.

Experimental analysis [139] of the parameters of welds produced by laser welding of various types has shown that the best welds are produced by overlapping welding. Similar studies have been made [26] for welding of crossed and parallel nickel wires 0.38 and 0.5 mm in diameters. The welding of two parallel nickel wires 0.38 mm in diameter produces good welds with no marked oxidation. The molten zone is about 0.18 mm deep and 0.45 mm in diameter. In 0.5-mm nickel wire the molten zone diameter is also about 0.45 mm, and the zone is 0.25 mm deep. The best melting conditions are obtained with the focussed spot size about 0.4 mm; a decrease in the spot size to 0.18 mm results in intense vaporization of the metal and formation of a cavity.

Laser welding of parallel wires is facilitated because multiple reflections increase the absorption of the beam energy by the metal. Therefore, welding of parallel wires takes less energy than welding of crossed wires.

The following main conditions must be satisfied in laser spot welding.

1. Highly accurate (radial) positioning of the component being welded. For instance, for an 0.025-cm wire the accuracy of positioning must be about 0.05 mm.

2. Accurate positioning of the component in the focus of the laser to prevent large variations in the power density. The variation in the area of the focussed spot due to the displacement l from the focus can be estimated as (for long-focal-length lens)

$$S' \approx S_f(1 \pm 2l/F) \tag{244}$$

where S_f is the area of the focussed spot and F is the focal length. Using Eq. (244) we can estimate the permissible variations in the laser power density which do not result in deterioration of the weld parameters.

3. Creating of a protective atmosphere in welding. If the materials to be welded are easily oxidized in the air welding must be done in protective, for instance, argon, atmosphere. Thus, it has been found [70] that argon protection of the welding zone increases the plasticity of the weld in laser welding of tungsten-rhenium thermocouples.

4. Symmetrical laser heating. For instance, for laser welding of two wires the laser beam must be focussed on the contact of the wires to improve the parameters of the weld.

When a wire is welded to a massive workpiece experiments show that melting of the massive body must be more advanced. To provide for a high-quality weld the massive body must be melted by about half the radius of the wire. If the wire is melted earlier the drop of the liquid metal rapidly travels along the metal surface hindering heat exchange and welding. Therefore, the following conditions must be satisfied in welding a massive body and a wire.

1. The time interval τ_{m1} before melting the massive body must be shorter than the time interval before melting the wire τ_{m2}:

$$\tau_{m1} < \tau_{m2} \tag{245}$$

2. The time interval τ_{b1} before the vaporization of the massive body must be longer than the laser pulse duration τ_i

$$\tau_{b1} > \tau_i \tag{246}$$

If we neglect the heat exchange between the wire and the massive body the temperature distributions in the massive body and in the wire can be found by analyzing independently the heating of a semi-infinite body and the heating of a rod by a constant heat flux. Using Eqs. (245) and (246) and the solutions of the above problems we can find the permissible radiation power densities and pulse durations for the welding of wire and massive body, which will vary in different ranges for each pair of metals to be welded. Different pulse durations permit different deviations of the radiation intensity; hence, the optimum pulse duration is the one which allows for the widest range of intensity variation. For instance, for the welding of the kovar and copper wires with the diameters of about 60 μm the optimum pulse duration is 1-3 ms and the permissible power density deviation is about 30% (hence, using Eq. (244) we can determine the permissible error in positioning the wire with respect to the laser focus). This operational regime provides for high reproducibility of the welding results, and the welds have a high tensile strength.

In welding two wires of different metals we must take into consideration the difference in their thermal properties. If the wire diameters are small and the contact area is close to the cross section of the wires the heating process can be analyzed in the framework of the one-dimensional model of two wires with ideal contact and heat absorption at the contact surface [70].

The temperature distribution in this model is given by

$$\frac{T_1(x, \tau) - T_0}{T_0} = \frac{2q_0\sqrt{a_1\tau}}{\lambda_1 T_0}\,\frac{K_\varepsilon}{1 + K_\varepsilon}\,ierfc\,\frac{x}{2\sqrt{a_1\tau}} \tag{247}$$

$$\frac{T_2(x, \tau) - T_0}{T_0} = \frac{2q_0\sqrt{a_2\tau}}{\lambda_2 T_0}\,\frac{1}{1 + K_\varepsilon}\,ierfc\,\frac{x}{2\sqrt{a_2\tau}} \tag{248}$$

where T_0 is the initial temperature and $K_\varepsilon = \sqrt{\frac{\lambda_1 c_1 \rho_1}{\lambda_2 c_2 \rho_2}}$.

The power density q_0 in Eqs. (247) and (248) can be defined as the radiation power averaged over the pulse duration divided by the cross-sectional area of the wires.

The laser power density can be determined from the condition that the temperature in the welding zone must be slightly higher than the boiling temperature. For instance, for welding of nichrome and brass wires the reflectivity averaged over the pulse duration is $R_1 \approx R_2 = 0.8$, and we find the required power density $q_0 \approx 1.35 \times 10^6$ W/cm².

Table 6 [70] presents the experimental results for ruby-laser welding of the tungsten-rhenium thermocouples 0.1 and 0.35 mm in diameters (free-generation laser operation).

In welding of 0.1-mm wires a bead 0.25-0.27 mm in diameter is formed at the end of the wire. The bead is asymmetrical with respect to the horizontal plane owing to the gravity effect.

TABLE 6. Welding of Tungsten and Rhenium Wires

Diameter, mm	Pulse energy, J	Focal distance, cm	Pulse duration, ms	Weld type	Note
0.1	1.12	18.8	3	Butt	Welding in air and in argon atmosphere
0.35	5.2	5.0	1	Parallel (overlapped)	

In welding of 0.35-mm wires the focal spot is somewhat displaced (by 0.1-0.2 of its diameter) towards the wire having a higher rhenium content. If no protective atmosphere is used when welding in the air the heat-affected zone becomes larger, which can be visualized from the annealling colours (its size is up to 1.3 mm for 0.1-mm wire and 1.0 mm for 0.35-mm wire). X-ray diffraction analysis results for the weld produced in the air show that laser pulses with a duration of 3 ms give rise to tungsten oxides in the welds. Therefore, welding is subsequently done in the inert-gas atmosphere (commercially pure argon) with the pressure of not more than 0.05-0.1 kgf/cm^2. Under such conditions the resulting thermocouple bead has a bright nonoxidized surface, and the wires are uniformly welded into it. The heat-affected zone is smaller by a factor of 3-4, and its size is not more than 0.22-0.25 mm. The protective argon atmosphere increases the strength of the weld by 20-25% and the bending angle by 40-45° compared to the welds produced in the air.

Let us consider the welding of a wire to a film [76]. The period before melting of the film material τ_{m3} must be longer than the period before melting of the wire material τ_{m2} with the same radiation power density:

$$\tau_{m3} > \tau_{m2} \tag{249}$$

The pulse duration τ_i must be longer than the melting time τ_{m2} for the wire and shorter than the melting time τ_{m3} for the film:

$$\tau_{m3} > \tau_i > \tau_{m2} \tag{250}$$

Analysis of heat conduction yields the following forms of the conditions (249) and (250):

$$\frac{1}{R_2} \frac{T_{m3}^2 \lambda_3^2 a_2}{T_{m2} \lambda_2 a_3} > q_0 \frac{A_3^2}{A^2} \tag{251}$$

$$\frac{T_{m3}^2 \lambda_3^2}{a_3} \frac{1}{q_0^2 A_3^2} > \tau_i > \frac{T_{m2} \lambda_2}{a_2} \frac{R_2}{q_0 A_2} \tag{252}$$

where R_2 is the reflectivity for the wire, A_2 is the absorption capacity of the wire and A_3 is the absorption capacity of the film.

Equation (251) gives the maximum permissible power density which would not damage the film, and Eq. (252) gives the permissible range of variations in the pulse duration for a given pair of materials. The welding process is more stable when longer laser pulses are used because in the latter case variations in the power density are smaller than in the case of shorter pulses.

Let us consider some features of laser welding of thin plates. Generally, thin plates are joined by direct welding of the surfaces or by coating the surfaces with low-melting metal and subsequent welding. Application of a low-melting metal reduces the effect of the instability of thermal resistance at the contact [128]. The thickness of the low-melting metal coating can be determined by analyzing the temperature distribution in the two-layer plate model. Two factors affect the optimum thickness of the low-melting metal layer; on the one hand, this layer must be sufficiently thick to eliminate the adverse effect of the variation in the gap width on the welding process; on the other hand, the layer thickness must be limited to prevent critical dependence of the welding process on the variation in the power density.

Let us analyze the kinetics of weld formation in laser pulse welding. This can be conveniently done by high-speed filming of the heat-affected zone with the laser power density of 10^4-10^6 W/cm^2 [5]. A high-speed camera is used to study the laser-affected zones in tungsten-rhenium alloys and platinum (Fig. 35). In some experiments a mirror filter is used to cut off the long-wavelength radiation from the affected zone.

As the speed of filming is very high the welding process cannot be recorded for one specimen from the beginning of the pulse and ending with weld formation. Therefore, successive stages of the process are recorded in 3-4 identical specimens while welding 0.1- and 0.35-mm wires of tungsten-rhenium alloys and

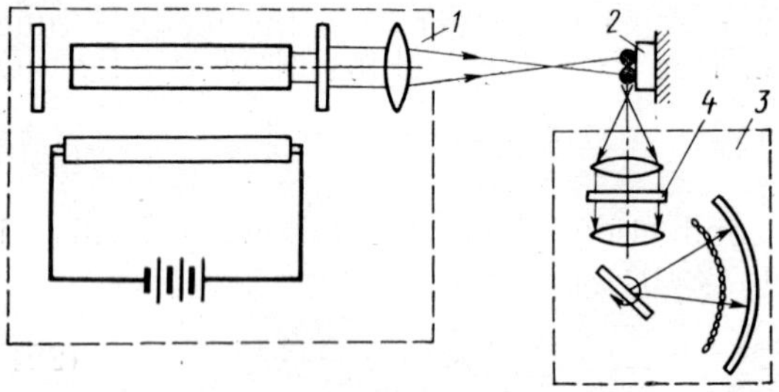

FIG. 35. Welding of W-Re wires
1—ruby laser; *2*—wires; *3*—high-speed camera; *4*—filter (100% reflection mirror for radiation with $\lambda = 6943$ Å)

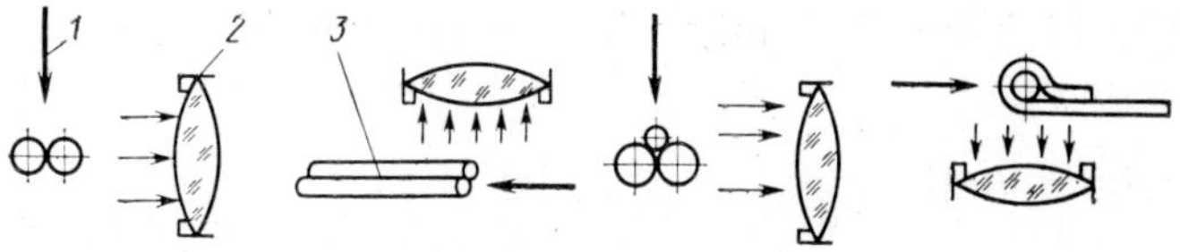

FIG. 36. Various types of welds, orientation of laser radiation and arrangement of recording of welding of W-Re wires
1—laser beam; 2—high-speed camera; 3—wires

welding such wires to the 32-μm platinum foil. Figure 36 shows the types of welds, orientation of laser radiation and the arrangement of recording of welding with a high-speed camera.

The experimental results show that melting of the welded metals starts 50-200 μs after the beginning of the laser heating; the subsequent processes are the formation of the liquid phase, vaporization and ejection of the molten particles. Variations in the size of these particles can be seen from their tracks. Though the laser heating is aimed at melting ($q_0 \approx 10^5$ W/cm^2) and the velocity of ejection of particles is negligible their trajectories are straight. This fact suggests, similar to other results [74], that the contribution of gravity is insignificant. The particles are ejected into the solid angle 2π and can be ejected parallel to the incident beam. The velocity of the ejected particles depends on their mass; it varies in the 5-10 m/s range. The factor responsible for ejection of particles is the spike structure of the laser pulse. The photographic evidence suggests also that the ejection of the liquid phase particles can be due to a high gas concentration in the melt [119].

Experiments show that when the laser power is close to the minimum required for welding the weld can be formed only if there is no gap between the wires being welded (the gap width varying from 0.1 to 1.0 of the wire diameter). The laser radiation of optimum power produces welds of sufficient strength even when there is a gap between the wires (which can be as wide as the wire diameter). Experiments are done with shifting the focussed spot from the ends of the wire by 0.5-3 diameters of the wire; for the weld to be of a satisfactory strength the displacement of the focussed spot cannot be more than 1.5-2.0 diameters. With a larger displacement the welded bead can break off from the wires. Experiments show that a high-quality weld can be produced even if the ends of the wires are not aligned and the radiation is focussed on both ends. The films show that both ends are gradually trimmed and aligned by melting, and finally a regular welded bead is formed.

In laser welding of large-diameter wires fusion can be incomplete resulting in a low strength of the weld. Therefore, welding of two 0.35-mm W-Re wires is done with the addition of 0.1-mm

W-Re wire. The high-speed filming shows that a weld is formed when the gap between 0.35-mm wires is filled with the molten added metal.

Welding of a platinum wire to a platinum foil by the laser pulse with $q_0 = 1.4 \times 10^3$ W/cm^2 is accompanied by boiling of the melt and accomplished in 500-800 μs.

High-speed filming shows that radiation drives the liquid phase to the edges of the pool so that a cavity is formed in the centre of the molten pool. A decrease in radiation power density results in a levelling off of the molten metal and its subsequent bulging in the centre of the pool. No ejection of the molten metal is observed with the above laser pulse parameters. The welding of metal-ceramic alloys is accompanied by ejection of drops of the liquid metal from the melting zone owing to a high concentration of dissolved gases in the bulk of the alloy (up to 28 mg/g for W-Re alloys).

Let us consider deep-penetration or "keyhole" laser welding. In the process of deep-penetration melting the laser pulse produces a sort of a "keyhole" in the material through which the radiation energy is absorbed by the material. The processes of deep-penetration laser melting are similar to those in melting of metals with the electron beam at the power density about 10^6 W/cm^2 [48, 124].

To produce deep-penetration melting the power density of the source must be higher than the critical value $q_c^{(2)}$ corresponding to the beginning of disintegration of materials. The critical value $q_c^{(2)}$ has been estimated to be smaller than the critical value $q_c^{(3)}$ corresponding to significant vaporization by the end of the pulse. If the radiation power density is higher than $q_c^{(2)}$ but lower than $q_c^{(3)}$

$$q_c^{(2)} < q_0 < q_c^{(3)} \tag{253}$$

then deep-penetration melting is feasible.

The following qualitative description of the deep-penetration melting can be given. Absorption of radiation by the material surface results in its heating, formation of the melting zone and beginning of vaporization from the surface. The vapour pressure which is sharply nonuniform pushes out a part of the liquid phase, and the heat source moves inside the material. The heat from the source is transferred to the deeper material layers. When the laser pulse ends the keyhole is filled mostly by the liquid metal (on its walls and on the heated surface) which then solidifies.

When the heat source is moving it continuously produces a keyhole in the liquid phase owing to the vapour pressure effect on the molten pool surface.

Let us find the shape and size of the cavity in the molten pool when the surface heat source has the gaussian intensity distribution [114].

The problem can be described by the system of equations of motion for the melt:

$$\frac{\partial \vec{v}}{\partial \tau} + (\vec{v}\nabla)\vec{v} = -\frac{1}{\rho}\nabla p + \nu\Delta\vec{v} + \vec{g} + \vec{F}_S \qquad (254)$$

where $\vec{v}$ is the velocity of liquid particles, p is the liquid phase pressure, ν is the kinematic viscosity, $\vec{g}$ is the bulk force vector and $\vec{F}_s$ is the surface force vector.

Equation (254) and equations of continuity, conservation of energy and conservation of mass give a full description of deformation of the molten pool surface. As these equations are obviously too complex we shall limit ourselves to solving them for $\vec{v} = \text{const}$. If we take the components of the resulting simplified Eq. (254) along the normal to the liquid phase surface we obtain for the profile $Z(r)$ of the deformed pool surface (for axially symmetrical pool):

$$\frac{Z''}{(1+Z'^2)^{3/2}} + \frac{Z'}{r(1+Z'^2)^{1/2}} - \frac{Z}{a_k^2} = \frac{F_S(Z, r)}{\sigma} \qquad (255)$$

where $a_k = (2\sigma/\rho g)^{1/2}$ is the capillary constant, F_s is the surface pressure and σ is the surface tension. Equation (255) does not take into account the temperature variation of the parameters in it. Numerical technique can be used to find the solution of Eq. (255) with a given dependence of the surface pressure on Z and r.

We shall consider below the limiting cases of Eq. (255) which can be solved in the analytical form.

1. Small deformations of the surface correspond to weak evaporation from the surface. In this case $Z'^2 \ll 1$. Then Eq. (255) can be written as

$$Z'' + \frac{1}{r}Z' - \frac{1}{a_k^2}Z = \frac{F_S(Z, r)}{\sigma} \qquad (256)$$

Generally, the surface forces $\vec{F}_s$ depend on Z and r for the deformed surfaces. If the surface forces are produced by the reactive pressure owing to vaporization the function $F_s(Z, r)$ can be written as

$$F_s(Z, r) = \frac{A(Z, r)}{T^{1/2}(Z, r)} \exp[-B/T(Z, r)] \qquad (257)$$

where B is a constant, $A(Z, r)$ is determined by the properties of the material and deformation of the surface and $T(Z, r)$ is the temperature distribution over the deformed surface of the molten pool.

The solution of Eq. (256) for an arbitrary form of F_s is

$$Z(r) = C_1 Z_0(r) + C_2 Z_0(r) \int \frac{dr}{r Z_0^2(r)} + Z_0^2(r) \int \frac{1}{r Z_0^2(r)} \left(\int r Z_0(r) \frac{F_s(r)}{\sigma} dr \right) dr \tag{258}$$

where integration is over the total range of variation of r; C_1 and C_2 are constants and $Z_0(r)$ is the solution of the homogeneous equation (256)

$$Z_0(r) = J_0(ir/a_k) \tag{259}$$

If the function F_s is approximated by a constant independent of r Eq. (256) is transformed into

$$Z(r) = \begin{cases} -\dfrac{F_s}{g\rho} & \text{at } r < |\delta| \\ 0 & \text{at } r > |\delta| \end{cases} \tag{260}$$

where δ is the radius of the area where the force F_s is effective, δ being smaller than the radius of the molten zone.

Since the resulting deformation of the surface should also have the stepwise character which is not realized in practice because derivatives of Z with respect to r must be continuous and $Z'^2 \ll 1$, we shall write the solution of Eq. (256) in the following form:

$$Z(r) = -\frac{F_s}{g\rho} + C_1 \left[1 + (r/2a_k)^2 + \frac{1}{4}(r/2a_k)^4 + \ldots \right] \tag{261}$$

where C_1 is a constant found from the boundary conditions.

When $r < 2a_k$ (for typical metals $a_k \approx 0.5$ cm) we can assume $\frac{1}{4}(r/2a_k)^4 \ll 1$ and retain only two terms of the series in Eq. (261). Then we obtain

$$Z(r) = -h + (F_s/g\rho - h)(r/2a_k)^2 \tag{262}$$

where the depth of deformation

$$h = \frac{F_s}{g\rho} \frac{\mu^2}{1 + \mu^2} \tag{263}$$

and $\mu = \delta/2a$. For $F_s = 10^3$ dyne/cm², $\rho = 10$ g/cm³ and $\mu = 0.5$ we find $h \approx 0.02$ cm.

Let us make a more real approximation of F_s writing it in the following form:

$$F_s = F_0 (1 - r^2/\delta^2) \exp(-kr^2) \tag{264}$$

where F_0 is the maximum pressure and δ is the effective size of the zone where the pressure is different from zero; F_s is assumed to be independent of Z.

For F_s given by Eq. (264) Eq. (256) has the following exact solution:

$$Z(r) = -\frac{F_0 \exp(-kr^2)}{4\rho g a_k^2 \delta^2 k^2} \tag{265}$$

at $r < \delta$ and

$$k = \frac{1}{2\delta^2}\left(1 - \sqrt{1 - \delta^2/a_k^2}\right) \tag{266}$$

Equation (266) shows that no steady-state solution exists for $\delta > a_k$ since k becomes a complex quantity. The physical meaning of this is that the capillary phenomena become insignificant for large deformation zones. When $\delta = a_k$ the deformation of the molten pool surface is

$$Z = -\frac{F_0}{g\rho} \tag{267}$$

2. Large deformations of the surface correspond to intense surface vaporization of the material and we assume $Z'^2 \gg 1$.

Then Eq. (255) yields

$$Z(r) = -a_k^2\left(\frac{F_s(Z, r)}{\sigma} - \frac{1}{r}\right) \tag{268}$$

Equation (268) gives implicitly the shape of the molten pool surface as a function of r. The maximum depth of deformation is determined from Eq. (268) with an accuracy of about $1/Z'^2$, and it can be used for estimating the melting depth. The limiting depth of the keyhole h (for $r = \delta$) can be found from

$$\sigma^{-1} a^2 F_s(h, \delta) + h - a_k^2/\delta = 0 \tag{269}$$

To find h from Eq. (269) we have to know the function $F_s(h, \delta)$, that is, the character of the dissipation of the beam energy in the metal vapour in the keyhole. Since $F(h, \delta)$ decreases with h Eq. (269) has the unique solution.

The value of deformation of the molten pool surface can be used for estimating the depth of the heat source. For deep-penetration melting ($Z'^2 \gg 1$) the internal surface of the keyhole can be regarded in the first approximation as a linear heat source in the bulk of the body. With small deformation ($Z'^2 \ll 1$) of the surface the heat source is distributed over a surface area of a large curvature radius. In this case the shape of the molten zone is different from that produced by a heat source which does not distort the surface of the molten pool.

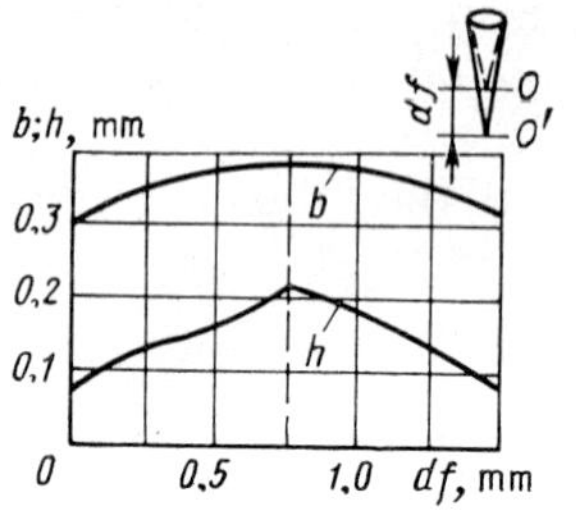

FIG. 37. Variation in width b and depth h of molten pool produced by laser pulse in steel (У8А) plate as a function of position of focussed spot

O—focal plane of focussing system; O'—focussed spot of laser beam

Now we shall mention some features of the seam welding by pulsed laser radiation. Laser butt welding of sheets using pulsed radiation with a high pulse repetition rate is, in practice, spot welding with the spots overlapping by, for instance, their radius. In pulsed laser welding the thickness of the metal sheet is limited by the energy and duration of the pulse; the permissible thickness of the sheet to be welded increases with the power and duration of the laser pulse [94, 174]. The pulse energy needed for welding of sheets must exceed the pulse energy for welding of wires of the same size and material by 50%. To obtain a complete-penetration weld the spot welds must overlap to a large extent (the distance between the centres of individual spot welds must not exceed the diameter of the molten zone from the back side of the sheet). A square-butt seam can be produced only by welding on both sides of the sheet.

Let us consider some technological features of pulsed laser welding. The energy incident on the surface of the workpiece depends on the losses in the optical system which shapes the laser beam (defocussing, diaphragming, and filtering) and the conditions of focussing. The output laser energy E_1 minus losses in the optical system is given by [4]:

$$E_1 = E(1-r)^n(1-\alpha)^m \tag{270}$$

where m is the number of optical elements, r is the reflectivity for glass ($r = 0.045$), n is the number of the glass surfaces without the mirror reflection surfaces, and α is the absorption coefficient of glass which depends on its grade (for instance, for the glass of grade 0 with the thickness of 1 cm $\alpha = 0.006$).

The reflection and absorption light losses predicted from Eq. (270) are in a good agreement with the calorimetric results [4] and amount to 10-12%.

Measurements of the reflectivity for steel with surfaces of varying roughness [4] irradiated by the ruby laser beam have shown that the reflectivity decreases with decreasing angle of incidence. Intense mirror reflection occurs when the roughness of the surface is comparable to the wavelength of the laser ($\lambda = 0.694$ μm); then it steadily decreases with increasing roughness of the surface. With the surface roughness of about 0.694 μm the diffuse reflection is slight (about 5.5%); it increases to the maximum at roughness of 2.4 μm. Then the reflectivity is

$$R = R_1 R_2 \tag{271}$$

where R_1 is the mirror reflectivity and R_2 is the diffuse reflectivity.

From Eqs. (270) and (271) we can find the energy absorbed by the material:

$$E_2 = E(1-r)^n (1-\alpha)^m (1-R) \tag{272}$$

If the system includes filters a term must be added to Eq. (272) to take into account the transmittance of the filters. The effect of diaphragms decreasing the radiation energy can be evaluated by introducing the diaphragming factor ω:

$$\omega = (D_d/D_l)^2 \tag{273}$$

where D_d is the diaphragm diameter and D_l is the diameter of the laser active element.

For instance, at $D_l = 6.3$ mm the 3-mm diaphragm reduces the radiation energy by a factor of 2.2. But diaphragming changes all the parameters of radiation including the pulse duration (slightly) as the generation processes are not simultaneous in the laser active medium, and the laser beam is not uniform.

The radiation power density on the surface of the material can be controlled by defocussing the beam, that is, by drawing the workpiece nearer or farther from the objective of the optical system. The radiation power density of the heated spot decreases with defocussing, and the size of the molten zone is changed. Figure 37 presents variation in the width and depth of the molten zone in steel plates (steel У8A 2.6 mm thick initially) produced by laser pulses with the duration $\tau_i = 2.2$ ms (the focal distance $F = 20$ mm). The depth of the molten zone can be seen to decrease when the workpiece is moved away from the focussed spot.

The quality of welds can be considerably improved by preparation of the contact surfaces. It has been mentioned above that nonideality of the contact has a significant role in welding of plates of different materials. The surfaces for welding must be degreased and cleaned. However, insulated wires can be welded without removing insulation; this is an advantage of welding compared to soldering.

4
LASER DRILLING

Laser machining, in particular drilling, of materials is now widely used in technology. Laser drilling was one of the first practical applications of laser technology and there is a high demand for effective techniques for microhole drilling.

In industry laser drilling is widely used for rough drilling, for instance, in watch jewels, diamond dies and other machine parts in various industries where high precision is not needed.

4.1. DISINTEGRATION OF METALS AND OPAQUE MATERIALS DRILLED BY LASER BEAMS

As no comprehensive theory of laser effects has been developed it is very important to analyze the experimental results of laser radiation effects in various materials.

Laser radiation with the power density exceeding $q_c^{(2)}$ produces melting and vaporization of material and ejection of solid particles, which is typically due to thermal stresses if they are higher than the cleavage strength of the material.

With the radiation power densities of 10^6-10^9 W/cm^2 formation and ejection of disintegration products are usually observed at the very beginning of the laser pulse. The high-speed photorecording (Fig. 38*a*) for the beginning of the laser effect shows a jet-like ejection of vapour, which repeats the spike structure of the laser pulse and starts practically with the pulse generation.

With increasing diameter and depth of the hole the increasing part of the disintegration products consists of the molten material which is melted on the walls and bottom of the hole and expelled by the vapour pressure. Photorecordings show the distinct traces of drops of the molten metal and red-hot solid particles. By the end of the pulse the ejection of the disintegration products decreases owing to the reduction of the radiation power density. With decreasing power density the disintegration products consist mostly of the molten material (Fig. 38*b*), and when the power

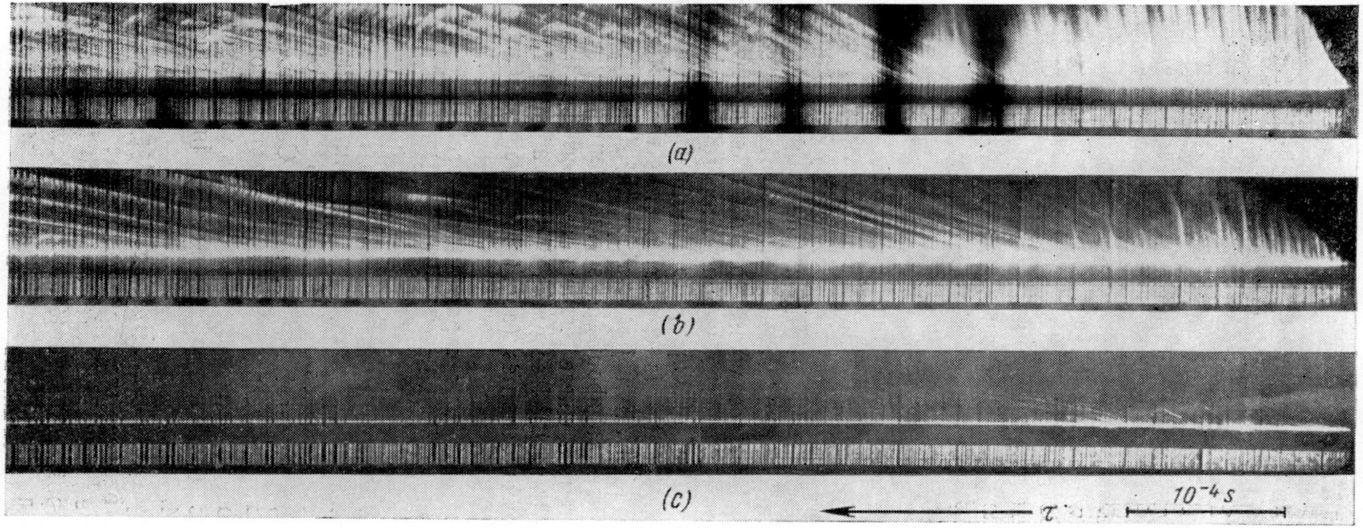

FIG. 38. High-speed photorecordings of ejection of disintegration products and laser beam intensity when drilling steel 45 by laser pulses with duration about 10^{-3} s and various energies

(*a*) 42 J; (*b*) 9 J; (*c*) 0.4 J. The focal distance of lens is 50 mm, the diaphragm diameter is 100 mm density is close to $q_c^{(2)}$ only the surface of the material is fused. The respective photorecording (Fig. 38*c*) shows only isolated traces of metal drops.

The surface temperature of the laser-irradiated zone in the material T_0 is typically higher than the boiling temperature T_b at the normal pressure. It depends on the radiation power density, for instance, the surface temperature of iron-carbon alloys under laser radiation with a power density of 10^6-10^8 W/cm^2 is higher than 4000 K [102]. The upper temperature limit reported in [102] seems to be overestimated (about 9000 K) since this temperature is higher than the critical point for these alloys.

The metal disintegration products produced by laser radiation with a power density of up to 10^8 W/cm^2 are ejected outside as a jet perpendicular to the irradiated surface.

When the vaporized material freely expands into vacuum or ambient atmosphere and the vapour pressure is considerably higher than the ambient pressure the velocity of escaping vapour is equal to the velocity of sound in it (the Jouguet condition). The vaporized material is accelerated to the velocity of sound in the thin layer at the surface whose thickness is a few free paths of the vapour particles. The vapour temperature sharply drops causing the transition of vapour to the unstable supersaturated state which subsequently is converted into the saturated state owing to the condensation jump while the temperature of the plasma jet increases to about the surface temperature. The condensation jump region where temperature increases has the width of about or less than 10^3 free paths of the vapour particles; with low radiation power densities (about 10^6 W/cm^2) it can be about 10^{-1} cm.

The vapour jet further expands and accelerates in the rarefaction wave; owing to cooling the metal vapour condenses into drops up to 10^{-4} cm in size; the temperature and velocity of these drops are close to those of the vapour [34].

Energy liberation with condensation can result in acceleration of the ejected particles and changes in some of their parameters (density, temperature, pressure). Condensed particles about 10^{-5}-10^{-4} cm in size produced with expansion of the ejected vapour jet have been found in the experiments with laser radiation of some metals (Al and other metals) in vacuum (about 10^{-2} mm Hg) [34, 73].

The velocity of the vaporization boundary v_0 is determined by the difference between the vaporization rate v_b and the condensation rate v_c, the latter being due to the back flow of vapour molecules to the vaporization front; the back flow is produced as a result of collisions of the vapour molecules and the atmosphere ones. The velocity of the vaporization front $v_0 \approx 0.8\ v_b$ [1].

When vapour freely expands into vacuum the back flow of particles is small and no distinct boundary can be found between the stage of heating and the stage of vaporization of material. When material is heated in the air the stage of heating is replaced by the stage of vaporization, and the surface temperature becomes T^* (T^* being the temperature of the phase transition from the condensed state into vapour at a given pressure) [20, 34].

A part of the incident laser radiation is absorbed and dissipated by the disintegration products. The major contribution to radiation absorption is made by the vapour of the material, which is a low-temperature weakly ionized plasma whose transmittance depends on its concentration and temperature (in case of free-generation laser pulses with the intensity in the above range). The main radiation absorption processes in gas (vapour) are absorption by atoms and absorption by electrons (inverse bremsstrahlung); the condensate particles in the vapour also absorb and dissipate radiation [34]. The absorbed radiation energy heats the disintegration products; the increase in brightness of the ejected products shows a correlation with the spikes of the laser pulse. It has been found [2, 34] that radiation absorption in the plasma produced by laser vaporization of the material in vacuum becomes significant at power densities $q_{0\,\max} \gtrsim 10^8\text{-}10^9$ W/cm^2. This is, approximately, the upper limit of the operating range of the power densities used for laser drilling. The lower limit of the operating power density range is the power density level at which the formation of a cavity starts at the end of the laser pulse, that is, it corresponds to the threshold of metal disintegration $q_c^{(2)} \approx \approx 10^5\text{-}10^6$ W/cm^2.

The liquid-vapour jet melts and washes off the walls of the cavity and increases its diameter. Depending on the radiation power density, pulse duration and the properties of the irradiated material the content of the liquid phase in the disintegration products varies from 30 to 80%; it increases with increasing thermal conduction of the material and the difference between T_m and T_b. The molten metal generally moves along the walls of the cavity forming a characteristic ridge around its entrance; if a large amount of metal is melted it is expelled outside and freezes forming a flattened rim slightly above the metal surface (Fig. 39).

With increasing velocity of the liquid metal large drops of the liquid metal are increasingly ejected along the periphery of the vapour jet. A large number of small drops ejected from the end of the cavity is observed also in the central region of the jet. The drops of the molten metal ejected outside are decelerated by the evaporation of the metal from their surface. Individual particles with the size about $10^{-3}\text{-}10^{-2}$ cm travelling with the velocity $v \approx (2\text{-}5) \times 10^2$ cm/s can reverse the direction of their motion

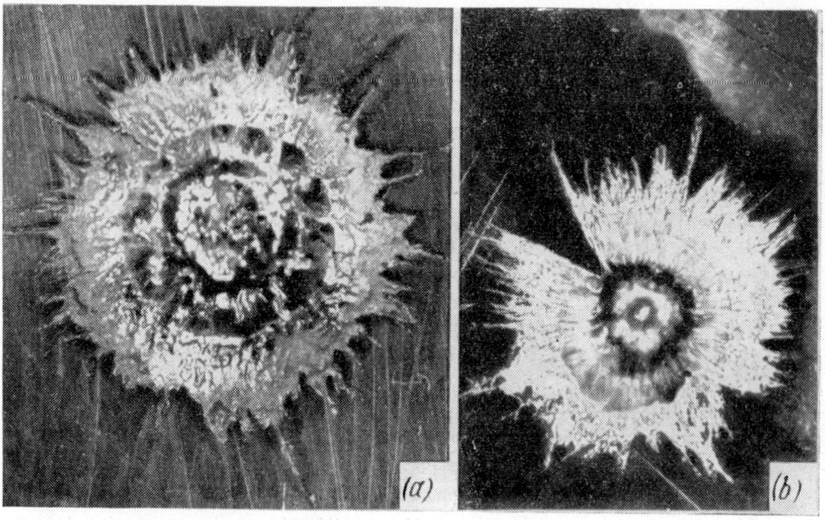

FIG. 39. Laser-affected zones
(a) steel ХВГ (×4); (b) steel 45 (×4)

and return to the cavity (see Fig. 38*a*) during the time of generation of several radiation spikes at a distance about 1 cm or more from the surface of the material.

There can be seen concentric bands of the annealling colours due to different rates of cooling in the metal on the outer surface of the rim (see Fig. 39).

The metal surface outside the rim is covered with a layer of oxides produced by metal combustion in the air. When steel specimens are irradiated in vacuum no oxide layer is found. The bottom and the walls of the cavity are covered with small, frequently deformed, crystallized metal drops about 10^{-5}-10^{-4} cm in size which are produced by condensation of the metal vapor on the surface.

The cracks or mechanical defects in materials are produced by the recoil momentum transferred to the bottom of the cavity and by the development of thermal elastic stresses due to high temperature gradients of the order of 10^5-10^6 C/cm (Fig. 40). The material is disintegrated by the effect of thermal elastic stresses rather than by vaporization if the brittle strength for the material is reached before the temperature in the heat-affected zone reaches the boiling point. The brittle strength of the material in the region of the hole when melting its walls can be estimated from the following equation [78]:

$$\sigma_{\varphi\varphi} = \frac{\beta E^* \Delta T}{3(1-2\sigma_1)} \frac{r^2(t)}{\rho^2} \tag{274}$$

FIG. 40. Cavity surface with microcracks in steel ХВГ (×600)

where β is the temperature coefficient of volume expansion of the material, E^* is Young's modulus, σ_1 is Poisson's ratio and $\Delta T = T_m - T_0$ is the variation in temperature at the melting front.

Equation (274) implies that the thickness of the heated layer is considerably smaller than the hole diameter and that the temperature in the layer is constant. The tangential stresses $\sigma_{\varphi\varphi}$ stretch the material and if $\sigma_{\varphi\varphi} > \sigma_r$ (σ_r is the tensile strength of the material) a radial crack is formed in the material. The length of the crack can be found from Eq. (274):

$$l = \left[\frac{\beta E^* \Delta T}{3(1 - 2\sigma_1)\sigma_r} r(\tau)\right]^{1/2} \tag{275}$$

Here the length l is reckoned from the centre of the hole *.

If we substitute the reported parameter values into Eq. (275) we can see that the length l can be of the order of a few hundred

* These equations for thermal elastic stresses are applicable only to thin plates when the crack length l is considerably larger than the thickness of the plate. However, even if $l \gg h$ the estimates have only the order-of-magnitude accuracy since the use of statical moduluses of elasticity for analyzing pulsed effects leads to considerable errors in estimates.

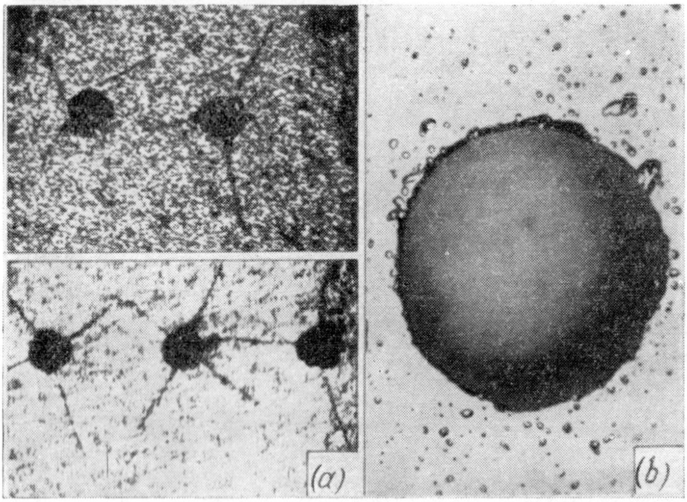

FIG. 41. Laser-drilled holes of 100 μm diameter in ferrite produced by pulses of various duration
a) 1.5 ms (radial cracks); (b) 0.1 ms (no cracks)

micrometres for some materials (for instance, steel among metallic materials and ferrite among nonmetallic materials). The resulting values of l are in good agreement with the experimental data. For example, for ferrite $\beta = 2.3 \times 10^{-5}$ K^{-1}, $\sigma_1 \approx 0.1$, $E^* = 2 \times 10^{11}$ dyne/cm^2 and with the hole radius $r = 50$ μm the crack length l is about 400 μm. Figure 41 shows holes in ferrite plates produced by laser pulses of various duration [86]. The longer laser pulses produce larger thermal stress zones in the material and give rise to cracks (Fig. 41*a*). The recoil momentum produces additional disintegration of the bottom of the hole [17, 66]. At the radiation power density of about 10^7-10^8 W/cm^2 the resulting pressures on the bottom of the hole can be as high as 10^2-10^3 kgf/cm^2 [34]. The estimates using the velocities of the vaporized material show that the resulting tangential stresses can be higher than the strength of the material as confirmed by the experimental results. The boundary of the material ejected from the hole bottom due to strain produced by the recoil forces can be distinctly seen in Fig. 42.

The final profile of the hole depends to a great extent on the conditions of focussing of the laser radiation. Holes of various shapes — conical, cylindrical, bulb-shaped, and others — can be produced by changing the focussing conditions in the laser optical system.

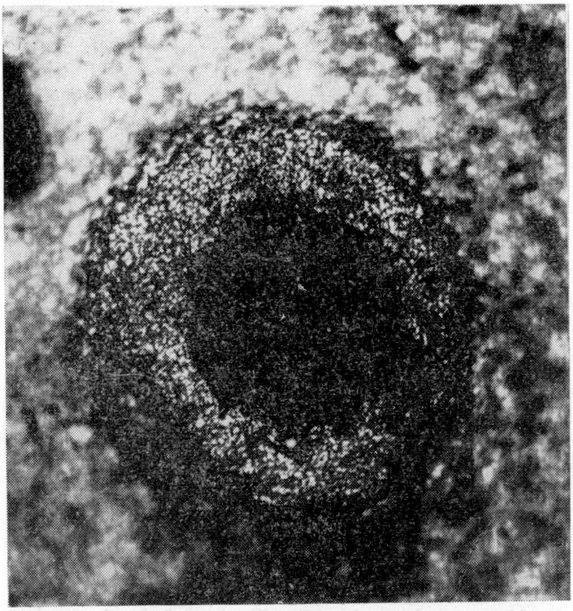

FIG. 42. Laser-drilled hole (from the back side) in steel specimen (×600)

The longitudinal sections of various steel specimens exhibit white fringes around the hole contours which represent a (poorly etched) layer of the phase and structural changes. The width of this layer increases from the bottom of the hole to its entrance; it depends on the thermal properties of the material, the parameters and shape of the laser pulse and other factors (for the specimens shown in Figs. 41 and 42 the width of this layer is not more than 10^{-3}-10^{-2} cm).

In the cross section of the white layer the temperature varies from T_m or more (the liquid metal can be considerably superheated) to a temperature below T_h in the metal layer of the initial (unchanged) structure. Hence we can find the temperature gradient which can be as high as 10^5-10^6 C/cm by the end of the pulse providing for rapid heating and cooling and structural and phase transformations of the metal in the laser-heated zone.

The quasistationary laser radiation (spikeless generation) produces the effects in materials, which differ from the effects produced by spiked laser pulses. For the quasicontinuous-wave pulse the instantaneous power at any moment is practically identical to the mean power. For instance, for a pulse of the total energy of 50 J focussed on the area about 10^{-2} cm^2 with the duration about 10^{-3} s the peak power density is 5×10^6 W/cm^2. For the spiked pulse of the same total energy focussed on the same area with the duration about 5×10^{-7} s the instantaneous power density is about 5×10^8 W/cm^2 [40, 41]. This large difference in the power density (by two orders of magnitude) has a significant effect resulting in a difference in the character of material disintegration between the quasistationary laser pulse and the spiked pulse discussed above. In experiments with laser heating no liquid phase is found in the products of disintegration of most metals up to the power densities of about 5×10^6 W/cm^2 but a noticeable vaporization of the material starts at lower intensities and rapidly increases with increasing radiation intensity [40]. The peak velocity of the vapour jets in the vicinity of the specimen surface is 10^6 cm/s, and the high-speed photorecordings show, apart from the vapour jets, the traces of molten particles with the velocities (perpendicular to the specimen surface) about 10^3-10^4 cm/s [40].

Thus, with the same laser pulse energy the radiation power density for the quasistationary radiation can be considerably lower (by two orders of magnitude) than the power density for the spiked radiation pulse owing to the longer time of effect and more uniform spatial structure. The rate of energy absorption by the material for the quasistationary laser pulse differs but slightly from the rate of energy dissipation which is mainly due to heat conduction.

In the metals irradiated with quasistationary laser pulses a system of concentric waves of the solidified metal is found on the surface of the molten zone; the distance between the ridges of the waves is steadily increasing with increasing distance from the centre of the heat-affected zone. Formation of these waves has been attributed to the reactive pressure of the vapour jets flowing from the heat-affected zone of the material [41].

4.2. FORMATION AND MOTION OF LIQUID PHASE IN LASER-DRILLED METALS

In analysis of drilling of deep holes in opaque materials it is assumed, typically, that all the molten material is expelled from the hole by the excess pressure. However, the experimental studies of the disintegration processes in metals and metallographic analysis of the heat-affected zone produced by laser radiation show that not all the molten metal is expelled from the hole. A considerable part of the melt is left in the hole, and when the laser pulse ends the melt solidifies on the walls and bottom of the hole. This is explained by decreasing power density on the bottom of the hole with increasing depth of the hole due to defocussing of

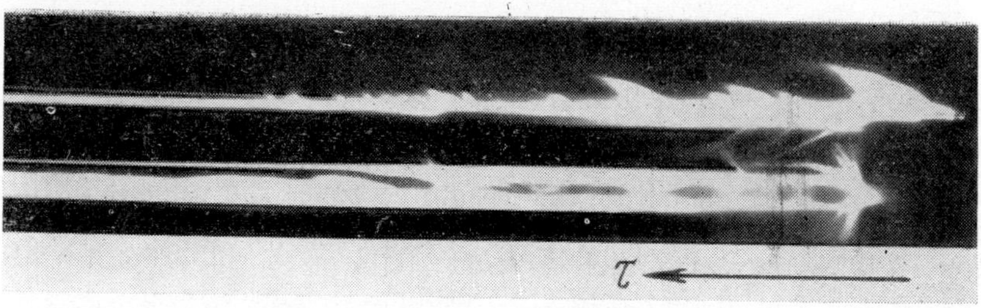

FIG. 43. High-speed photorecordings of laser drilling of thin tungsten plates 50 μm thick (below) and ejection of disintegration products (above). The focussed spot diameter is 250 μm, $\lambda = 1.06$ μm

the laser beam, absorption of radiation by the disintegration products and decreased laser power by the end of the pulse. Under such circumstances the vaporization rate and the vapour pressure decrease so that only a part of the melt is expelled and the hole is filled with melt; this is observed for the laser pulses of about 1 ms or longer.

Figure 43 presents high-speed photorecordings of laser drilling of thin metal plates. The hole can be seen to be sealed with the molten metal either during the pulse or at the end of the pulse when the laser power density decreases or just after the end of the pulse. Figure 44 [86] illustrates sealing of a hole with the liquid metal.

Laser radiation with the power density $q_0 < q_{\min}$ melts the surface layer of the material to a negligible depth, and the motion of the melt has no importance (Fig. 44*a*). When radiation density

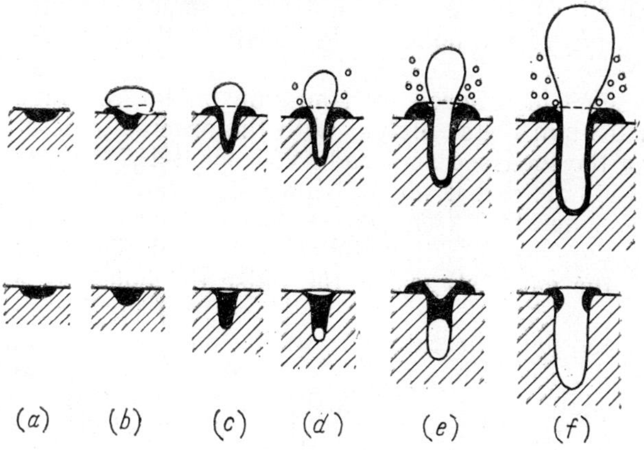

FIG. 44. Profiles of laser-drilled holes at various radiation power densities during the pulse (above) and after the end of the pulse (below)

$q_0 > q_{min}$ the vaporization of the material begins in the centre of the focussed spot where the temperature is higher and the vapour pressure pushes the melt boundary inside the material (Fig. 44*b*). Radiation of higher power density results in considerable deepening and widening of the hole owing to the vaporization and displacement of the melt from the irradiated zone. The velocity of the laminar motion of the melt is relatively low. The melt is not ejected with the vapour jet in drops; it forms a ridge due to the surface tension on the surface of the material directly around the entrance of the hole (Fig. 44*c*). The melt which has not solidified flows into the hole after the end of the pulse or at the end of the pulse when the radiation power density decreases; the crystallized metal seals the hole. When the radiation power density becomes higher the motion of the liquid metal becomes partly turbulent, and drops of metal can be ejected from the hole (Fig. 44*d*). Such holes are sealed only partially after the end of the pulse; they contain voids which can be also formed by sealing the hole at some depth by the melt that subsequently solidifies during the pulse. When the radiation power density is still higher the melt motion becomes completely turbulent, and a large part of the melt is ejected in drops with the vapour jet. Nevertheless, the remaining amount of the melt can be sufficient for sealing the hole near its entrance mainly with the pulses about 1 ms with long tails (Fig. 44*e*). Only very high power densities (about 5×10^7 W/cm^2) produce holes which are not sealed with the molten material (Fig. 44*f*). However, even at such power densities the initial profile of the hole can be considerably distorted by redistribution of the melt after the end of the pulse. Below, we shall discuss the experimental results of laser drilling of 0.3-mm metal plates (the power density is about 10^6-10^7 W/cm^2) and make some estimates for quasiregular sealing of the holes on the basis of the concepts developed in [146].

Figure 45 presents high-speed transmission photorecordings of laser drilling of 0.3-mm metal plates. With the ordered-generation laser pulse and the focussed spot radius of about 180-200 μm the diameter d_1 of the hole exit is about 1 μm after irradiation period of about 10^{-5} s. When the laser beam is focussed on the front surface of the plate the entrance diameter d_2 of the hole is larger than d_1, that is, at the early stages of the pulse the hole has a conical shape.

Now we must estimate the thickness of the molten layer on the walls of the hole. The time period required to produce a through hole in the plate can be estimated as $\tau_0 = h/v_0$ where h is the thickness of the plate and v_0 is the rate of disintegration of the material. From the quasistationary vaporization theory we find $v_0 \approx q_0/\rho L_b$ [34]. At $q_0 \approx 10^6$-10^7 W/cm^2 and $h = 0.3$ mm

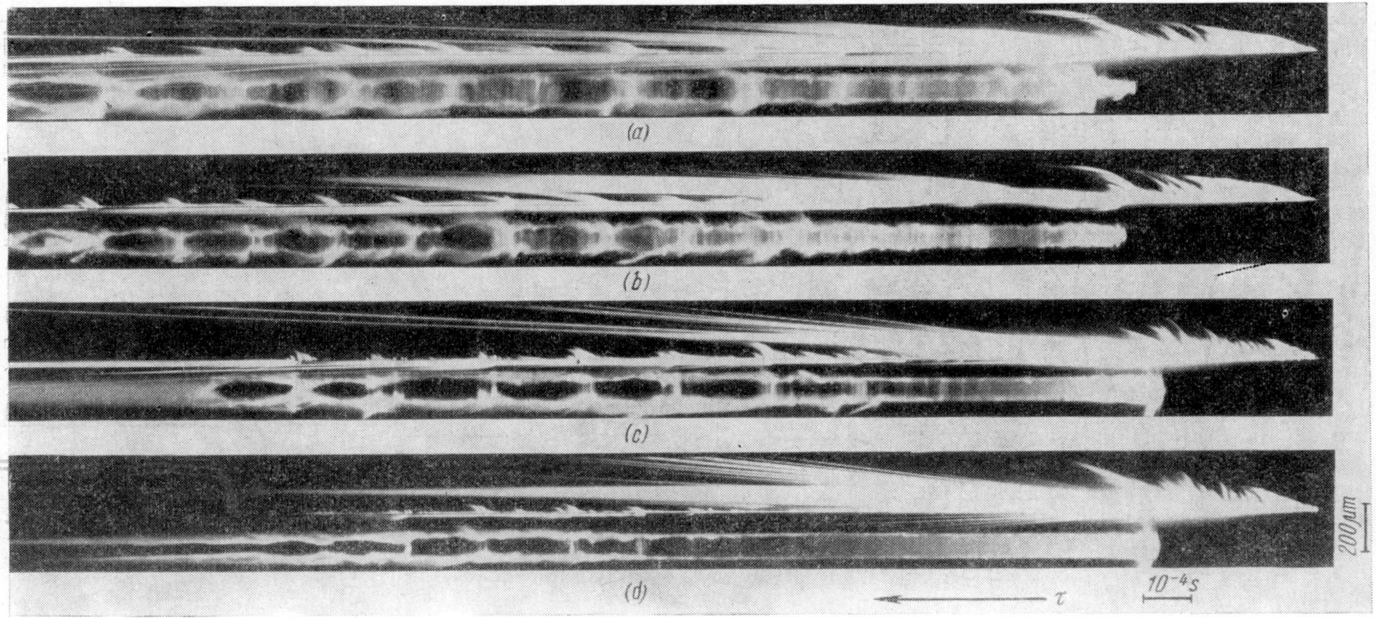

FIG. 45. High-speed photorecordings of ejection of disintegration products (above) and variation in hole cross section (below) for laser-drilling of 0.3-mm plates of molybdenum (*a*), tantalum (*b*) and tungsten (*c* and *d*)

we have for iron $v_0 \approx 10^2$ cm/s and $\tau_0 \approx 3 \times 10^{-4}$ s. As the rate of metal disintegration is higher than the vaporization rate predicted from a one-dimensional model the real period τ_0 is somewhat shorter.

The thickness of the molten layer on the inner surface of the hole as a function of time can be found by solving the system of hydrodynamical equations and Stefan's temperature distribution problem in the cylindrical coordinate system with the heat source whose intensity varies with time and along the axis of the hole. We shall limit ourselves to separating the analysis of the thermal problem and the hydrodynamical problem. The photorecordings show that the hole is sealed by the melt for the first time approximately in 90 μs after formation of the through hole. To make a qualitative estimate of the thickness of the molten layer on the walls of the hole in about 200 μs we shall use the results reported in [34, 86].

The boundary of the melting zone has the coordinate

$$\xi = 2p_0\sqrt{a\tau} \tag{276}$$

where p_0 is the root of the transcendental equation

$$\exp(p_0^2)\mathrm{Ei}(-p_0^2)[M - p_0^2] = N \tag{277}$$

Here $M = q_l/4\pi a\rho L_m$, $N = (T_m - T_0)c/L_m$, T_m is the melting point, T_0 is the starting temperature, c is the specific heat, L_m is the melting heat, and q_l is the intensity of the linear heat source.

For the optimum laser pulse parameters the roots of Eq. (277) can be found from a simpler equation derived by expansion of $\mathrm{Ei}(-p_0^2)\exp(p_0^2)$:

$$p_0^2 = (M/N + 1) \tag{278}$$

For the iron with the hole radius $R = 20$ μm, $a = 0.1$ cm²/s, $\rho = 7$ g/cm³, $L_m = 270$ J/g, $\tau = 100$ μs and $q_l = 2\pi Rq_s$ we obtain $\xi \approx 80$ μm. The result is somewhat overestimated since Eqs. (276) and (277) have been derived for a one-phase model, that is, assuming that the molten material is immediately removed and the hole diameter is the same throughout the plate. The same conclusion follows from the estimates of the molten layer thickness made below for the quasistationary process. If the parameters of the metal vapour are assumed to be constant throughout the hole ξ can be found according to [36]:

$$\xi = \frac{q_s H}{2\omega_l \rho_l L_m}\left[\left(1 + \frac{4Da_l\omega_l\rho_l L_m \ln C/B}{q_s^2 H}\right)^{1/2} - 1\right] \tag{279}$$

where H is the depth of the hole, ω_l is the rate of the flow of the melt from the bottom to the hole, and the constants are

$$B = \rho_s c_s (T_m - T_0) + \rho_s L_m$$
$$C = B + c_l \rho_l (T_b - T_m) \qquad (280)$$
$$D = C + \rho_l L_b$$

Here the subscript s denotes the solid phase, l denotes the liquid phase and T_b is the boiling temperature which is a function of the radiation power density.

Equation (279) can be simplified for large H:

$$\xi = \frac{Da_l}{q_s} \ln C/B \qquad (281)$$

To use Eqs. (279)-(281) we must know the rate ω_l of the flow of the melt from the bottom of the hole to the hole. If no condensation on the walls occurs Bernoulli's equation yields

$$\omega_l = [2(p_c - p_v)\rho_l^{-1}]^{1/2} \qquad (282)$$

where p_c is the pressure inside the hole and p_v is the outside pressure.

In most cases $p_c - p_v \approx p_c \approx 0.5\ p_0$ where p_0 is the saturated vapour pressure, and Eq. (282) transforms into

$$\omega_l \approx (p_0/\rho_l)^{1/2} \qquad (283)$$

The pressure p_0 which depends on temperature can be found by successive approximations. The estimates according to Eqs. (280)-(283) give the thickness of the molten layer about 50 μm, that is, smaller than the estimate from Eq. (276).

As shown by theoretical estimates and experimental results, laser radiation for a few tens of microseconds produces a narrow hole with the diameter about 10-20 μm in the plate; the hole is surrounded by a molten layer having the thickness of the same order of magnitude. The pressure gradient in the hole gives rise to the motion of the molten metal. If the hole in the plate is blind the recoil pressure on the melt on the hole bottom pushes it out from the bottom to the molten layer on the walls and then out of the hole. We do not consider here explosive ejection of the molten metal due to bulk vaporization [120].

The formation of a through hole changes the conditions in the hole as the pressure in it drops sharply tending to the outside pressure. This results in a change in the flow of the molten layer on the hole walls. Some time after opening of the hole on the lower side of the plate the molten metal flows down owing to gravity and surface tension, and at a certain hole diameter the melt seals the hole. This results in increased radiation absorption in the lower part of the hole leading to higher temperature and recoil pressure and, after a certain period, to next opening of the hole, and the process repeats itself.

The process of sealing and opening of the hole is not purely periodic owing to inertia of thermal phenomena. For each cycle of the process the diameter of the hole increases and a part of the molten metal is removed. Therefore, when the exit diameter of the hole reaches a certain value d_1 the hole cannot be sealed any longer and from that moment the hole diameter increases steadily until the intensity of the heat source on the hole walls drops below the threshold for the beginning of melting.

Let us estimate a possible contribution of screening of the incident laser radiation by the disintegration products expelled from the heat-affected zone. The estimates according to [34] show that under described conditions the laser radiation can be screened by the high-temperature iron vapour at the radiation power density of 10^7 W/cm^2 starting from the moment about 0.3×10^{-3} s after the beginning of the pulse (the pulse duration in the experiments is $1.5\text{-}1.8 \times 10^{-3}$ s). At the averaged power densities of 10^6-10^7 W/cm^2 used in the experiments the screening by plasma, condensate drops [34] and large metal drops [98] ejected in explosive vaporization cannot be very significant. Simple numerical estimates show that vapour leaves a hole 0.3 mm deep with the velocity of about 10^4-10^5 cm/s in about 1-10 μs; this period is much shorter than the period of pulsations of the liquid phase volume observed in the experiments. Hence, at radiation power densities below 10^7 W/cm^2 screening of laser radiation makes no significant contribution to the observed process of sealing the hole with the molten metal.

Let us now analyze sealing the hole by the molten metal flowing down the walls of the hole. The down flow becomes possible after the formation of the through hole and drop in pressure in the hole. Since the diameter of the capillary in the lower part of the hole is small about 10^{-5} s after the formation of the opening (in experiments d_1 is about 10-20 μm) the conditions of sealing the hole are not greatly affected by the presence of the capillary.

We shall analyze sealing a narrow blind hole owing to the effect of the capillary and gravitational forces [96]. It is assumed that a pressure gradient which remains after the opening of the bottom of the hole has no significant effect on the motion of the molten material in the hole since the characteristics time of pressure equilibration after opening of the hole is about 10^{-6} s. Then the melt motion in the case of axial symmetry can be described by the Navier-Stokes equation in the following form:

$$\rho \frac{\partial v_z}{\partial \tau} - \mu \left(\frac{\partial^2 v_z}{\partial r^2} + \frac{1}{r} \frac{\partial v_z}{\partial r} \right) = -\rho g - \frac{\sigma}{H(\tau)} \left(\frac{1}{R_u} + \frac{1}{R_l} \right) \qquad (284)$$

where v_z is the vertical component of the velocity of the melt, μ is the dynamical viscosity, σ is the surface tension, $H(\tau)$ is the time-dependent thickness of the molten layer on the bottom of the hole which seals the opening in the bottom, and R_u and R_l are the radiuses of the melt in the upper and lower parts of the hole, respectively.

In Eq. (284) we have neglected the radial motion of the melt, that is, assumed that $v_r = 0$. We can assume also that $R_u = \xi$ which is the thickness of the molten layer, and R_l is equal to the radius of the hole.

Equation (284) must be supplemented by the mass conservation equation

$$2 \int_0^\tau \int_R^\xi v_z(r, t)\, r\, dr\, dt = R^2 [H(\tau) - H_0] \qquad (285)$$

where H_0 is the initial depth of the hole.

The system of equations (284) and (285) yields the function $H(\tau)$ which gives the time of sealing of the hole t (from the condition $H = 0$ at $\tau = t$). We can find the solution of the system of equations (284) and (285) for the two limiting cases — for high and low viscosities [96]. As temperatures in our problem are high the viscosity term can be omitted, and Eqs. (284) and (285) can be transformed into

$$\frac{\partial v_z}{\partial \tau} = -g \left[1 + \frac{\sigma}{\rho g H} \left(\frac{1}{\xi} + \frac{1}{R} \right) \right] \qquad (286)$$

$$2 \int_0^\tau \int_R^\xi v_z(r, t)\, r\, dr\, dt = R^2 (H - H_0) \qquad (287)$$

The boundary layer is not taken into consideration since for the times about 10^{-5}-10^{-6} s its thickness $\sqrt{\nu\tau}$ is about 10^{-4} cm for iron and it is small compared to the thickness of the molten layer (a few tens of micrometres).

Equations (286) and (287) yield the following equation for the time of sealing the hole [96]:

$$t = \frac{1}{\sqrt{2}} \int_0^{H_0} \frac{dH}{\sqrt{\beta_1 (H_0 - H) - \alpha_1 \ln (H/H_0)}} \qquad (288)$$

where

$$\alpha_1 = \frac{\sigma (\xi^2 - R^2)}{\rho R^2} \left(\frac{1}{\xi} + \frac{1}{R} \right), \quad \beta_1 = \frac{g (\xi^2 - R^2)}{R^2}$$

Neglecting the capillary forces we can transform Eq. (288)

$$t_g = \frac{R}{\xi}\sqrt{\frac{2H_0}{g}} \tag{289}$$

and ignoring the gravitational forces we obtain

$$t_s = H_0 \frac{R}{\xi}\sqrt{\frac{\pi\rho R}{2\sigma}} \tag{290}$$

Numerical estimates according to Eqs. (289) and (290) show that under our conditions the capillary forces dominate ($t_g \approx 10^{-4}$ s, $t_s \approx 10^{-6}$ s at $R = 10^{-3}$ cm, $H_0 = 10^{-3}$ cm, $\xi = 10^2$ cm, $\sigma = 2 \times 10^3$ dyne/cm and $\rho = 10$ g/cm^3) the process of sealing the hole.

Let us analyze the equilibrium of a molten metal drop on the bottom of the hole in a metal plate and find the relationship between the radius of the hole and the vapour recoil pressure $p = p(R)$ which is needed to estimate the time necessary for the formation of an opening in the bottom of the hole. The equilibrium equation is

$$2/3\,(\rho g R^2) - pR - 2\sigma = 0 \tag{291}$$

Hence the equilibrium radius is

$$R = -\frac{3p}{4\rho g} + \sqrt{\frac{9p^2}{16\rho g} + \frac{3\sigma}{\rho g}} \tag{292}$$

Figure 46 presents the relationship between the equilibrium radius R and the pressure p for a typical metal melt ($\rho = 10$ g/cm^3 and $\sigma = 10^3$ dyne/cm). When R is about 10^{-3} cm the pressure p is more than 1 atm; this means that the vapour recoil pressure disturbs the equilibrium and produces the opening of the bottom of the hole after a time interval with the duration of the order of the interval required for the development of the steady-state vaporization process τ_s ($\tau_s \approx a/v_s^2 \approx 10^{-5}$ s). These results show a qualitative agreement with the experimental data.

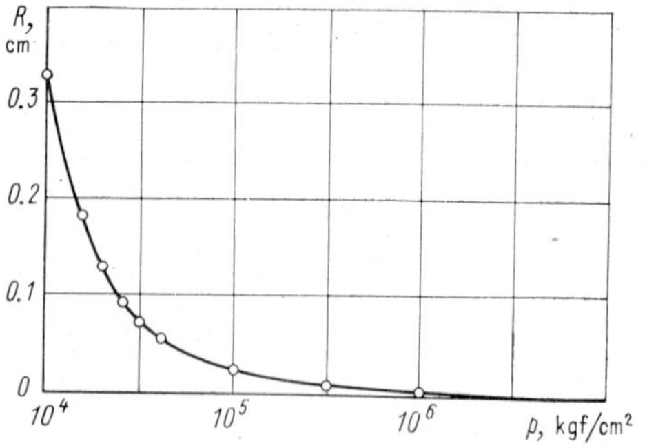

FIG. 46. Equilibrium hole radius R as a function of vapour recoil pressure p

The spike structure of the laser pulse does not have a significant effect on the motion of the molten layer in the hole since pulsations of the melt have lower frequency (about

10^4 Hz) than spike pulses (about 10^6 Hz); hence the above effects must occur also for the quasistationary laser pulses.

The qualitative estimates of the motion of the molten metal in the hole made above show that high-quality laser drilling requires power densities of at least 10^7 W/cm^2. However, the laser power density should not be too high since the disintegration products can screen the laser beam and the hole will be sealed by the molten metal. Hence, the optimum ranges must be determined for the laser intensity and pulse duration to provide for effective laser drilling of holes of a given diameter and shape.

4.3. FORMATION OF HOLES IN OPAQUE MATERIALS BY LASER DRILLING

A one-dimensional model can be used for analyzing the formation of a hole by the laser beam with a not-too-high intensity only when the depth h of the hole is smaller than the size d of the focussed spot. In this case the absorbed energy E^0 is $L_b \rho d^3$ [34].

When the absorbed energy E is higher than E^0 and $h > d$ the process of hole development is much more complicated since heating, melting and disintegration of its walls are affected by many factors. Among these factors are vapour condensation, radiative and convective heat exchange between the disintegration products and the hole walls, absorption of defocussed laser beam and others.

The heat flux intensity F_v due to vapour condensation and absorption of the defocussed laser beam energy on the walls is 0.01-0.05 of the intensity q_0 of the radiation incident on the surface. For instance, at $q_0 \approx 10^7$-10^8 W/cm^2 the intensity F_v is about 10^5-10^6 W/cm^2. During the laser pulse there is observed an increase not only in the depth of the hole but also in its diameter which by the end of the pulse is four times that of the focussed spot on the surface of the material [9, 34, 62]. The threshold heat flux intensity F_0 sufficient for heating the walls of a cylindrical hole with the radius r_0 to the melting temperature T_m is given by [20]

$$F_0 = \frac{2\lambda T_m}{r_0 \ln (4a\tau / C_E v_0^2)} \quad \text{at} \quad r_0 \ll \sqrt{a\tau}$$

$$F_0 = \frac{\sqrt{\pi}}{2} \frac{\lambda T_m}{\sqrt{a\tau}} \quad \text{at} \quad r_0 \gg \sqrt{a\tau} \qquad (293)$$

where C_E is Euler's constant.

For most metals at $\tau \approx 10^{-3}$ s and $r_0 \approx 10$-50 μm the threshold intensity F_0 is about 10^5-10^6 W/cm^2.

The liquid metal which comprises a considerable part of disintegration products (up to 80%) is expelled from the hole by the vapour pressure [34, 62, 63]. By the end of the pulse with a not-too-high radiation density the main mechanism of disintegration of the material is its melting and washing out [62]. Ejection of the liquid metal can be also due to its boiling as a result of considerable superheating produced in short periods compared to the laser pulse duration [12, 119, 120]. The factors contributing to this effect are high concentration of dissolved gases in the metal and the presence of bubbles and pores containing gases in the bulk of the metal.

The spiked structure of the free-generation laser pulses also contributes to the growth of bubbles.

A simplified phenomenological model of development of deep holes due to laser radiation with not-too-high intensities has been put forward by the authors of references [9, 34] on the basis of experimental results. The model assumes that the material of the hole bottom can be only vaporized and the molten material from the hole walls is effected, the heat flux intensity on the walls F_v being approximately equal to r^{-2} where r is the hole radius. The vaporization and melting processes are assumed to be quasistationary. Moreover, the model ignores the temperature dependence of the vaporization heat L_b of the material and screening of the incident radiation by disintegration products. Then the rate of hole deepening is

$$\frac{dh}{d\tau} = \frac{r_f^2 q_0}{(r_f + h \tan\varphi)^2 L_b \rho} \tag{294}$$

where $r_f = \alpha F$ is the radius of the focussed spot, $\tan\varphi = (D - \alpha l)/2F$, D is the diameter of the emitting end of the laser and l is the distance from the laser to the front focus of the optical system with the focal distance F.

Assuming the hole to have a cylindrical shape and using the solution of Eq. (294) we can find how the radius and the depth of the hole, the liquid-vapour ratio and the specific disintegration heat depend on the pulse duration and the total radiation energy [34]. The radius $r(\tau)$ of the growing hole and its depth $h(\tau)$ are related by the light cone equation for the apex angle 2φ [20]

$$r(\tau) = r_f + h(\tau)\tan\varphi \tag{295}$$

For quasistationary vaporization and melting processes the energy balance in the hole at a given moment is described by the following equation:

$$P_s(\tau)\,d\tau = L_b \pi r^2(\tau)\,dh + L_m 2\pi r(\tau)\,h(\tau)\,dr \tag{296}$$

The system of equations (295) and (296) with the initial conditions $h(0) = 0$ and $r(0) = d_f/2$ has the following solutions for $h(\tau)$ and $r(\tau)$ when $h(\tau) < r_f$:

$$h(\tau) \approx P_s \tau / \pi r_f^2 L_b \tag{297}$$

$$r(\tau) = r_f + P_s \tau \tan\varphi / \pi r_f^2 L_b \tag{298}$$

and at $h(\tau) \gg r_f$:

$$h(\tau) = \left[\frac{3P_s\tau}{\pi \tan^2\varphi\,(L_b + 2L_m)}\right]^{1/3} \tag{299}$$

$$r(\tau) = h(\tau)\tan\varphi = \left[\frac{3P_s\tau}{\pi (L_b + 2L_m)}\right]^{1/3} \tag{300}$$

Table 7 presents the values of the depths and diameters of the laser-drilled holes on the front surface in aluminium, steel, molybdenum, tungsten and ferrite predicted from Eqs. (299) and (300).

TABLE 7. Values of Depths (h) and Diameters (d) of Holes (in millimetres)

E, J	Al		Steel		Mo		W		Ferrite	
	h	d	h	d	h	d	h	d	h	d
0.1	0.58	0.14	0.45	0.11	0.40	0.105	0.40	0.10	0.45	0.11
0.5	1.10	0.23	0.85	0.18	0.80	0.18	0.70	0.16	0.85	0.18
1.0	1.37	0.29	1.10	0.24	1.00	0.22	0.90	0.20	1.10	0.24
2.0	1.75	0.37	1.40	0.30	1.35	0.29	1.20	0.26	1.40	0.30
5.0	2.40	0.50	1.90	0.40	1.85	0.39	1.65	0.35	1.90	0.40

Notes: 1. $\tau_i = 10^{-3}$ s, $\tan\varphi = 0.1$, $r_f = 10$ μm.
2. The specific vaporization energies of metals at 0 K (J/cm³): Al—3.1×10^4, steel—5.7×10^4, Mo—6.9×10^4, W—8.8×10^4, ferrite—5.1×10^4.

Equations (299) and (300) show a nonlinear energy dependence of h and d; for instance, at $d \gg r_f$ the hole depth $h \approx E^{1/3}$, and the hole diameter $d \approx E^{1/3}$ where $E = P_0\tau$ is the total pulse energy. At the initial stage of laser drilling the variation in the hole diameter is small while the hole depth increases with time as the material is vaporized from all the surface of the hole. This process requires high energy consumption per unit mass corresponding to the heat of sublimation of the material. As the hole diameter increases with time owing to intense melting of the walls and washing out of the melt the rate of deepening of the hole decreases and in the limiting case ($\tau \to \infty$) the depth and radius of the hole vary as $\tau^{1/3}$, while the shape of the hole remains practically the same. This drilling process is known as the quasistationary regime. The features of this regime are the constant liquid-vapour ratio in the products of disintegration (the liquid phase amounts to about 2/3 of the disintegrated material [34]) and the decrease of specific disintegration heat to

$$L_d \approx \frac{(L_b + 2L_m)}{3} \tag{301}$$

The time required for the development of the quasistationary regime $\tau_{qs} \geqslant 2r_f/v$. After the period τ'_{qs} the hole depth h is $v(\tau'_{qs} - \tau_{qs}) = 2r_f$. It should be noted that τ'_{qs} is always higher than τ_{qs}. Since $\tau'_{qs} = 2r_f/v + \tau_{qs}$ we obtain $\tau'_{qs} \approx 10^{-5}$ s at $\tau_{qs} = 10^{-9}$-10^{-7} s ($q_0 \approx 10^8$ W/cm^2), $2r_f = 10^{-2}$ cm and $a = 0.1$ cm^2/s. The time required for the development of the hole shape is of the order of the time required for the development of the constant (peak) rate of the material removal equal to τ'_{qs}.

Equations (299) and (300) show that the relationship between the depth and diameter of the hole produced by quasistationary disintegration of the material is determined only by $\tan\varphi$ which is a parameter of the optical system (2φ is the apex angle of the light cone after the local plane of the optical system). The smaller $\tan\varphi$ the larger the ratio h/d:

$$h/d = \frac{1}{2\tan\varphi}\left[1 - \frac{1}{\sqrt[3]{1 + \dfrac{3E\tan\varphi}{\pi r_f^3 L_b}}}\right]$$

The optimum condition for laser drilling of deep small-diameter holes is $\tan\varphi = 0$; it is satisfied in the optical system which provides for what is called cylindrical light pipe between the focal plane and the image plane of the laser emitting end (here $l \approx D/\alpha$). Then we obtain

$$h = E/\pi r_f^2 L_b;\quad d = d_f = 2r_f,\quad h/d = E/2\pi r_f^3 L_b \tag{302}$$

and in the first approximation the disintegration of the material is quasistationary vaporization [20]. When $\tan\varphi \neq 0$ the power density on the bottom of the hole decreases due to defocussing of the laser beam; hence, there exist some limiting diameter and depth of holes produced by laser pulses with the energy E:

$$h_{\lim} = \frac{\sqrt{\dfrac{E}{\pi q_1}}}{\tan\varphi},\quad d_{\lim} = 2\sqrt{\frac{E}{\pi q_1}} \tag{303}$$

where q_1 is the threshold power density for which the quasistationary vaporization regime is established by the end of the pulse.

Experiments have been done to verify the predicted results. For instance, the variation in the depth and diameter of the hole

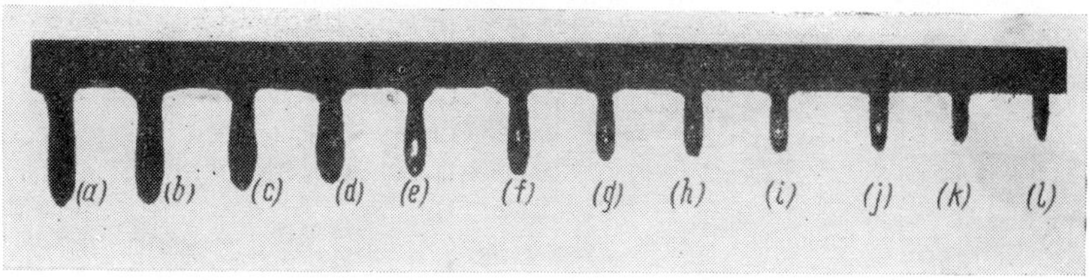

FIG. 47. Profiles of laser-drilled holes in steel specimens at pulse duration varying from 1.25 ms (*a*) to 0.2 ms (*l*) (×6)

with time is determined in the following way. A number of holes are drilled side by side in a specimen by laser pulses of different durations but with the same pumping power. To vary the laser pulse duration continuously from 0.1 to 1.3 ms, that is, to the duration of the initial laser pulse a rotating disk is placed in the path of the beam. Then the sections of the specimen are prepared, and the hole sizes are measured (see Fig. 47) to find the variation in the depth and diameter of the hole with time taking into account some differences in the pulse energies.

Figure 48 presents such experimental results for steel 45. The total pulse energy is about 25 J. The laser beam passes a 10-mm diaphragm and is focussed by the lens with $F = 65$ mm on the surface of the steel specimen. For a certain time interval the experimental plots $h(\tau)$ and $d(\tau)$ show a qualitative agreement with the predicted variation, and the depth and diameter of the hole in the first approximation vary as $\tau^{1/3}$. But after a period of about 0.8 ms the hole diameter remains practically constant; at any rate its time dependence is much weaker than $\tau^{1/3}$. This is, apparently, due to decreased power density during the last third of the pulse duration (see Chapter 2).

The plots in Fig. 48 show that an increase in the depth and diameter of the hole is most rapid at the beginning of the laser pulse. For instance, the depth of the hole reaches a half of its maximum value in 0.3 ms, that is, a fourth of the pulse duration. By the midpoint of the pulse the depth of the hole is about 70% of the maximum and the diameter is about 90% of the maximum. Figure 49 presents the variation in the rates of deepening and widening of the hole with time derived from the plots in Fig. 48. It can be seen that the deepening rate v_1 sharply decreases during the pulse. This effect is attributed to decreasing power density on the bottom of the hole with its growth caused by defocussing of the laser beam and increased radiation absorption by the disintegration products. Plotted in Fig. 49 is also the ratio between the deepening rate v_1 and the widening rate v_2 for the hole as a function of time. It can be seen that the predominance of v_1 com-

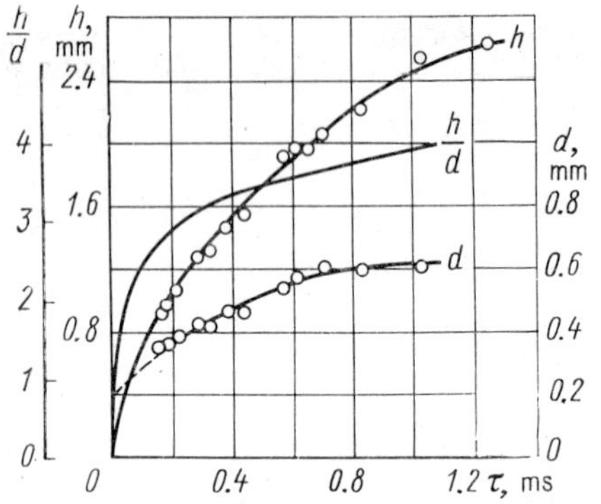

FIG. 48. Variations in depth h, diameter d and ratio h/d of blind holes as functions of radiation time in steel 45 irradiated by laser pulse with energy about 25 J and duration about 1.3 ms

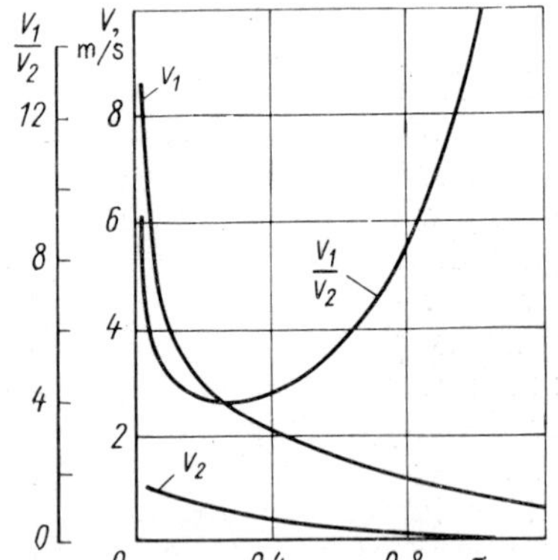

FIG. 49. Rates of hole depth increase (v_1) and hole diameter increase (v_2) as functions of radiation time in steel 45 irradiated by laser pulse with energy about 25 J and duration about 1.3 ms

pared to v_2 decreases from the beginning of the pulse owing to a more rapid decrease of v_1. Starting from a certain moment the predominance of v_1 again increases and by the end of the pulse v_1 is much higher than v_2 as the diameter of the hole remains practically constant.

When the laser power density is higher than the threshold for the disintegration of the material the development of the hole starts from the beginning of the pulse irrespective of the initial state of the specimen surface. Only when the laser power density is close to the threshold the roughness of the surface can delay the vaporization of the material and the development of the hole which start some time after the beginning of the pulse. Apparently, such a delay can occur when the laser pulse front is not steep enough. In this case the power density can be close to the threshold value as in the quasi-continuous-wave pulse in which the mean power density is lower than the power density of individual spikes in a spiked pulse with the same total energy and duration.

In these experiments the laser pulses have an almost rectangular shape with sharply increasing radiation power density at the onset of the pulse. Therefore, the vaporization of the material in laser drilling starts practically with the first spikes of the pulse as shown by the photorecordings of the vaporization processes. The targets used in the experiments are ground or polished up to ∇13. The pulse energy is

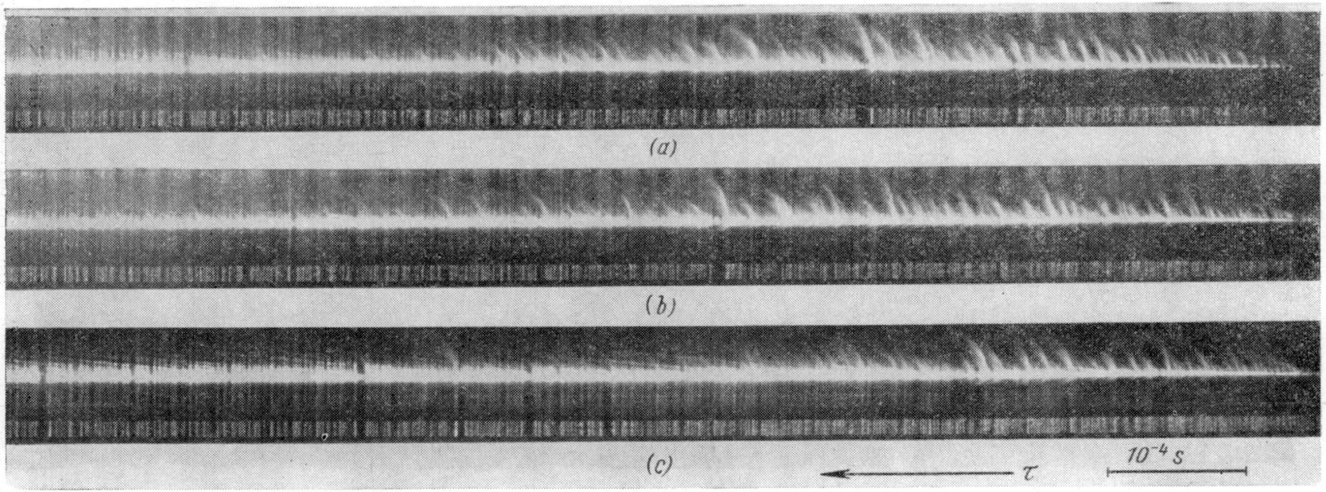

FIG. 50. High-speed photorecordings of vaporization of material from specimens of varying roughness irradiated by laser beam
(a) ∇7; (b) ∇10; (c) ∇13

9 J and the pulse duration is about 1.3 ms. The laser beam is focussed on the specimen surface with a 50-mm lens.

Figure 50 presents the photorecordings of the vaporization processes observed in these experiments.

4.4. RELATIONSHIP BETWEEN ENERGY PARAMETERS OF LASER PULSE AND PARAMETERS OF LASER-DRILLED HOLES

For given focussing conditions the ultimate parameters of the laser-drilled hole are determined by the energy and duration of the laser pulse. Many laser drilling installations produce pulses of constant duration. Under such circumstances to produce a hole of a certain size we have to select the energy of the pulse. Thus, the relationship between the pulse energy and the size of the hole has a fundamental importance in laser drilling.

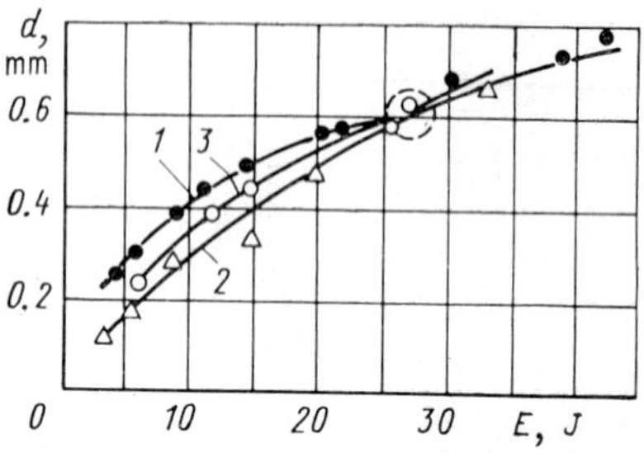

FIG. 51. Hole diameter d as a function of laser pulse energy E

1—energy variation by filters with constant pumping voltage V_p=4.2 kV, diaphragm diameter D=7 mm; 2—energy variation by pumping, D=7 mm; 3—energy variation by diaphragming with constant pumping voltage V_p=4.2 kV

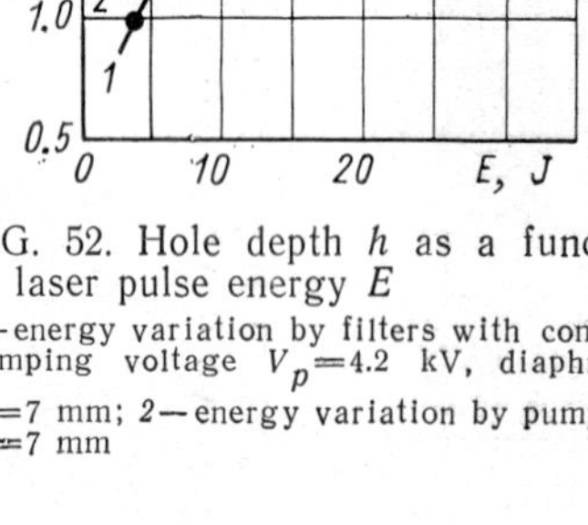

FIG. 52. Hole depth h as a function of laser pulse energy E

1—energy variation by filters with constant pumping voltage V_p=4.2 kV, diaphragm D=7 mm; 2—energy variation by pumping, D=7 mm

We should take into account that there are three techniques for varying the laser pulse energy and producing variations $h(E)$ and $d(E)$: (1) variation in the pumping energy (the most popular technique); (2) use of filters; (3) diaphragming of the beam. Figures 51 and 52 show the $d(E)$ and $h(E)$ plots for experimental drilling of steel. A 60-mm lens is used for focussing the laser beam on the specimen surface.

For the pulses of the same energy the diameters of the laser-drilled holes are smaller when the pulse energy is varied by pumping or diaphragming than when the energy is varied by filters (see Fig. 51). The diameter difference is large, especially for lower pulse energies. For instance, for the pulse energy of a few joules the hole diameters differ by a factor of 2-2.5.

The smaller hole diameters are explained, with diaphragming of the beam, by decreasing divergence and with pumping variation, by decreasing divergence and duration of the pulse. Variation in the hole depth with the pulse energy (see Fig. 52) shows that, as should be expected, the plot for pumping variation in the pulse energy is above the plot for filter variation in the pulse energy owing to decreased divergence of the beam in the former case.

An important feature of diaphragming with constant pumping is that the variation in the diaphragm diameter within a wide range has only a small effect on the depth of the laser-drilled hole while the hole diameter directly depends on the diaphragm diameter. Table 8 presents the sizes of the laser-drilled holes and the diaphragm diameters for the constant pumping of the active medium.

TABLE 8. Hole Sizes for Various Diaphragm Diameters and Constant Pumping of 4.2 kV

Diaphragm diameter, mm	Pulse energy E, J	h, mm	d, mm	h/d
12	42.2	2.2	0.80	2.3
10	38.9	2.3	0.75	3.1
8	30.2	2.2	0.70	3.1
6	21.4	2.3	0.60	3.8
5	14.7	2.5	0.45	5.6
4	11.4	2.5	0.40	6.3
3	6.0	2.1	0.25	8.4

Diaphragming the beam does not affect the hole depth because it does not change the power density distribution in the cross section of the beam and changes only the beam divergence and diameter. As the rate of the deepening of the hole is determined by the power density the diaphragming of the laser beam must not change significantly the depth of the hole. For some comparatively small diaphragm diameters the hole depth is even larger by 10-12% (see Table 8) owing to the fact that the diaphragm cuts out the peripheral regions of the beam where the power density is lower and thus leads to a decrease in the proportion of the liquid phase in the disintegration products of the material and a decrease in the energy expenditure for the removal of the liquid phase from the hole.

The above experimental results for the variation in the depth and diameter of the laser-drilled hole with changing pulse energy show, at least, a qualitative agreement with Eqs. (299) and (300)

which relate the hole depth and diameter to the laser beam parameters, in particular, the apex angle of the light cone after focussing.

4.5. EFFECTS OF TEMPORAL STRUCTURE OF LASER PULSE ON HOLE PARAMETERS

As can be seen from Eqs. (299) and (300) for the quasistationary disintegration of material [20] the depth and diameter of the laser-drilled hole are determined by the product of the laser power by the pulse duration, that is, by the total energy of the laser pulse. This implies that the final size of the hole does not depend on the power and duration of the laser pulse if its total energy remains constant.

But Eqs. (299) and (300) have been derived [20] for a simplified model which neglects, for instance, screening of the laser radiation by the disintegration products; hence, these formulas are not accurate enough for describing some practical applications. Special tests have shown a significant dependence of the hole size on the power and duration of the pulse with the constant total energy. The pulse duration is varied by means of a rotating disk. The laser beam after a 10-mm diaphragm is focussed by a 35-mm lens on the surface of a steel 45 specimen. The laser pumping remains constant to maintain the laser output parameters constant while the intensity-varying filters are used to produce pulses of a constant total energy.

Figure 53 presents a longitudinal section of a specimen with holes drilled by laser pulses of different durations. The diameters and depths of these holes are given in Table 9. The pulse energy E remains constant within 5-10%. Table 9 presents also the radiation power densities q_0 calculated from the following equation: $q_0 = E/S\tau$, where τ is the pulse duration and $S = \pi\left(F \tan \frac{\varphi}{2}\right)^2$, φ is the divergence angle of the laser beam, and F is the focal distance of the lens.

TABLE 9. Depth h and Diameter d of Laser-Drilled Holes for Various Pulse Durations

E, J	τ, ms	q_0, W/cm²	h, mm	d, mm	h/d
5.4	0.25	2.2×10^8	1.2	0.42	2.9
5.1	0.35	1.5×10^8	1.3	0.39	3.3
5.9	0.55	1.1×10^8	1.5	0.38	3.9
5.7	0.75	7.6×10^7	1.6	0.36	4.4
5.4	0.85	6.3×10^7	1.8	0.30	6.0
5.0	1.15	4.4×10^7	1.6	0.26	6.1

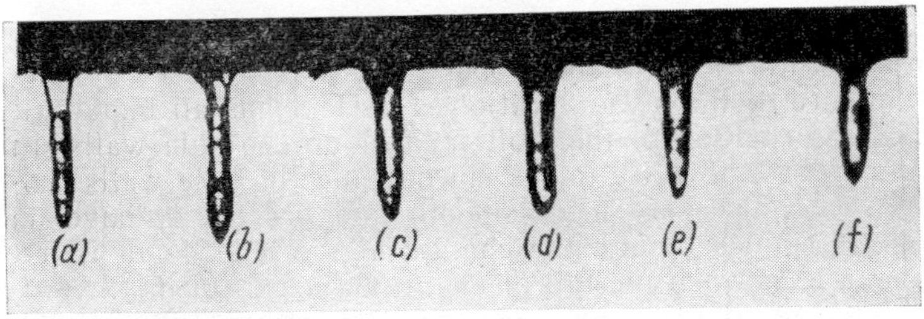

FIG. 53. Profiles of holes drilled in steel 45 by laser pulses of the same pulse energy and duration (×10)
(*a*) 1.15 ms; (*b*) 0.85 ms; (*c*) 0.75 ms; (*d*) 0.55 ms; (*e*) 0.35 ms; (*f*) 0.25 ms

An increase in the pulse duration from 0.25 to 0.85 ms results in a 50% increase of the hole depth and a 30% decrease in the hole diameter. The variation in the diameter and depth of the hole with pulse duration for the constant pulse energy can be attributed to two factors. First, the power density is higher in the central region of the beam, and as filters decrease the power density the peripheral regions of the beam where the power density becomes lower than the critical values for melting and vaporization of the target material can be increased. Thus, with the constant total pulse energy the size of the affected zone of the target is reduced. This is equivalent to a decrease in the apex angle of the light cone containing the effective flux. Hence, we can see from Eqs. (299) and (300) that the use of filters must lead to a certain increase in the hole depth and decrease in the hole diameter, and the higher the optical density of the filters the larger the changes.

The second factor is more important; it is the decrease in radiation screening by the disintegration products when the power density is reduced by longer pulse duration. This decrease in screening is due, on the one hand, to lower vapour density inside the hole and, on the other hand, to the increased proportion of the liquid phase which remains on the walls of the hole and cannot be expelled from the hole by lower vapour pressure. Thus, the liquid phase is practically not involved in screening of radiation. Indeed, the metallographic sections of the laser-drilled specimens show that short pulses (of higher power) produce holes with smooth walls, and increasing pulse durations result in ripply hole walls owing to thicker molten layers. For relatively long pulses with low power density melting becomes the predominant process, the hole depth decreases and the hole can be sealed by the molten metal (see Fig. 53*a*).

If a laser pulse has a long tail it can make a significant contribution to the formation of the molten layer on the walls of the hole. The laser used in the above drilling experiments has produced pulses with the tails about 350-400 μs long, which amounts to about a third of the pulse duration. This fact has an important effect on the formation of the molten layer on the hole walls. In some experiments the pulse tail is cut off, and the hole walls produced by such pulses are more smooth and practically have no molten metal ripples though the change in the hole depth is not large (about 15% when the tail of about 300 μs is cut off).

4.6. EFFECTS OF LASER BEAM FOCUSSING ON SHAPE AND SIZE OF LASER-DRILLED HOLES

The shape and size of the laser-drilled hole are considerably affected by the conditions of focussing of the laser beam, that is, by the focal distance of the focussing system and by the position of the focal plane of the system in respect to the surface of the workpiece.

Let us analyze how defocussing of the laser beam affects the geometry of the laser-drilled holes for lenses with the focal distances of 39, 60 and 91 mm. The laser beam in the experiments is focussed successively above the surface of the workpiece, on it and below it. Figure 54 presents the paths of the light beams and the resulting holes in steel Cт. 3 for the three lenses with different focussing of the beams. The laser pulse energy in the experiments is constant (about 35 J) and the diaphragm diameter is 8 mm.

When the beam is focussed near the surface of the workpiece the diameters of the hole entrance are a few times that of the focussed spot. For instance, in case of the most accurate focussing on the surface the entrance diameter of the hole is 4.8, 3.5 and 2.6 times that of the focussed spot for the lenses with $F = 38$, 60 and 91 mm, respectively (the focussed spot radius is found, as usual, from the formula $d_f = 2F \tan(\varphi/2)$ where φ is the angle of divergence of the beam). When the beam is focussed accurately on the workpiece surface the maximum power density is obtained on the surface resulting in the most intense vaporization of the material and in the predominance of the vapour in the disintegration products. The high vapour pressure gives rise to intense ejection of the molten metal from the walls of the hole increasing its size.

The ratio of the entrance diameter of the hole to the focussed spot diameter decreases with increasing focal distance of the lens as the radiation power density in the focussed spot is reduced by the larger focussed spot area owing to increased focal distance of the lens.

When the beam is highly defocussed, which is seen in Fig. 54 for the 39-mm lens, the entrance size of the hole becomes smaller;

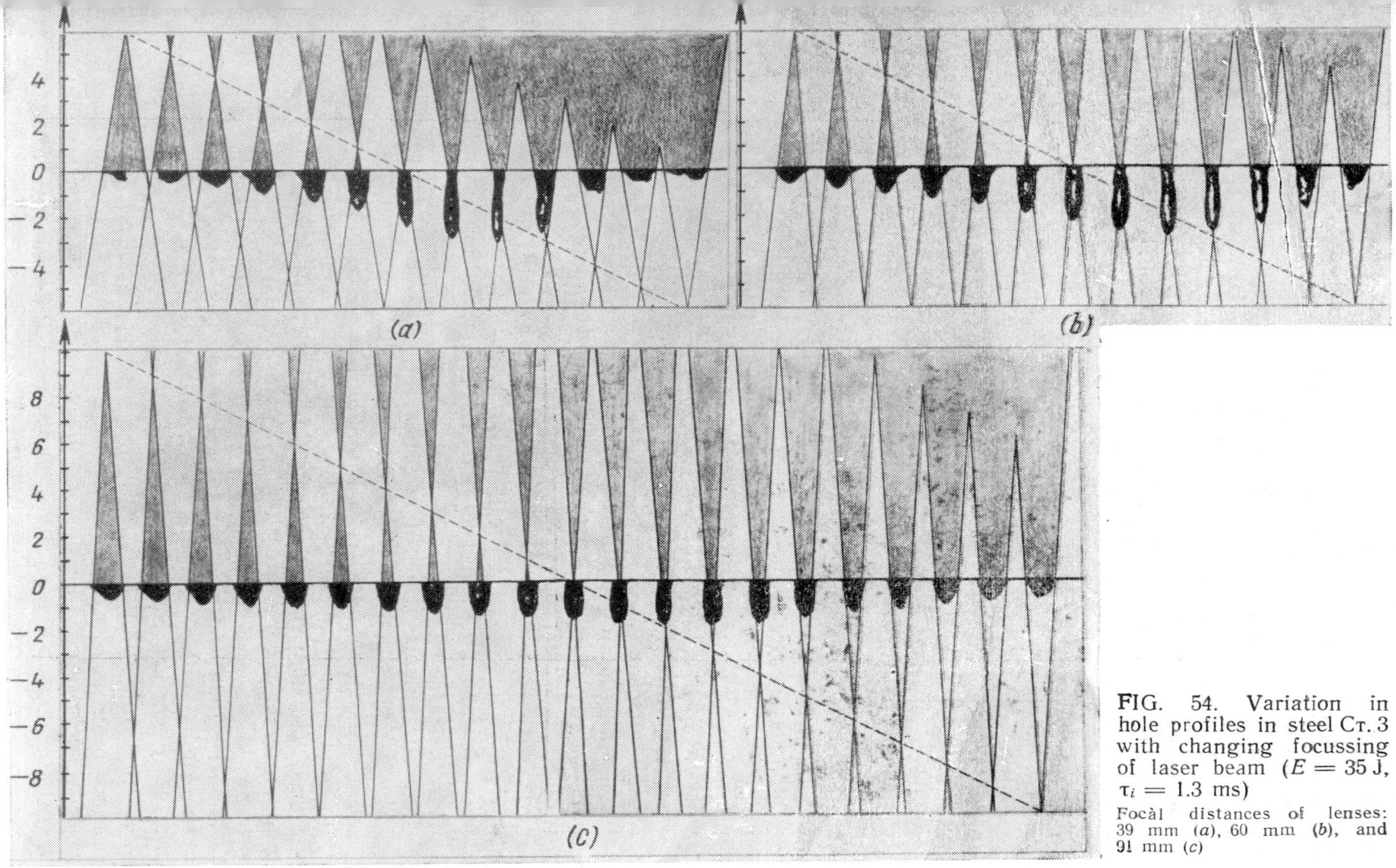

FIG. 54. Variation in hole profiles in steel Cт. 3 with changing focussing of laser beam ($E = 35$ J, $\tau_i = 1.3$ ms)

Focal distances of lenses: 39 mm (*a*), 60 mm (*b*), and 91 mm (*c*) this is due to the collection of the molten metal at the entrance because of low power density. Changing the focussing of the beam we can produce holes of various shapes — parabolical, conical (with highly defocussed beam), cylindrical (with the beam focussed on the surface) and flask-like which is a combination of a cone and a cylinder (when the beam is focussed below the workpiece surface). The hole sizes after focussing experiments are measured on the sections of the samples, and results are presented in Fig. 55 where the entrance diameter d and the depth h of the hole and the area S of the entrance are plotted as functions of defocussing of the beam.

Figure 55*a* indicates that the largest depth of the hole and the smallest diameter are obtained with various degrees of defocussing of the beam. The deepest hole is drilled by the laser beam focussed on some plane below the surface of the workpiece by the distance about 2/3 of the depth of the hole, that is, in our case about 1.5-2 mm depending on the focal distance of the lens. To drill deeper holes of smaller diameters the lens with shorter focal distances are needed.

Topography of the laser-drilled hole walls. The specific technological applications of laser drilling depend on the state of the laser-treated surfaces. The laser-drilled holes in technological workpieces are too small (tens or hundreds of micrometres) for easy evaluation of the state of the hole walls. Therefore, such evaluations should be done for holes produced by laser pulses of large enough energy and duration so that the topography of the hole walls can be recorded using standard profilographs and profilometers.

The holes to be studied are drilled in steel 45 specimens by 40-J laser pulses with the duration about 1.3 ms, the numbers of pulses used for drilling being 1, 2, 3, 4 and 6 for holes Nos. *1*, *2*, *3*, *4* and *8*, respectively (see Fig. 56).

The surface finish (roughness) of the walls of laser-drilled holes is measured on the sections of the specimen which is cut along the plane parallel to the hole axis at the distance about 0.1 mm from the centre. Then the specimens are ground with abrasive disk to remove barbs and to obtain precisely the axial plane of the hole. A profilograph with a diamond needle designed by K. Ammon is used to record profilograms and measure the surface finish of the walls of the holes. The accuracy of the profilographs is 0.20 μm, the horizontal magnification is ×30. The class of surface finish is determined by comparing the measured profilograms with the profilograms for the steel standards of machining.

The drilled specimen is placed into a special holder with the hole axis precisely parallel to the direction of the motion of

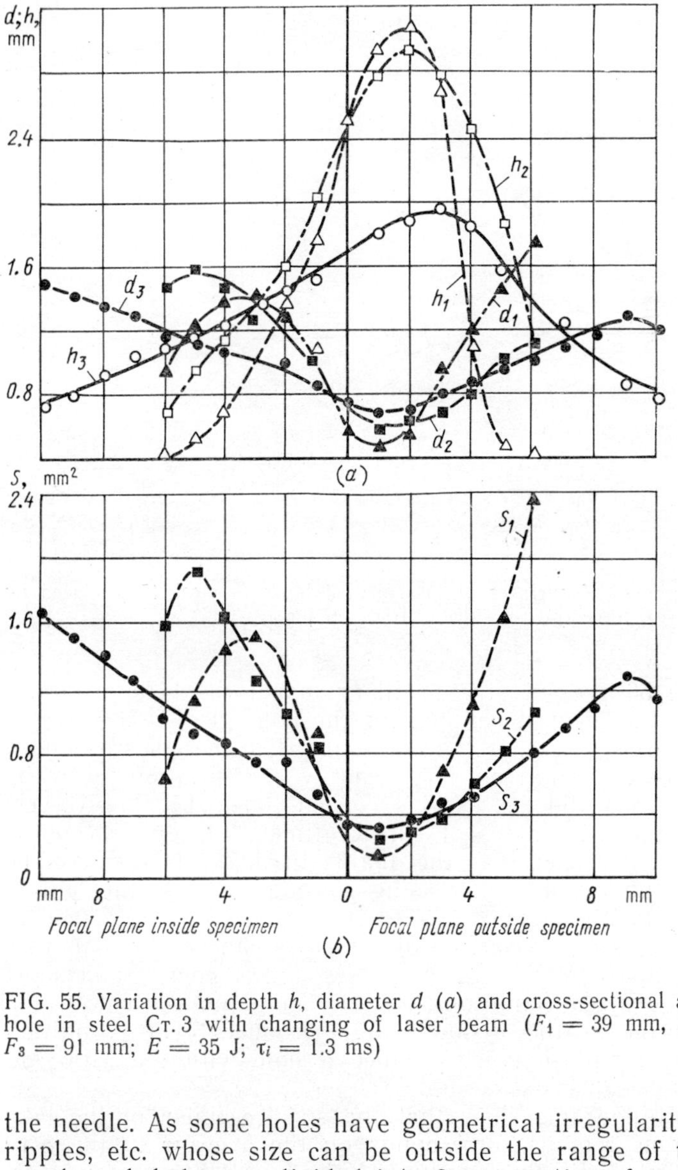

FIG. 55. Variation in depth h, diameter d (a) and cross-sectional area S (b) of hole in steel Ст. 3 with changing of laser beam ($F_1 = 39$ mm, $F_2 = 60$ mm, $F_3 = 91$ mm; $E = 35$ J; $\tau_i = 1.3$ ms)

the needle. As some holes have geometrical irregularities such as ripples, etc. whose size can be outside the range of the profilograph such holes are divided into 3-mm portions for which separate profilograms are recorded. These profilograms for the parts of the hole are later combined to form one profilogram for the hole in total, which demonstrate the macro- and microgeometry of the hole.

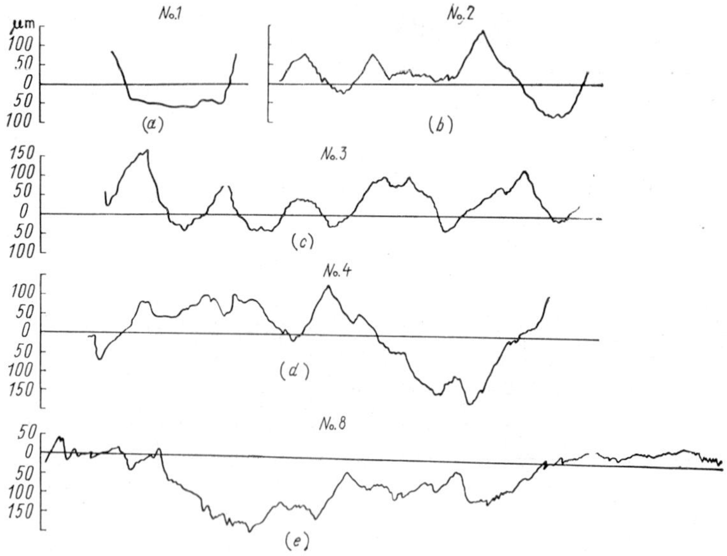

FIG. 56. Roughness of surfaces of laser-drilled holes in steel 45 produced by one (*a*), two (*b*), three (*c*), four (*d*) and six (*e*) pulses

Figure 56 presents the profilograms of the hole walls recorded with the above technique for the holes drilled by various numbers of laser pulses. It can be seen that in the holes produced by laser pulses of relatively large energy and duration the walls have large peaks and depressions — from 100 to 250 μm, corresponding to the 3-1 classes of surface finish. In contrast to the results reported earlier [64] the quality of the surface cannot be significantly improved by increasing the number of drilling pulses with the same parameters.

The defects of the surface of the holes can be eliminated by drilling the holes by a series of pulses of low energy and short duration to provide conditions for the minimum content of the liquid material in the disintegration products. This condition means that one pulse must produce a hole with the ratio h/d about 1.

The profilograms show that the surface finish of the hole walls on some regions corresponds on the average to the 8-10 classes of surface finish according to the USSR State Standards. These results suggest that the optimum regime of laser drilling can be successfully employed for drilling at least rough holes, for instance in draw plates which can then be sized with one or several finishing operations.

The state of the walls of laser-drilled holes depends on the properties of the material [64]. For instance, in laser-drilled carbon steels (the carbon content being from 0.2 to 0.6%) the roughness of the hole walls decreases with increasing carbon content (Fig. 57). This fact is attributed to the decreasing melt viscosity with increasing carbon concentration in steel which provides for easier melt removal from the hole.

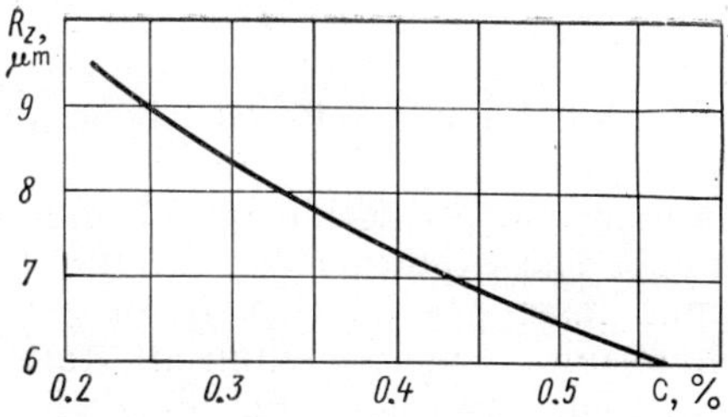

FIG. 57. Size R of microprojections on surface of laser-drilled hoes in carbon steels as a function of carbon content

4.7. FACTORS AFFECTING ACCURACY AND REPRODUCIBILITY OF LASER DRILLING

The analysis of materials drilling with laser beams producing power densities of 10^6-10^9 W/cm^2 shows that this process cannot produce sized holes of better than the 3 grade of accuracy [86]. However, laser techniques are successfully used for rough drilling of watch jewels (rubies) and diamond die plates in respective industries. Development work is underway to use laser technology for drilling of die plates for manmade fibres, ferrite plates of computer memories, sprayer heads and in such machine parts as nozzles, blades, filters, etc. made of hard and high-melting materials.

As can be seen from the analysis of the physical processes in laser drilling the poor accuracy and quality of the laser-drilled holes are mainly determined by the high content of the liquid phase in the disintegration products (30-80%) and uncontrollable redistribution of the molten material on the walls and bottom of the hole by the end of the laser pulse and after it. The thickness of the molten layer on the walls and bottom of the hole is determined by the energy and duration of the laser pulse. The pulses of longer duration produce disintegration products with higher liquid phase content thus hindering the control of the drilling process and increasing variability of the hole sizes. Too long pulses (over 1 ms) produce large zones of virtual structural changes in the material owing to large sizes of the heat-affected zones around the holes. In the brittle materials long pulses can produce cracking owing to large thermal stress zones (see Fig. 41). The probability of cracking of the material is much lower with shorter laser pulses. For instance, no cracking is found near the laser-drilled holes in ferrite if the laser pulse duration is not more than 100 μs.

We have discussed above the effect of the pulse duration on the accuracy of the laser-drilled holes (or, more exactly on the formation and motion of the molten material which can seal the hole). The experimental results can be analyzed to find the optimum laser pulse duration for drilling high-quality holes in such materials as molybdenum and tantalum.

The accuracy and quality of laser drilling is considerably affected by variation in the power density during the pulse with random spiked structure, that is, by the temporal structure of the pulse and by the shape of its envelope. The laser optical systems with proper alignment and diaphragming make possible the drilling of holes with sufficiently accurate shape and size [86]. On the other hand, the lack of coaxial alignment of the laser beam in the optical focussing system can distort the power density distribution in the focussed spot or make the drilled hole nonperpendicular to the workpiece surface despite its proper alignment with respect to the optical system.

The angular distribution of radiation within the directivity pattern and the intensity distribution over the beam cross section have a significant effect on the accuracy of the size and shape of the laser-drilled hole. Irregular distribution of intensity over the beam cross section which is typical of the random-generation spiked laser pulses makes an important contribution to drilling errors (distortion of the cross section shape of the hole). Angular asymmetry of the beam also results in distortion of the focussed spot even when the beam is diaphragmed. Asymmetric distribution of intensity can be determined by the inhomogeneity of the active laser medium or the light field of pumping, or inaccurate alignment of the resonator elements and resonator itself with respect to the optical system.

The laser radiation with the symmetric but nonuniform distribution of intensity which decreases from the centre to the periphery of the beam (as is typical for free-generation laser pulses with random spiked structure) produces holes of regular shape but sometimes with molten edges and considerable entrance conicity.

G. Suslov has analyzed the measured intensity distributions over the beam cross section and recommended the following optimum configurations of the spherical resonator: $R/L = 0.6$ for single-pulse mode of laser operation and $R/L \geqslant 6$ for pulse-repetition operation (R is the curvature of the resonator mirrors and L is the distance between the mirrors). As the pumping pulse repetition rate increases and becomes higher than the generation threshold by a factor n_W the optimum configuration approaches the confocal configuration while an increase in pumping rapidly produces a beam with the most uniform intensity distribution obtainable for the given resonator configuration, which can be produced even for $n_W \geqslant 2$. For the laser with a flat resonator the extent of uniformity of the intensity distribution over the beam cross section slightly depends on the threshold and is determined mainly by the uniformity of pumping over the cross section of the active element.

The profile of the laser-drilled hole is greatly affected by such factors as the caustic shape of the optical focussing system, the position of the workpiece with respect to the caustic, variation in the intensity distribution along the beam axis with increasing hole depth, etc.

A variety of factors determine the accuracy and reproducibility of laser drilling which are essential for some technological laser applications, for instance, for producing identical holes in parts of different materials. These factors can be classified into three groups [86]. The factors of the first group which affect mainly the reproducibility of drilling results (for instance, the sizes or shapes of holes) stem from the instability of the laser radiation parameters (energy, pulse duration, divergence angle, spatial and temporal structure of the laser beam).

Laser drilling makes use of pulsed laser operation providing for sufficiently high productivity of the process. The experiments with neodymium-glass and ruby lasers have shown that the laser radiation parameters vary significantly until a steady-state thermal regime of the active element is established (what is called transient conditions). The variation in the laser radiation parameters is caused by heating of the laser active element, which changes the spectral and emitting parameters of the active medium and by thermal deformation of the resonator because of non-uniform distribution of temperature over the cross section of the active medium.

The transient conditions duration is determined by the size of the active element, its thermal diffusivity and the efficiency of the cooling system; typically it is not more than a few tens of seconds. By the end of the transient conditions period the pulse energy can increase by a factor of 5-10 compared with the first pulse, the pulse duration can increase by a factor of 5-6 and the beam divergence by a factor of 2-3. These changes in the radiation parameters must result in changes in laser drilling results, for instance, the sizes of the laser-drilled holes must be increased under the transient conditions.

The parameters of laser radiation (energy, pulse duration and the divergence angle) in the steady-state operational mode are determined by the power of pumping and the efficiency of the cooling system which provide for a certain steady-state temperature of the active medium and resonator. Hence, only the steady-state operation of laser can make it possible to produce identical holes (or, more exactly, holes with close shapes and sizes). With the single-pulse operation the variability of the hole size is due primarily to the instability of the laser pulse parameters in the steady-state regime. The instability of a laser radiation parameter (and, respectively, of a laser-drilled hole parameter, for instance its size) is gauged by the average relative deviation $\varkappa_x = \Delta x/x$; with the gaussian distribution it is proportional to the root-mean-square deviation $\sigma_r = 1.25|\overline{\Delta x}|$ [86]. The quantity $\varkappa_x$ is convenient for describing the instability of the parameters because it can be readily found from the experimental results. Moreover, this measure of instability enables us to analyze separately the effects of various factors on the reproducibility of laser drilling results.

The most variable parameter in the pulse-repetition steady-state operation is the radiation energy whose instability $\varkappa_W$ is much higher than the instability $\varkappa_\tau$ of the pulse duration and the instability $\varkappa_\varphi$ of the beam divergence. For instance, for the neodymium-glass laser with the pulse energy of 2-3 J we have $\varkappa_W = 5\%$, $\varkappa_\tau = 0.5\%$ and $\varkappa_\varphi = 1\%$ [95]. An increase in the instability of the pulse energy (as the main factor) results in increasing instability $\varkappa_d$ of the diameter of through holes; the thicker the material being drilled the higher the pulse energy instability. This can be seen from the results presented in Table 10 which gives the instability of the pulse energy of the neodymium-glass laser and the instability of the laser-drilled hole diameter in materials with different thickness [95].

TABLE 10. Instability of Pulse Energy of Neodymium-Glass Laser and Instability of Hole Diameter in Materials of Different Thickness

Material	Thickness, mm	Instability, %	
		energy	diameter
Ferrite	0.5-0.6	2.0	6.6
		5.0	11.5
Aluminium foil	0.1	1.5	3.4
		6.5	7.5
Chromium film on polycrystalline glass	Not measured	2.0	1.7
		5.0	2.0

The stability of the radiation parameters in the steady-state pulse-repetition operation mode is determined primarily by surpassing the generation threshold n_W. For instance, when the threshold surpassing increases from 1.1 to 3.5 the instability of the laser pulse from the neodymium-glass laser drops from 18 to 2% and is then determined by the pumping instability and temperature fluctuations in the active medium. The pulse energy instability $\varkappa_W$ is related to the pumping instability $\varkappa_{W_p}$ by the following equation:

$$\varkappa_W = \frac{n_W}{n_W - 1} \varkappa_{W_p} \tag{304}$$

Much weaker effect on the stability of the (neodymium-glass) laser parameters is produced by the repetition rate of the pumping pulses and the characteristics of the cooling system such as the type of the cooling liquid, its flow rate and its temperature.

In the ruby laser, on the contrary, all the factors leading to temperature increase in the active medium exert a considerable effect on the stability of the laser parameters since the generation threshold in ruby markedly increases with temperature (and the surpassing of the threshold thus decreases). The instability of the radiation energy is related to the temperature by the following equation (with constant surpassing of the generation threshold):

$$\varkappa_W = \frac{2\varkappa_T}{n_T^2 - 1} \tag{305}$$

where $\varkappa_T$ is the relative instability of the temperature, $n_T = T_{max}/T$, T_{max} is the maximum temperature at which generation is still possible (the temperature of generation collapse), and T is the operational temperature for which $\varkappa_W$ and $\varkappa_T$ are determined.

Equation (305) shows that at T_{max} we have $n_T = 1$ and $\varkappa_W \to \infty$ that is, the generation threshold increases with increasing T and the temperature T_{max} corresponds to the threshold conditions of excitation of the active medium irrespective of pumping.

The instability of the pulse duration $\varkappa_\tau = \frac{\Delta\tau}{\tau}$ is related to the duration of the pumping pulse τ_p and n_W by the following equation:

$$\varkappa_\tau = \frac{\varkappa_{W_p} t_\mu [1 - \exp(-t_p/t_\mu)]}{\left(n_W + e^{-\tau_p/\tau_\mu} - 1\right)\left(\tau_p - \tau_\mu \ln \frac{n_W}{n_W - 1 + e^{-\tau_p/\tau_\mu}}\right)} \tag{306}$$

where τ_μ is the lifetime of the metastable atomic level.

Equation (306) indicates that in the four-level active mediums (for instance, in the neodymium glass) the instability of parameters is highly dependent on surpassing the generation threshold, and the main contribution to instability is made by the pumping instability. However, the energy instability is always higher than the instability of the pulse duration.

The stability of the pulse duration for the three-level active mediums (for instance, ruby) depends also on the thermal regime of the active medium; it increases with decreasing temperature and temperature fluctuations. The stability of the laser radiation parameters is also highly dependent on the wear and ageing of the resonator mirrors, reflectors of the lamp and elements of the optical system and on variation in the properties of the active medium owing to the effect of the ultraviolet radiation of the pumping lamp. These factors determine the life of the laser or its elements and the stability of the laser parameters for long-term operation of the laser. The wear and ageing of the laser components result in a gradual decrease in the pulse energy, pulse duration and increase in divergence of the beam and hence in variation of the hole diameter and hole depth in laser drilling. The nature of the wear and ageing of the optical system components determine the reproducibility of diameters and shapes of the laser-drilled holes in technological laser applications.

The statistical computer results for a large number of laser-drilled holes in ferrite plates (multiple diameter measurements and radiation parameter measurements for each pulse) show for the steady-state laser operation that laser parameters and hole diameters have gaussian distribution [86] (Fig. 58). The measurements of the diameters of the laser-drilled holes in the draw plates for synthetic fibres have also demonstrated the gaussian distribution of the mean diameters [86].

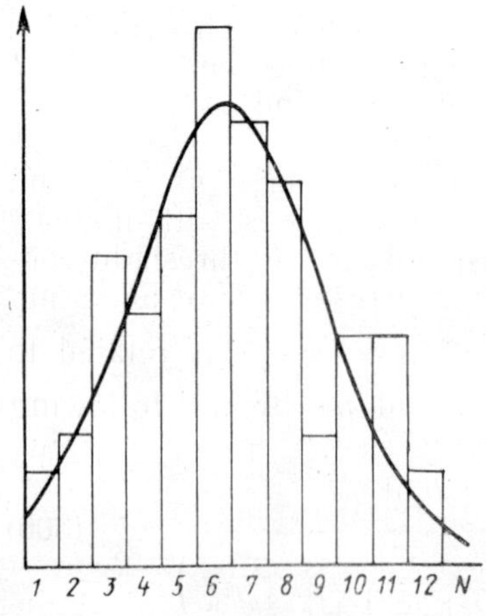

FIG. 58. Distribution of diameters of holes drilled in ferrite by single laser pulses

The factors of the second group which also affect the reproducibility of laser drilling results are the errors in positioning and motion of the workpiece with respect to the laser beam.

For instance, if the workpiece is moved so that the perpendicular to the surface is not parallel to the optical axis of the focussing system for a series of drilled holes the laser beam will be increasingly defocussed, which results in the variation in the hole diameter and shape. The effect of these factors can be minimized.

The factors of the third group are due to the inhomogeneity of the properties and structure of the material; these factors also affect the reproducibility of laser drilling results. These factors are significant when the hole size is comparable to the size of inhomogeneities or when their concentration is high. If the material contains dissolved gases or impurities which can act as nucleating centres in boiling the liquid phase can boil at lower temperature than the chemically pure material. Laser-drilling experiments have been done with specimens of various grades of copper (cathode, anode, blister, vacuum-melted, industrial and porous) which have different gas concentrations and porosity, and it has been found that the hole shape in porous specimens differs significantly from that in nonporous specimens [12]. The copper specimens of different grades exhibit different relationships between the hole depth and the laser pulse energy; the largest deviation of results is found for the blister copper specimen; this can be attributed to higher degree of inhomogeneity of this material. Below we shall discuss the stability of the main parameters of a technological laser installation and measurements of the series of holes drilled with this installation.

This installation is used for drilling draw plates for synthetic fibres; drilling is done by several laser pulses with the repetition rate varying from 1 to 8 Hz. It has been assumed that the following parameters determine the stability of the individual components of the installation: (1) the electric power of the pumping power supply; (2) the light power of the pumping light; (3) the laser pulse energy; (4) the duration and shape of the light pulse of the pumping lamp and the duration and shape of the laser pulse; (5) the divergence of the laser beam; (6) the thermal regime of the active medium.

The experimental installation for simultaneous measurement of the above parameters is shown in Fig. 59. The first three parameters are measured with a loop oscillograph *24*; the signal from photocell *23* which receives the light from a pumping lamp *1* from the side of the "total-reflection" mirror *18* of the resonator is fed to the first input terminal of the oscillograph, the signal from the inductance coil wound on the discharge wire of the pumping lamp is fed to the second input terminal, and the signal from photocell *10* which receives a part of the laser beam from a beam-splitting glass plate *8* is fed to the third terminal. The signals fed to the loop oscillograph are recorded on the photographic paper as straight sections whose speed varies according to the pulse repetition rate. Since the duration of the analyzed

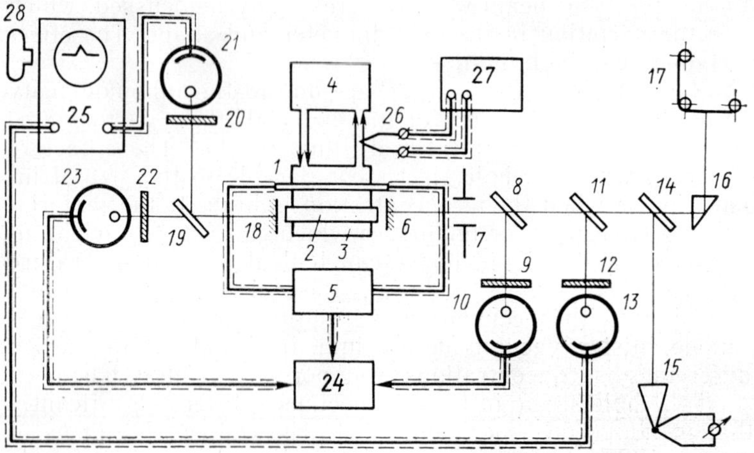

FIG. 59. Experimental installation for measuring stability of laser parameters
1—pumping lamp; *2*—laser active element; *3*—reflector; *4*—cooling unit; *5*—power supply unit; *6*—semiopaque resonator mirror: *7*—diaphragm; *8*, *11*, *14*, *19*—beam-splitting glass plates; *9*, *12*, *20*, *22*—filters; *10*, *13*, *21*, *23*—photocells; *15*—calorimeter; *16*—prism; *17*—photographic paper and drive; *18*—"total-reflection" resonator mirror; *24*—loop oscillograph; *25*—two-channel CRT oscillograph; *26*—thermocouple; *27*—recorder; *28*—camera

pulses is 200 μs the length of the sections recorded on the photographic paper is proportional to the pulse energy as the oscillograph operates in the ballistic regime when τ is in the 10^{-3}-10^{-7} s range. A two-channel cathode-ray oscillograph *25* records the shape and duration of the pumping pulses and the laser pulses simultaneously or separately by receiving signals from photocells *13* and *21* which receive a part of the laser beam and a part of the pumping light split by plates *11* and *19*, respectively. A CRT oscillograph with a memory tube can record on the same screen several successive laser pulses (up to 5-7) by vertical displacement of the oscillograph ray; this is convenient for analyzing the laser operation under the transient conditions when a pulse series with a high repetition rate must be recorded. The filters placed before the photocells separate the laser radiation or the pumping lamp light.

The laser output energy is measured with calorimeter *15* which makes it possible to determine the mean laser power for high pulse rates (over 4 Hz).

The cross-section shape of the laser beam emitted from the resonator is found from the trace left by the beam on the photographic paper *17*. As the motion of paper is synchronized with laser pulses a large number of beam images are recorded making it possible to observe variations in the shape and divergence of the beam during operation. The divergence angle φ is found from the standard equation

$$\tan\frac{\varphi}{2} = \frac{D - D_0}{2L}$$

where D_0 is the diaphragm diameter, D is the diameter of the beam image on the paper and L is the optical path of the beam from the diaphragm to the paper.

The thermal regime of the active medium is studied using thermocouple *26* placed into the pipe of the cooling system through which the heated water from the laser is fed to a cooling unit *4*. The signal from the thermocouple is fed to recorder *27*.

The pumping powers in the experiments vary from the generation threshold (the capacitor voltage is about 750 V) to the peak power of the pumping lamp (the capacitor voltage is about 1350 V). The laser pulse rates vary from 1 to 6 Hz.

As the loop oscillograph records simultaneously the laser radiation energy, the light energy of pumping and the electric supply power of the pumping lamp we can determine the stability of these parameters and the relationship between them for various regimes of laser operation. Figure 60 presents a typical oscillogram recorded for intermediate pumping power and laser pulse rate. The oscillograms show that in all operational regimes for some time after the beginning of the periodic pumping the laser operates under the transient conditions with sharp variation in the radiation parameters, namely, the duration, energy and shape of the pulse and the divergence of the beam.

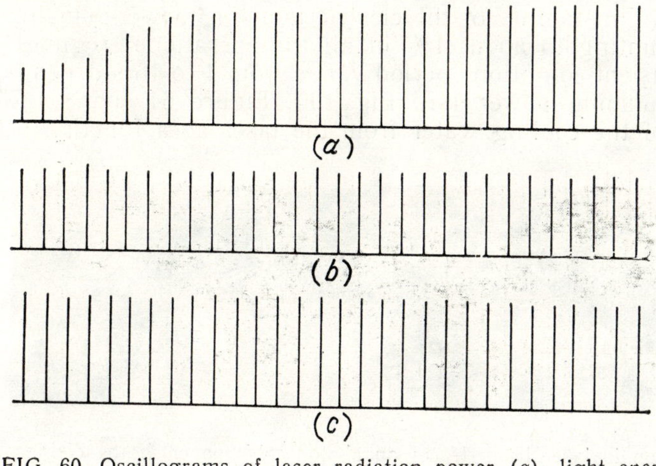

FIG. 60. Oscillograms of laser radiation power (*a*), light energy of pumping lamp (*b*) and pumping power supply (*c*) (frequency is 3 Hz, pumping voltage is 1000 V)

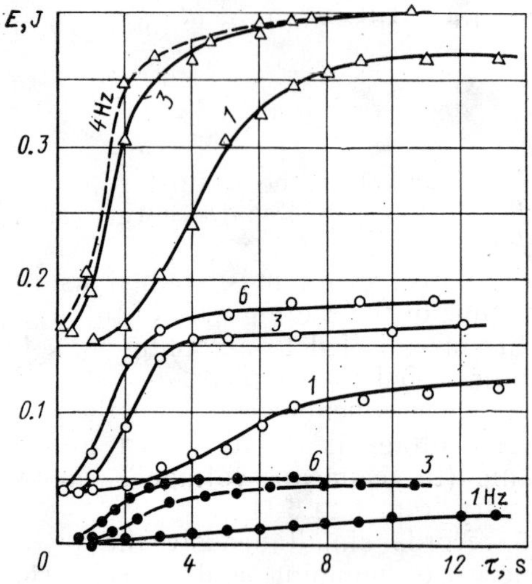

FIG. 61. Pulse energy as a function of time (from the beginning of pumping) for various frequencies and various pumping voltages

●—750 V; ○—1000 V; △—1350 V. The dashed line represents the collapse of the frequency operational regime of the power supply for pumping voltage over 1000 V from 6 to 4 Hz

The laser pulse energy sharply increases (by a factor of 3-10) by the end of the transient-conditions period and with increasing repetition rate of the laser pulses (Fig. 61). For instance, for 1000-V pumping the laser pulse energy at the pulse repetition rate of 6 Hz is 1.5 times that at 1 Hz. The divergence of the laser beam also increases several times under the transient conditions. With the constant pumping power the beam divergence increases 4.5 times when the pulse repetition rate increases from 1 to 3 Hz (Fig. 62).

Under the transient conditions the laser pulse duration decreases by a factor of 2-4 (Fig. 63) while the variations in the amplitude and shape of the pumping pulse are considerably smaller. From the moment the laser is switched on the instability of the electric pumping power remains constant amounting to about 1% in all the operational regimes.

The transient-conditions period varies from 4 to 15 s depending on the pumping power (see Fig. 61). Figure 64 shows the temperature of the cooling water from the laser as a function of

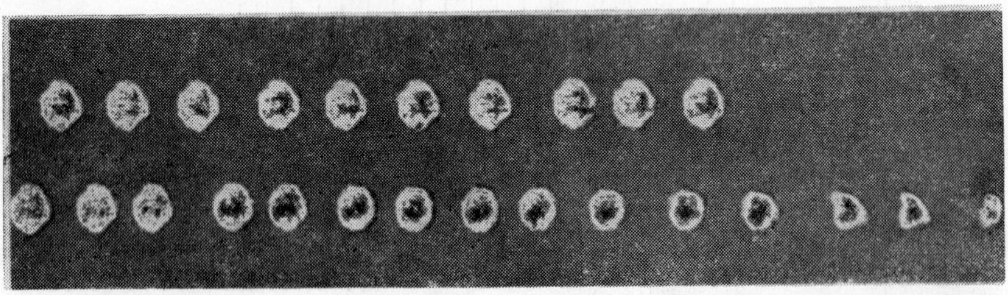

FIG. 62. Variation in laser beam traces on photographic paper at the distance of 50 cm from diaphragm from the beginning of periodic pumping with $f = 3$ Hz and $V_p = 1350$ V

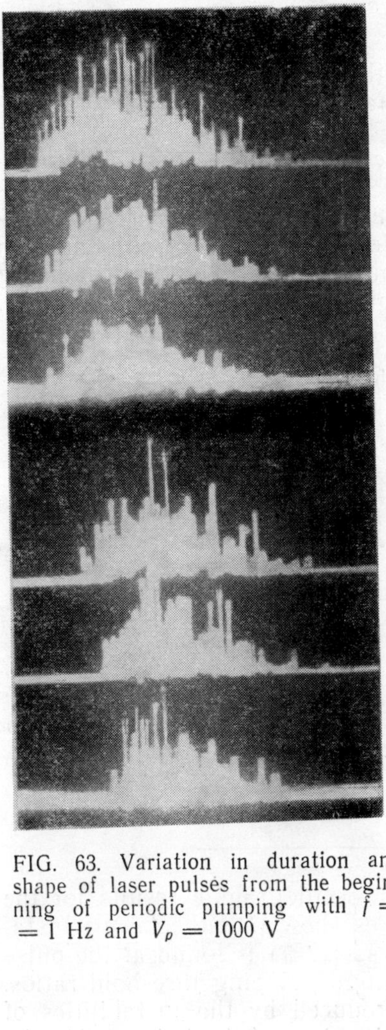

FIG. 63. Variation in duration and shape of laser pulses from the beginning of periodic pumping with $f = 1$ Hz and $V_p = 1000$ V

time in various operational regimes. It can be seen that the temperature of the active medium is stabilized much later than the output laser power. The transition to the steady-state thermal regime lasts from 20 s to 2 min depending on the pulse repetition rate and the pumping power. This difference in the transient-conditions period between the temperature regime and the laser radiation

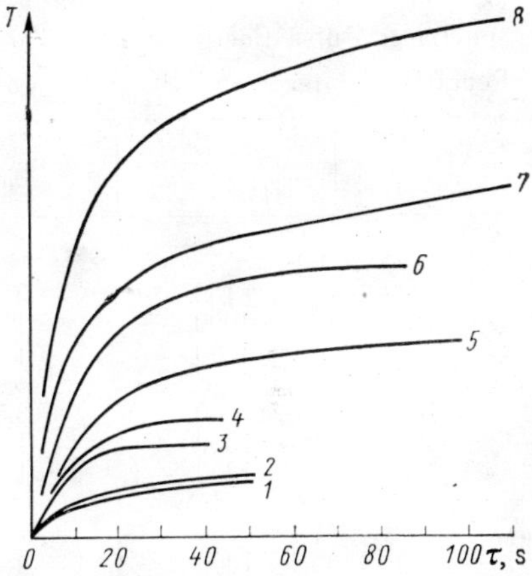

FIG. 64. Variation in temperature of water in laser cooling system with time from the beginning of pumping for various operational regimes
1 — V_p=750 V, f=1 Hz; 2 — V_p=1000 V, f=1 Hz; 3 — V_p=1350 V, f=1 Hz; 4 — V_p=750 V, f=3 Hz; 5 — V_p=1000 V, f=3 Hz; 6 — V_p=750 V, f=6 Hz; 7 — V_p=1350 V, f=3 Hz; 8 — V_p=1000 V, f=6 Hz

parameters can be attributed to the following factors. An increase in the energy and duration of the laser pulses and the divergence of the beam under the transient conditions can be attributed to the fact that the flat parallel ends of the active element become spherical with heating of the active element following the beginning of the pumping. At a certain point in time the curvature of the ends stops growing though the temperature of the active element continues to increase. As the luminescent properties of neodymium in this temperature range show only a slight dependence on temperature the transition to the steady-state regime for the radiation parameters must occur faster than for the temperature regime of the active medium.

By the end of the transient-conditions period all the radiation parameters have steady-state values, and their variation is determined by the instability of the steady-state laser operation. Statistical processing of the oscillograms recorded with the loop oscillograph has yielded the instabilities of the electrical pumping power, the light power of the pumping lamp and the pulse energy. The results are given in Table 11.

TABLE 11 Instabilities of Pumping Light Power $\varkappa_{E_p}$ and Laser Pulse Energy $\varkappa_{E_l}$ for Various Pulse Repetition Rates f and Pumping Voltage V_p

f, Hz	V_p, V	$\varkappa_{E_p}$, %	$\varkappa_{E_l}$, %
1	750	3	4
	1000	1	1.5
	1350	1	1
3	750	3.5	4.5
	1000	1.5	1.5
	1350	1	1
6	750	4	4 5
	1000	2	2

The data of Table 11 and the oscillographic results for the shape and divergence of laser pulses show that the lowest instability of all the radiation parameters (1-2%) is found at the pulse repetition rate of 1-3 Hz and with high pumping threshold ratios.

Let us analyze the effects produced by the instabilities of the laser radiation parameters on the sizes of the holes drilled by the above laser installation. The experimental holes are drilled in steel plates of standard thickness (50 and 100 μm) to eliminate the potential effects of thickness variation.

Series of holes (80-100 holes) are drilled in various operational regimes with various pulse repetition rates and pumping powers. The exit hole diameters are carefully measured, and the statistical processing of the results yields the instabilities of the hole sizes. The results show that the instability of the hole sizes decreases with increasing pumping power (that is, increasing

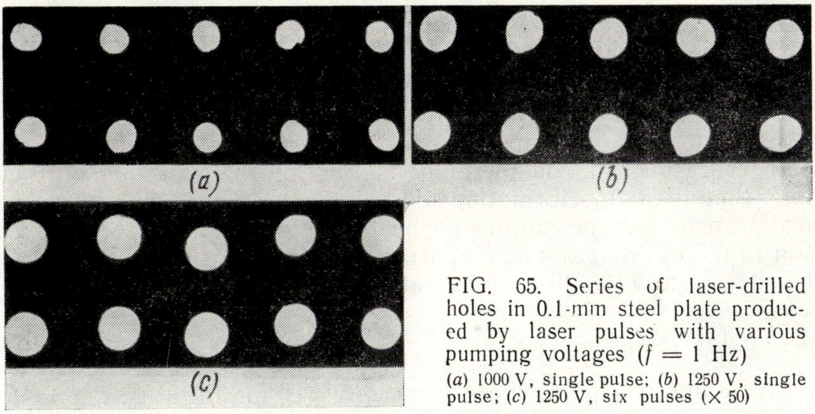

FIG. 65. Series of laser-drilled holes in 0.1-mm steel plate produced by laser pulses with various pumping voltages ($f = 1$ Hz)

(*a*) 1000 V, single pulse; (*b*) 1250 V, single pulse; (*c*) 1250 V, six pulses (× 50)

pumping threshold ratio). For instance, the through holes drilled in 0.1-mm steel plate by single pulses with the repetition rate of 1 Hz have the diameter instability of 14% at the pumping voltage of 750 V while at the pumping voltage of 1350 V the diameter instability drops to 4%. The instability of the hole diameters is 2-3 times the instability of the respective laser pulse energy. This is illustrated in Fig. 65*a* and *b* which shows a series of holes drilled by single laser pulses in the 0.1-mm steel plate (the pulse repetition rate of 1 Hz) with various pumping powers.

The accuracy and reproducibility of laser drilling can be improved by multipulse technique. A comparison of the quality of holes drilled using single-pulse and multipulse techniques is given in Fig. 65*c* which shows a series of holes drilled with six laser pulses in the same laser installation as shown in Fig. 65*b*. It can be seen that multipulse laser drilling produces holes of better quality and reproducibility. The statistical processing of the results gives the instability of the hole diameters as low as 2-3%.

4.8. TECHNIQUES FOR IMPROVING ACCURACY AND REPRODUCIBILITY OF LASER DRILLING

The techniques used for improving the accuracy and reproducibility of laser drilling can be classified into two groups [86]. The techniques of the first group involve directly the laser operation, for instance, the selection of drilling conditions providing for minimized heat transfer to the walls of the hole, the regulation of the shape and structure of the pulse, the selection of an appropriate laser drilling procedure (multipulse drilling, cylindrical light pipe technique, etc.) and others. The techniques of the second group include various calibration procedures, chemical etching, blowing of the holes with compressed air, etc.

(a) SELECTION OF DRILLING CONDITIONS

To decrease the content of the liquid phase which makes a major contribution to deterioration of the quality and accuracy of the laser-drilled holes laser drilling can be done by pulses of short duration when heat transfer to the walls is low. The radial heat transfer results in considerable melting of the entrance of the hole, that is, in the formation of the entrance cone. The effect of heat transfer can be minimized by limiting the time τ_v of transition to the quasistationary vaporization (for a certain laser power density q_0) to r_f^2/a, that is, by satisfying the condition [86]:

$$\tau_v(q_0) \leqslant r_f^2/a \tag{307}$$

Taking an approximate equation [34] for the rate $v(\tau_v)$ of nonstationary vaporization as a function of time and assuming that $v(\tau_v) = 0.9\, v_0$ we obtain the following condition for the necessary laser power density [86]:

$$q_0 \geqslant 30 \frac{L_b a}{r_f} \left(\ln \frac{s r_f}{30 a} \right)^{-1} \tag{308}$$

For many metals for which the vaporization heat $L_b \approx \approx (1\text{-}5) \times 10^4$ J/cm³, $a \approx 1$ cm²/s and the sound velocity $s \approx \approx 10^5$ cm/s the minimum value of q_0 according to condition (308) varies from 10^7 to 5×10^7 W/cm² at $r_f = 100$ μm. When condition (308) is satisfied the entrance cone of the laser-drilled hole decreases in size but other factors which contribute to melting during the hole development remain unchanged.

(b) SELECTION OF SHAPE AND STRUCTURE OF DRILLING PULSE

In the analysis of the properties of laser radiation, the stability of its parameters and its effect on the errors in laser drilling we consider mostly the integral parameters of the laser radiation such as the pulse duration, pulse energy and the beam divergence. However, the temporal and spatial structures of the pulse and its shape produce an equally significant effect on the accuracy and reproducibility of laser drilling. Let us briefly discuss the effect of these features on laser drilling.

In most industrial laser-drilling installations the lasers have flat-mirror resonators which provide for random spiked generation of 0.5-1.5-ms pulses. The parameters of the resulting radiation are not stable leading to errors in laser drilling. Therefore, in precise laser machining (particularly, in drilling holes with the ratio $h/d > 2$) it is important to control the laser pulse parameters (to select and control the optimum pulse duration and to obtain a specified shape and structure of the pulse). For drilling deep holes the pulse duration must be increased since the drilling efficiency is considerably lower at the radiation power densities exceeding 10^8-10^9 W/cm^2. On the other hand, increasing pulse duration results in poorer accuracy because of increasing the molten phase content. Hence, to minimize the content of the molten phase drilling should be done with pulses of an intermediate duration (the experiments indicate that the duration must not exceed 0.1-0.2 ms) and high enough radiation power density. The hole of required depth is drilled by several successive laser pulses (the multipulse technique is discussed in detail below). There is practically no data on the contribution of the laser pulse shape (the durations of the front and tail and the pulse shape in general) to the process of laser drilling. The experimental results show that the process of material disintegration cannot be comprehensively analyzed only in terms of the duration and energy of the pulse as these parameters do not describe the character of the energy transfer to the affected zone of the material. But the character of energy transfer is an important contributing factor in laser drilling, which to a considerable extent determines the drilling result. The rate of heating of the material is largely dependent on the pulse shape while the threshold of disintegration of the material remains practically the same for the pulses of different shapes.

The duration of the pulse front determines the period τ_v of heating the material to the vaporization point. The period τ_v and the size of the heat-affected zone (and, hence, the diameter of the entrance cone) increase with increasing duration of the pulse front. When the pulse front is not steep the temperature of the material and, hence, its reflectivity increase slowly thus reducing the efficiency of laser drilling (this is most significant for the metals with high initial reflectivity). From the above discussion it follows that the duration of the pulse front should not exceed $0.1\,\tau_v$, i. e., $\tau_v = 0.1\, r_f^2/a$; at $r_f \approx 10^{-2}$ cm and $a \approx 1$ cm^2/s it amounts to 10 μs. The shorter the pulse tail the lower the probability that the hole will be sealed by the molten material as with the short pulse tail the vaporization of the material is abruptly discontinued at the moment the hole reaches the required depth.

As follows from the equation for the energy balance in the hole [Eq. (296)] the vaporization rate remains constant and equal to the initial rate v_0 if the radiation power is

$$P(\tau) = P_0 \left[\left(1 + \frac{\tau v_0 \tan\varphi}{r_f}\right)^2 + 2\left(1 - \frac{v_0 \tau \tan\varphi}{r_f}\right) \frac{v_0 \tau \tan\varphi}{r_f} \frac{L_m}{L_b} \right] \tag{309}$$

where P_0 is the initial radiation power.

For instance, to drill a hole with the diameter equal to two diameters of the focussed spot when $L_b/L_m = 10$ the laser power must increase 4.4 times by the end of the pulse [86]. The experimental results show that when the ratio $h/d \leqslant 10$ the pulse energy absorbed on the bottom of the hole decreases with the hole depth, and as the energy is lost owing to defocussing and absorption by the hole walls the laser power should increase during the pulse as τ^2 to provide for a constant vaporization rate.

In practically all existing industrial laser installations the laser pulses do not meet the above requirements. Even if the pulse front sometimes has a proper duration the pulse tail is too long: it is about 100-200 μs long for the pulses of 1.0-1.5 ms. For such pulses the accuracy and quality of drilling can be improved by introducing modulators into the system, for instance, rotating disks with slits installed in the cavity or outside it, which cut off the pulse tail. Such mechanical modulators can be used for shaping the pulse as a whole though this will lead to inevitable energy losses. The pulses of a specified shape can be produced by appropriate design of the pumping system. For instance, the power supply of the pumping lamp can include several discharge lines in series having a small time constant of discharge; these lines are charged to different voltages according to the required pulse power variation in the irradiated zone. Similar results can be produced by irradiating the material with a series of pulses whose energies vary in a specified way.

The spiked laser pulses are perfectly suitable for laser machining (including drilling) as they provide effective enough utilization of laser radiation for the disintegration of the material while the averaged pulse power remains comparatively low and the energy losses for heat conduction are negligible (these advantages are due to high instantaneous powers and short durations of spikes). The experimental data and theoretical results show that the disintegration products are ejected outside during the intervals between the spikes and, as a result, the screening of radiation by the products of the disintegration of the material is considerably reduced.

But some factors due to random generation of laser pulses impair the accuracy and quality of laser drilling. A considerably higher accuracy and quality of drilling can be obtained with ordered laser generation when all the spikes in the pulse have the same shape, energy and duration, and are separated by the same time interval. This laser operational regime is distinguished by uniform spatial distribution of radiation intensity across the laser beam.

There are several techniques for generating pulses with regular temporal and spatial structures, for instance, multimode generation with approximately constant Q-factor (with spherical resonator or resonator with lenses) or single-mode generation with mode selection. A pulse of a regular structure can be produced using low Q-switching; this is done by placing a telescopic system inside the resonator, a rotating disk being positioned in the focal plane of the system; the disk has slits with the width not larger than the focussed spot diameter (about 100 μm). The number and duty factor of spikes and the radiation power of a spike can be varied by changing the rotation speed of the disk, the number and spacing of the slits and the pumping/threshold ratio. For instance, a mechanical modulator can vary the number of spikes in a 1.5-ms pulse from a neodymium-glass laser from less than ten to 150 [86].

The multimode generation with constant Q-factor produces the radiation of comparatively high divergence (it is higher by approximately an order of magnitude than that of the random generation radiation) and therefore it cannot be focussed into sufficiently small spots. However, this technique for generating regular-shape pulses is the most handy one, especially for the ruby laser, and the resulting radiation has a highly uniform intensity distribution across the beam. For the neodymium-glass lasers the most convenient technique for producing regular-shape pulses is low Q-switching (with a rotating disk) since it is difficult to realize the multimode operation of the available lasers of this type with a spherical resonator.

Thus, the effect of a laser pulse with the regular structure is similar to the effect of multipulse radiation when the parameters of the spikes (duration τ_0, interval τ_1 and q_0) are such that each spike vaporizes the material melting a minimum amount of it. Such conditions have been shown [86] to exist if each spike gives rise to the quasistationary vaporization process (that is, $\tau_v = \tau_0$). If in the intervals between spikes the vaporization surface can cool down, for instance to the temperature when crystallization begins ($T(\tau_1) \geqslant T_m$), then a minimum amount of the melt is expelled from the bottom of the hole to the walls.

The optimum duty factor of the pulse can be found from Eq. (307) for τ_v and the equation which gives the temperature of the cooling surface:

$$T(\tau) = T_0 (\pi\tau)^{1/2} v_0$$

Then we find

$$\frac{\tau_1}{\tau_0} \geqslant 3.2 \times 10^{-3} \left(\frac{L_b}{\rho c T_m}\right)^2 \tag{310}$$

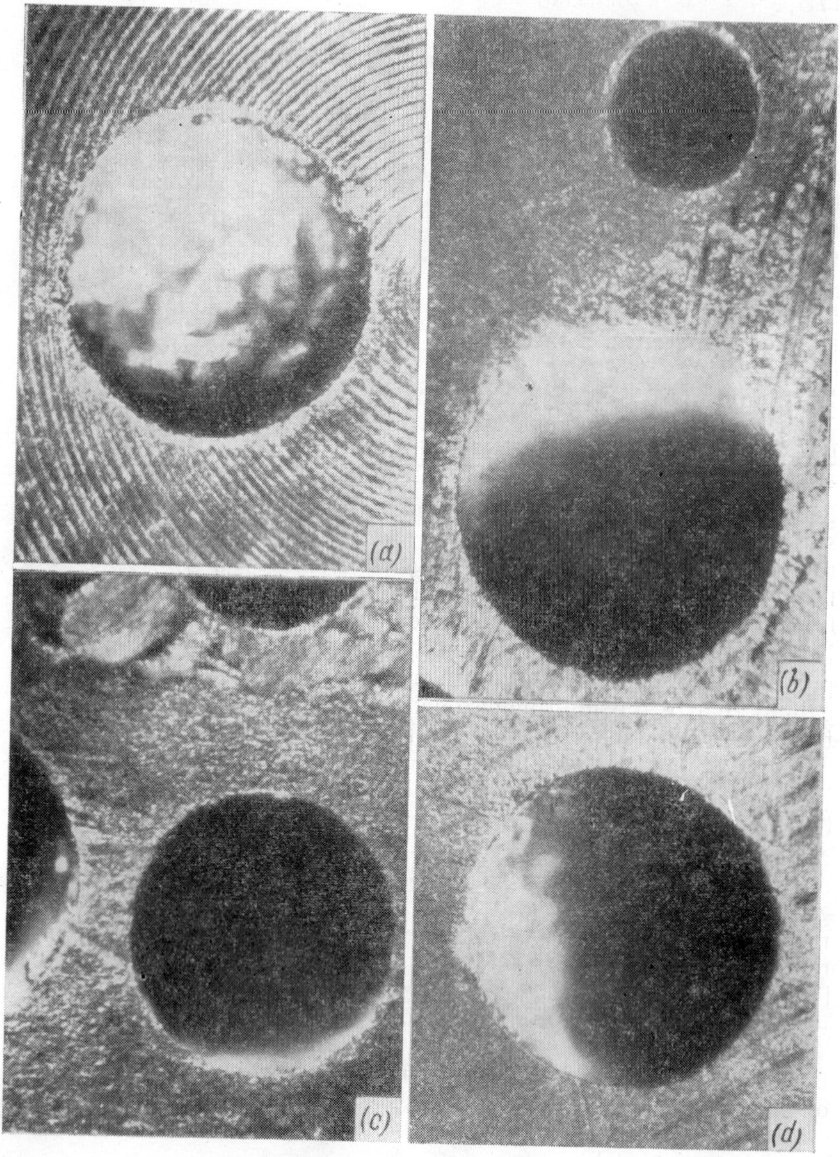

FIG. 66. Laser-drilled holes in copper and duralumin plates
(*a*) copper (×30); (*b*), (*c*), (*d*) duralumin (×50)

And we have

$$\frac{q_1}{T_0(q_1)} = \frac{10k}{\sqrt{a\tau_0}} \qquad (311)$$

where q_1 is the radiation power density in an individual spike.

In many metals where $L_b/\rho c T_m \approx 20$ the conditions (310) and (311) are satisfied by the pulses with the spike duty factor over 1, and the vapour produced by one spike has time to leave the hole of even 1-mm depth (the vapour velocity is about 10^5 cm/s) before the beginning of the next spike. Thus, laser pulses with regular spatial and temporal structures produce holes with better accuracy and quality. Figure 66 presents the photographs of the holes drilled in copper and duralumin specimens by regular-structure laser pulses (neodymium-glass laser with resonator of spherical mirrors having radius of curvature of 2 m in nonconfocal position). Pulses of various energies produce regular round holes of the 3 grade of accuracy; the reproducibility is good, the diameter varies from 0.1 to 2.0 mm and the depth/diameter ratio varies from 2.5 to 6. No ripples left by the molten metal are found on the walls of the hole and the metal surface.

(c) OPTICAL SYSTEM WITH CYLINDRICAL LIGHT PIPE (CLP)

The accuracy and quality of laser-drilled holes depend to a high extent on the geometry of the laser beam. One condition of accurate laser drilling is such a position of the workpiece with respect to the caustic of the optical focussing system that provides for the most uniform distribution of the radiation power density across the drilling zone. The caustic has two characteristic planes where the workpiece is positioned. For the far-field zone drilling (which is widely used for drilling round holes) the workpiece surface is positioned in the focal plane. For the near-field zone drilling the workpiece surface is positioned in the plane conjugate with the emitting laser end or the diaphragm. The spatial intensity distribution in the far-field zone is a superposition of the fields of all the generated modes; the distribution has typically the gaussian form, and the drilling zone has no distinct boundaries. For the near-field zone drilling the irradiated zone has distinct boundaries, and the spatial intensity distribution in it is identical to the field distribution on the emitting end of the active laser element (appropriately scaled). For such an arrangement of the laser and the optical focussing system when the sizes of the light spot in the focal plane and in the image plane of the emitting laser end (or the diaphragm) are identical the light pipe has a cylindrical surface [144]. In other words, the cylindrical light pipe is formed when the rays defining the beam emitted by the laser end (or from the diaphragm) intersect in the front focus of the optical system. Then the length of the cylindrical part of the laser beam is

$$x' = F^2\alpha/d \tag{312}$$

where d is the diameter of the emitting end (diaphragm), α is the angle of the divergence of the beam, and F is the focal distance of the optical system.

To control the diameter and length of the CLP between the focussing objective and the laser an additional telescopic system and two iris diaphragms are placed; one diaphragm is in the focus of the telescope objective and the other is near the telescope exit pupil. The CLP diameter is adjusted by simultaneously changing the diaphragm diameters, and the length is controlled by moving the second diaphragm along the optical axis and simultaneously changing its diameter. This system can form also conical light pipes with the parameters varying in a wide range. The CLPs make it possible to reduce considerably (and sometimes to eliminate completely) the direct radiation absorption by the hole walls (for the far-field zone technique when the beam is not exactly perpendicular to the workpiece surface) and thus to improve the accuracy of laser drilling. The CLPs also prevent a decrease in the radiation power density owing to the defocussing of the laser beam with increasing hole depth.

If the optical system of the laser installation provides for forming a CLP rectangular laser pulses can also be used in laser drilling. The experiments using CLPs [86] have shown that such beams can drill holes having the depth/diameter ratio of 15 with the deviation from the cylindrical shape of not more than 1/200. These results are much better (by a factor of 1.5 and 4) than the results of drilling by laser installations without CLP capacity. The CLP laser technique can be used to produce round holes of fine quality grade and complex-shape holes in plates up to 0.3 mm thick.

(d) ADDITIONAL OPERATIONS AND TECHNIQUES FOR IMPROVING QUALITY AND ACCURACY OF DRILLING

Let us discuss briefly some additional operations and techniques which can be used to improve the accuracy of laser drilling. Blowing the holes with compressed air makes it possible to enlarge the ejection of the molten material from the hole during drilling and thus to reduce uncontrollable redistribution of the molten material after the pulse, which impairs the accuracy and quality of the drilled holes. To provide for blowing the workpiece is placed as a wall in a compressed-air chamber, and the moment

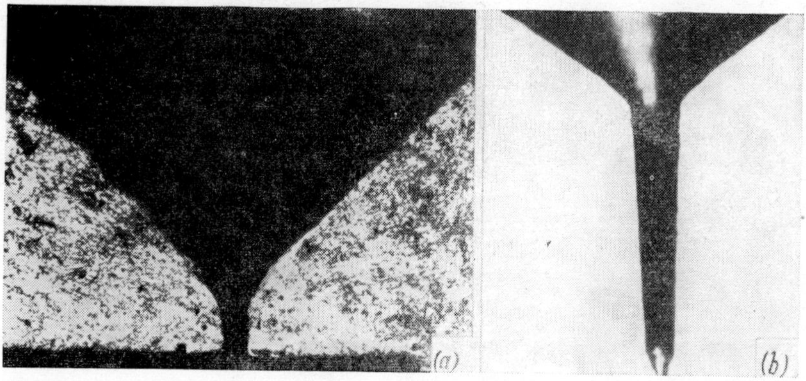

FIG. 67. Polished section of calibrated hole in draw plate for acetate fibre (*a*); punch for sizing holes in draw plate (*b*)

a through hole is drilled the escaping air pushes out the molten material off the hole walls. This technique reduces the roughness of the hole walls and provides for fine quality of drilling.

The disadvantages of this technique are the need for additional equipment and increased drilling time. The compressed air jet can be directed along the laser beam (or perpendicular to it) from a jet nozzle positioned between the laser and the workpiece. This arrangement ensures protection of the optical focussing system against condensed disintegration products and provides for the removal of the molten material and improved quality of through and blind holes.

Sometimes laser drilling can be usefully and profitably used for drilling rough holes which are subsequently sized using other techniques (diamond-dust grinding, sizing with punches, wires, etc.). This process has found successful industrial use in the USSR and other countries for drilling watch jewels (rubies), diamond draw plates and other components.

The laser-drilled holes in metal workpieces can be sized to a high accuracy with punches of carbon steel or hard alloys. The process for production of draw plates for acetate fibres has been developed. The holes in the draw plates are formed in two stages — first, a rough 35-45-μm hole is drilled by the laser beam, and second, the hole is sized to the diameter of 56 μm with a punch (Fig. 67). The resulting hole diameter is accurate to 2-3 μm, and the surface finish of the walls is of class 10. However, this sizing technique can be used to drill only relatively deep holes whose depth is comparable to the diameter; this limitation is due to the punch strength.

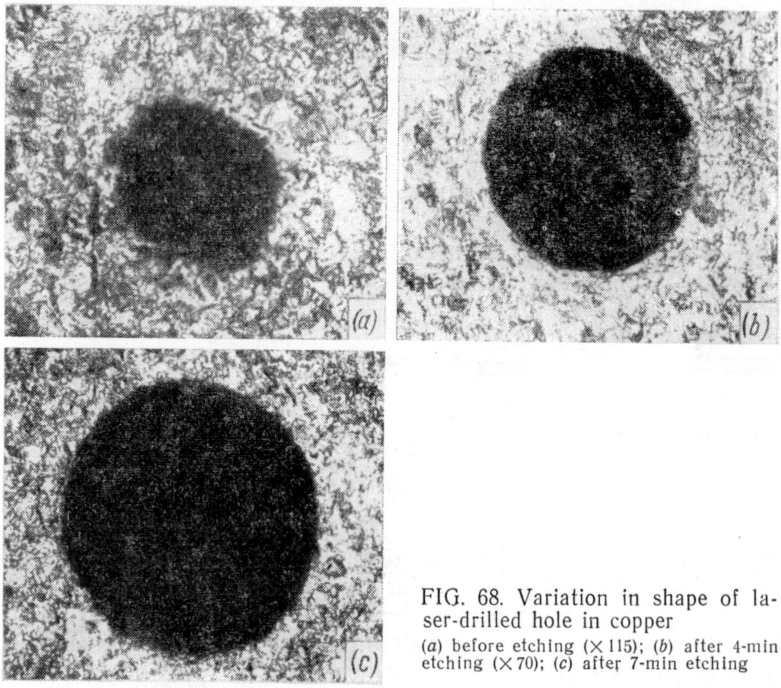

FIG. 68. Variation in shape of laser-drilled hole in copper
(a) before etching (× 115); (b) after 4-min etching (× 70); (c) after 7-min etching

The rough laser-drilled holes with the diameter about 40 μm in the 0.1-mm tantalum plate can be sized with a tungsten wire by passing the wire through the hole in both directions by the distance of not more than 2-3 diameters of the hole, the ratio between the free length of the wire and its diameter being not more than 15. The finished holes are of extra fine or fine quality grade and the walls have the surface finish of classes 11-12.

Chemical etching is another possible technique for sizing laser-drilled holes of practically any depth and diameter in metals.

Etching can remove any irregularities of the hole of round shape. Figure 68 shows a laser-drilled hole in a copper plate before etching and after 4 and 7 minutes of etching by 70% aqueous solution of nitric acid at the room temperature. Etching rapidly removes the laser-fused metal and the barbs inside the hole; the surface finish of the hole walls can be considerably improved (up to class 8 for copper).

There are other techniques for improving the accuracy and quality of laser drilling. To prevent sealing of the hole with the molten material the back side of the workpiece can be coated with a material whose vaporization temperature is below the melting point of the drilled material. The resulting vapour will push out the molten material from the laser-drilled hole. The materials for that purpose are chosen to have a high vapour pressure at the melting temperature of the drilled material (paraffin wax, glycerine, xylene, cedar oil).

To decrease the conicity of the laser-drilled hole and to prevent spattering of the molten material the workpiece surface is coated with liquid (the liquid must have a low surface tension) or a metal foil (for instance, aluminium foil). The liquid coating must be thin not to change the focussing conditions for the laser beam and have a low vapour pressure at the room temperature. For instance, oil coating decreases the molten layers at the exit and entrance of the hole while the silicone oil coating decreases the conicity of the hole.

(e) INCREASING REPRODUCIBILITY OF LASER DRILLING [86]

Laser drilling is frequently used to produce a series of identical holes, when it is essential to have a high accuracy and a good reproducibility of drilling. The reproducibility of drilling is determined by the stability of the laser radiation parameters, by the drilling conditions and other factors as the processes of melting and redistribution of the melt during the laser pulse and after its end are responsible not only for the accuracy but also for the reproducibility of the hole shape and size. The techniques that improve the accuracy of drilling improve also its reproducibility. It is important also to improve the stability of the laser pulse parameters (energy, duration, and spatial and temporal structures).

Two types of instabilities have been found in experimental studies of laser parameters. The instabilities of the first type are the instabilities found for the laser parameters when a relatively small number of pulses are emitted to drill a small number of holes in one or several workpieces, and the laser components experience no significant ageing and wear. The instabilities of the second type are found for long periods of laser drilling when there is significant ageing and wear of the laser components. In the latter case the following factors can improve the reproducibility of laser drilling: (1) steady-state operational regime of the laser, providing for the minimized parameter instability; (2) using high pumping/threshold energies to improve the stability of laser parameters; (3) using backfeed systems increasing the pumping level to compensate the stability decrease due to ageing and wear of the main laser components.

4.9. MULTIPULSE LASER DRILLING TECHNIQUE

The above analysis of the laser drilling processes and the factors affecting the accuracy and reproducibility of drilling shows that it is difficult to regulate the single-pulse drilling process owing to the complex character of the physical phenomena and the lack of control of the laser pulse parameters. The laser drilling processes which are the most difficult ones to control are the following: the development of the hole by the ejection of disintegration products, the redistribution of the molten material layer on the walls of the hole at the end of the pulse and after it, and the variation in the laser radiation power density at the vaporization front with increasing hole depth.

If these processes are not controlled the shape of the hole can be distorted, the heat-affected zones can be increased, the workpiece surface can be cracked, and so on. The best way to eliminate the above problems is to make use of the multipulse technique (MPT) which is the main method of improving the accuracy and quality of laser drilling. With MPT the hole is drilled not by a single laser pulse but by a series of pulses of specified energies and durations, which gradually produces a hole of the required diameter and depth.

Let us discuss the main features of the MPT. Each laser pulse vaporizes a layer of the material increasing the hole depth, the final depth being determined by the total energy of the pulses. Meanwhile, the hole diameter is determined by the averaged parameters of the laser pulse in the series; besides it depends on the laser beam diameter in the heated zone and the shape of the caustic.

Each pulse can remove a rather thin layer of the material so that in drilling of relatively deep holes (about 1 mm and more) the redistribution and removal of the liquid phase have a smaller effect on the hole shape than in single-pulse drilling. These factors determine also the microgeometry of the surface, and decreasing the thickness of the layer removed by one pulse makes it possible to decrease the roughness of the surface.

Using the MPT we can drill holes by a series of short pulses with the interval between them considerably longer than the cooling time for the material. Thus, the size of the heat-affected zone is determined by the duration of one short pulse. Using the short pulses of a high power density which is easy to achieve we can sharply reduce the content of the liquid phase in the disintegration products of the material.

Using the MPT with short pulses we can drill brittle materials without cracking owing to the more uniform (than in the single-pulse drilling) distribution of the residual stresses in the bulk of the material and increasing the dynamic strength with decreasing time of heat treatment. The radiation power density on the bottom of the hole decreases with increasing depth of the hole due to the defocussing of the laser beam. To eliminate this effect the workpiece must be displaced after each pulse by the thickness of the layer removed by the pulse. If the accuracy of the hole size is not essential the vaporization front can be aligned with the focal plane of the optical system focussing the laser beam. The accuracy of drilling of holes of round and other shapes can be considerably higher if the vaporization front is aligned with the image plane of the diaphragm, and the radiation power density is uniform across the beam.

Another factor contributing to increasing the reproducibility of MPT drilling is the decrease in the instabilities of various parameters. For instance, when instead of one laser pulse with the energy E drilling is done by n pulses with the energies E/n the standard deviation of the hole size decreases as $\sqrt{n}$.

The MPT is used generally for two drilling applications: (1) drilling the deepest holes of no better than plain quality and not too high accuracy of the shape; (2) high-precision drilling (of extra fine quality or better).

Different operational regimes are used for these applications of the MPT. In the first application the optimum MPT regime produces by each pulse the maximum ratio of the hole depth to its diameter $h/d > 1$ while in the second application the optimum regime produces by each pulse the hole depth increase smaller than the diameter ($h/d < 1$). Thus, in the first method the amount of disintegration products is relatively high, the hole diameter being determined by the first pulse, and the subsequent pulses only deepen the hole producing only a slight change in the diameter. In the second method the nonstationary regime of vaporization is used when the amount of the disintegration products is low and the molten layer can be minimized by eliminating or reducing many unfavourable factors, namely, heat transfer to the hole walls, defocussing of the laser beam and excessive duration of the pulse.

The MPT employing cylindrical light pipes also decreases considerably the amount of the molten material in the hole by decreasing the direct laser radiation absorption in the hole walls so that it can be used to drill high-precision holes whose sizes, profiles and shapes are determined by the parameters of the laser radiation and the optical focussing system.

(a) SOME FEATURES OF MULTIPULSE TECHNIQUE AND OPERATIONAL REGIMES FOR DRILLING DEEP HOLES

Figure 69 presents a microphotograph of a section of the specimen with holes drilled by various numbers of laser pulses. The pulse energy remains constant at about 40 J and the pulse duration is about 1.3 ms. The laser beam is focussed on the surface of the specimen by a 50-mm lens, and the beam is diaphragmed with a 10-mm diaphragm. The hole depth increases with increasing number of drilling laser pulses. The hole diameter remains practically constant and it is determined by the averaged pulse energy. Figure 70 presents a plot of the hole depth h as a function of the number n of the drilling pulses.

The final size of the MPT drilled hole here can be found from Eqs. (299) and (300) where the hole depth h is determined by the total energy of the series of pulses $E = \sum_{i=1}^{n} E_i$ while the diameter d is determined by the averaged pulse energy $\bar{E} = \frac{1}{n} \sum_{i=1}^{n} E_i$. Equations (299) and (300) in this case show that the

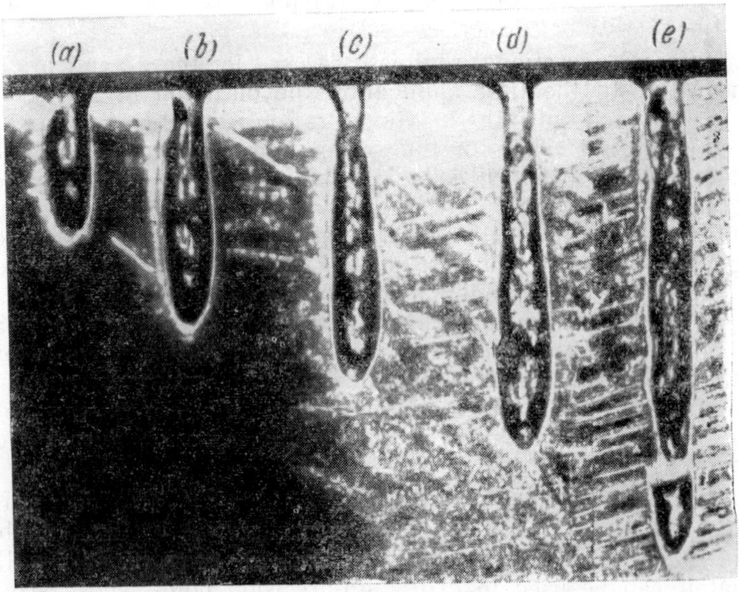

FIG. 69. Profiles of holes in steel 45 produced by 1(a), 2(b), 3(c), 4(d) and 6(e) laser pulses (×7)
Pulse energy=40 J, pulse duration=1.3 ms

main factors determining the depth and diameter of the hole are, as before, the value of $\tan\varphi$ describing the caustic after the focal plane of the optical system and the number of pulses drilling the hole. For $h > d$ the resulting depth/diameter ratio is

$$h/d = \frac{1}{2\tan\varphi} n^{1/3} \qquad (313)$$

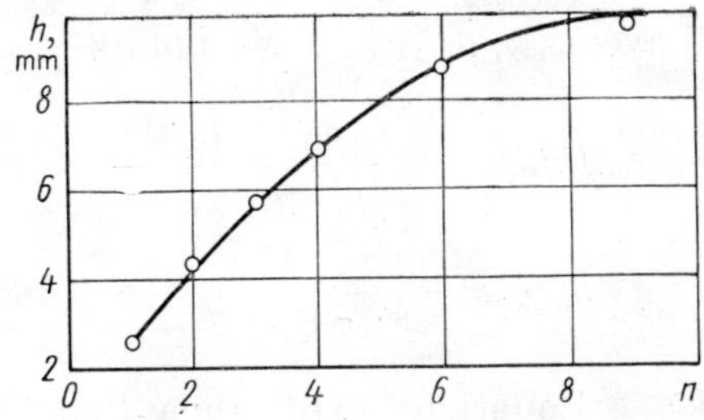

FIG. 70. Depth h of hole in steel 45 as a function of number n of drilling laser pulses with energy of 40 J and duration of 1.3 ms

It can be seen that MPT produces holes with the h/d value higher than the holes produced by single-pulse drilling by a factor of $n^{1/3}$. If the CLP is used for drilling ($\tan\varphi = 0$) the hole depth increases faster with the number of pulses. If we neglect the attenuation of the laser beam in the hole we find

$$h/d = \frac{n\overline{E}}{2\pi r_f^3 L_b} \qquad (314)$$

The experimental results show that the rate of increase in the depth of the MPT drilled hole reduces with increasing number of drilling pulses. This effect is due to decreasing radiation power density on the hole bottom as the laser beam is defocussed with increasing depth of the hole ($\tan\varphi \neq 0$). Clearly, the maximum hole depth is obtained when the radiation density on the hole bottom becomes equal to the threshold density for the disintegration of the material. Hence, the maximum parameters of the hole are

$$\begin{aligned} h_{\max} &\approx \frac{1}{\tan\varphi}\left(\sqrt{\frac{\overline{E}}{\pi Q^*}} - r_f\right) \\ d_{\max} &\approx 2\left(r_f^3 + \frac{3\overline{E}}{\pi L_b}\tan\varphi\right)^{1/3} \end{aligned} \qquad (315)$$

where Q^* is the threshold energy density corresponding to the development of the quasistationary vaporization by the end of the pulse.

From Eqs. (315) we can find $(h/d)_{\max}$ and the number $n_{\max}$ of pulses required for drilling the deepest hole. Since typically we have $h_{max}, d_{max} \gg r_f$ we obtain

$$(h/d)_{max} \approx \frac{0.28}{\tan^{4/3}\varphi} \frac{L_b^{1/3}\bar{E}^{1/6}}{Q^{*1/2}}$$
$$n_{max} \approx \frac{1.7}{\tan^{1/3}\varphi} \frac{L_b^{1/3}\bar{E}^{1/6}}{Q^{*1/2}} \quad (316)$$

Equations (316) show that the ratio $(h/d)_{max}$ and n_{max} only slowly increase with the pulse energy as confirmed by the experimental results. It should be noted, too, that $(h/d)_{max}$ and n_{max} depend on the pulse duration only through the parameter $E^{1/6}$.

Table 12 presents the maximum parameters of holes in steel ($L_b = 5 \times 10^4$ J/cm^3, $Q^* \approx 300$ J/cm^2) calculated for series of pulses of various energies with the duration of 100 μs at $\tan\varphi = 0.2$. If the laser drilling is done with CLP technique the maximum parameters of the pipe correspond to the hole size outside of which the above factors become effective.

TABLE 12. Maximum Hole Parameters and Maximum Numbers of Pulses Required to Drill Holes with Various Pulse Energies

$\bar{E}$, J	d, mm	h_{max} mm	$(h/d)_{max}$	n_{max}
0.1	0.15	0.5	3.3	4
1.0	0.33	1.5	4.5	6
10	0.73	5	6.9	9

The experimental plot of the vaporized mass as a function of the number of pulses (Fig. 71) reveals an earlier plateau than the hole depth plot (see Fig. 70). This means that after a certain number of pulses no more mass is vaporized. The apparent discrepancy is attributed to the fact that the mass of vaporized material is measured by weighing the specimen before and after drilling, that is, the amount of the material ejected outside is measured. But with increasing number of pulses an increasing part of the molten material remains on the surface of the specimen in the form of a rim or drops around the hole entrance. Hence, the mass of the specimen changes only slightly though the hole depth is increased.

When the multipulse technique is employed for drilling the hole diameters are determined by the parameters of an individual pulse and are only slightly affected by the number of drilling pulses. Starting from a certain hole depth the increasing part of the laser beam is incident on the hole walls owing to defocussing of the beam. It can be assumed that a part of this radiation will be reflected from the walls and make a contribution to the deepening of the hole. Then in addition to the laser beam geometry we must take this contribution into account in calculating the maximum depth of the MPT drilled hole. To analyze the effect of the hole walls a few pulses are used to drill a hole of such a depth that the marginal rays of the laser beam intersect the walls inside the hole. Then a diaphragm is placed after the focussing lens to cut off the marginal rays of the beam which are incident on the walls; the hole is drilled with one more pulse. A similar hole is drilled with a pulse without a diaphragm. The depth of the second hole is found to be always larger. The difference between the depths of the holes is a measure of the contribution of the wall-reflected radiation to the deepening of the hole, it amounts to 2-6% depending on the hole depth.

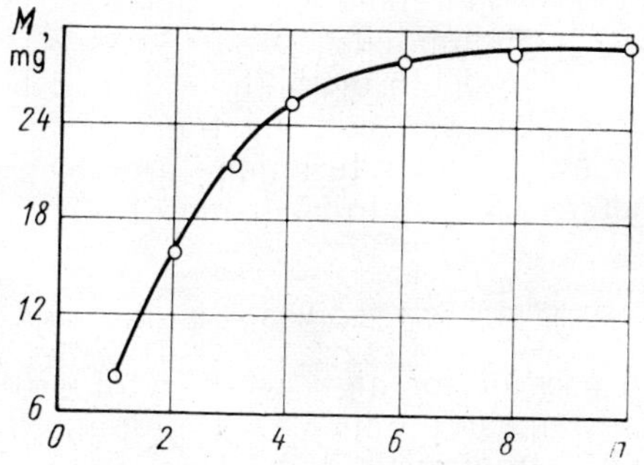

FIG. 71. Vaporized mass M as a function of number n of drilling laser pulses with energy of 40 J and duration of 1.3 ms for steel 45

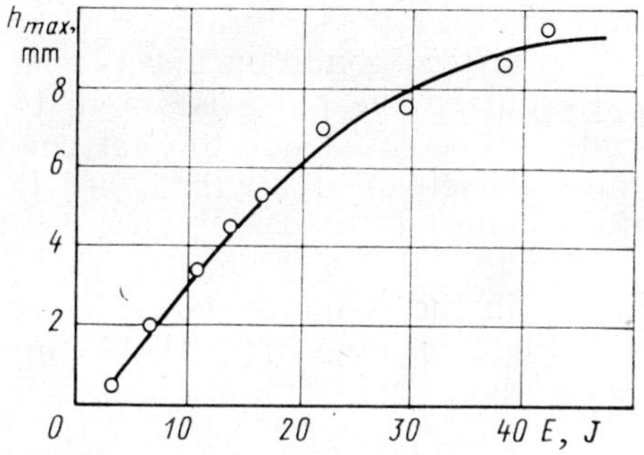

FIG. 72. Maximum hole depth h_{max} as a function of pulse energy E for multipulse drilling

The experimental results agree with the above analyses; they show that the MPT drilling can produce holes several times deeper than the single-pulse drilling can and for some materials ten times deeper if the proper radiation parameters and focussing conditions are chosen. For instance, the plot of h as a function of n shown in Fig. 70 demonstrates that the depth of the hole drilled by 9-10 pulses is 4 times the depth of the hole drilled by a single pulse. This plot shows also that the hole depth is about 90% of the maximum even after 6-7 laser pulses. Figure 72 illustrates the relationship between the maximum hole depth and the averaged pulse energy for steel 45. The holes are drilled by 6-9 pulses with the duration about 1.3 ms focussed on the surface of the specimen with a 50-mm lens.

The depth of the hole drilled by multipulse technique can be increased by displacing the lens focus into the specimen between the pulses. This technique produces holes with a large depth/diameter ratio (up to 25 or more).

(b) THE MPT REGIMES FOR HIGH-PRECISION DRILLING

To provide for high accuracy of drilling with the MPT each laser pulse must produce the minimum possible melting of the walls and bottom of the hole. This leads to the following two conditions for the operational drilling regime:

$$\tau_i \leqslant \frac{r_f^2}{a}, \; h(\tau_i) \leqslant 2r_f \tag{317}$$

Here τ_i is the pulse duration.

The first condition means that the heat transfer to the hole walls during the pulse is small, that is the melting of the hole walls is decreased due to heat conduction. The second condition limits the effect of all the other factors contributing to melting of the material.

If the CLP technique is used for laser drilling then the variation in the radiation power density on the bottom of the hole with the deepening of the hole can be assumed to be slight, and the depth of the layer vaporized by one pulse is

$$h(\tau_i) = v_0\left(\sqrt{\tau_i} - \sqrt{t_1}\right)^2 \tag{318}$$

Equation (318) is derived by integrating the approximate time function of the rate of nonstationary vaporization. Here t_1 is the time interval between the beginning of the pulse and the beginning of vaporization $\left[t_1 = \left(kT_0/q_0\sqrt{a}\right)^2\right]$ and v_0 is the steady-state vaporization rate. Substituting t_1 and the limiting values of $h(\tau_i)$ and τ_i from Eq. (317) into Eq. (318) yields

$$2r_f = \frac{v_0}{a}\left[r_f - \frac{kT_0(v_0)}{L_b v_0}\right]^2 \tag{319}$$

where $T_0 = T^* \ln^{-1} \frac{s}{v_0}$

Solving Eq. (319), for instance, by successive approximations and finding v_0 we can determine the radiation power density in the MPT drilling, which satisfies the conditions (317):

$$q_0 = v_0 L_b \approx \frac{2L_b a}{r_f}\left(1 + 3\ln^{-1}\frac{sr_f}{7a}\right) \tag{320}$$

Note that the specific energy of disintegration L_d for this drilling regime is higher than the vaporization heat:

$$L_d \approx L_b\left(1 + 3\ln^{-1}\frac{sr_f}{7a}\right) \tag{321}$$

The knowledge of q_0 readily yields other parameters of the drilling technique, namely, the pulse energy $E_i = \pi r_f^2 q_0 \tau_i$ (the pulse duration τ_i is already known: it is equal to r_f^2/a) and the number n of pulses required to drill a hole of the depth h $(n = h/2r_f)$. For instance, the following drilling parameters are required to drill a through hole with the diameter 50 μm in 0.5-mm steel plate ($a = 0.1$ cm²/s, $s = 2 \times 10^5$ cm/s, $L_b = 5 \times 10^4$ J/cm³): $\tau_i = 250$ μs, $q_0 = 2.6 \times 10^6$ W/cm², $E_i = 0.05$ J and $n = 10$.

Though the MPT drilling produces a smaller amount of the liquid (molten material) some material is necessarily melted as the temperature distribution over the bulk of the material is continuous and under the vaporization front there is a thin molten layer with the thickness $h_l \approx \frac{a}{\bar{v}} \ln T_0/T_m$ where $\bar{v} = 2r_f/\tau_i = 2a/r_f$ is the mean rate of vaporization of the material. If we assume that during each pulse the molten layer is partially expelled by the vapour pressure and uniformly distributed over the walls and the bottom of the hole by the end of the pulse the relative deviation $\Delta d/d_0$ of the diameter from the specified size which determines the quality of drilling is

$$\Delta d/d_0 = 0.1 \ln T_0/T_m \tag{322}$$

For holes drilled in the above plate we have $\Delta d/d_0 = 0.08$ with $d_0 = 50$ μm corresponding to the fine quality grade of machining. If the deviation of the diameter calculated from Eq. (322) is larger than the working tolerance the condition (317) should be made more strict by reducing the pulse duration.

It should be noted that the condition (317) for the optimum MPT regime is not universal, it is applicable mostly to metals. For nonmetallic materials in which the thermal conductivity is low the pulse duration can be limited owing to other factors. For instance, in drilling of brittle materials the pulse duration calculated from the condition (317) can be too long and lead to cracking of the material. Under such circumstances limitation to the pulse duration must be more strict.

Let us now discuss briefly multipulse drilling with pulses of varying duration. If we compare the sizes of the holes drilled by one long pulse and a series of short pulses with the same total energy we can find how the final size and the shape of the hole

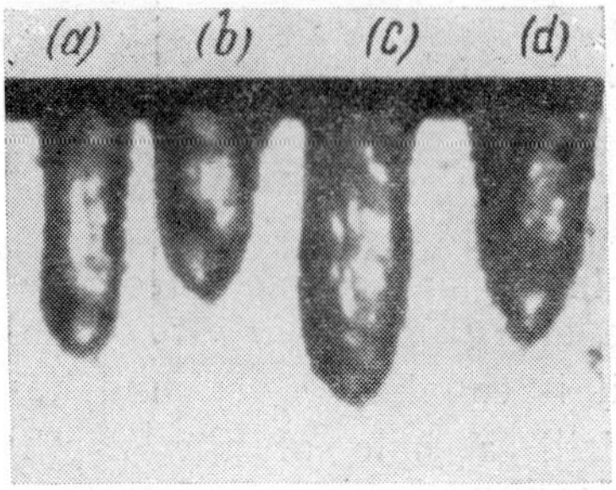

FIG. 73. Profiles of holes produced by one laser pulse (*b*, *d*) and five pulses of smaller duration (*a*, *c*) with the same energy consumption
(*a*) $E=3.8$ J, $\tau=0.15$ ms; (*b*) $E=19$ J, $\tau=0.65$ ms; (*c*) $E=6.8$ J, $\tau=0.25$ ms; (*d*) $E=34$ J, $\tau=1.2$ ms

are affected by the energy losses owing to the heat transfer to the walls and the absorption of radiation by the products of disintegration of the material.

When the hole is drilled by several pulses with the time interval between them, which is sufficient for complete removal of the disintegration products from the hole, the resulting hole is deeper than the hole drilled by a single pulse whose energy and duration equal the sums of energies and durations of the shortest pulses.

As can be seen from Fig. 73 the energy losses in single-pulse drilling decrease the hole depth by 30% compared to the multi-pulse drilling. The single-pulse drilled holes have an entrance cone indicating the presence of a large molten layer on the hole walls. A comparison of the microgeometry of the hole walls demonstrates also that in shorter-pulse drilling the molten layer on the hole walls is thinner.

An interesting approach would be to drill holes with a series of short pulses formed by splitting up a single long laser

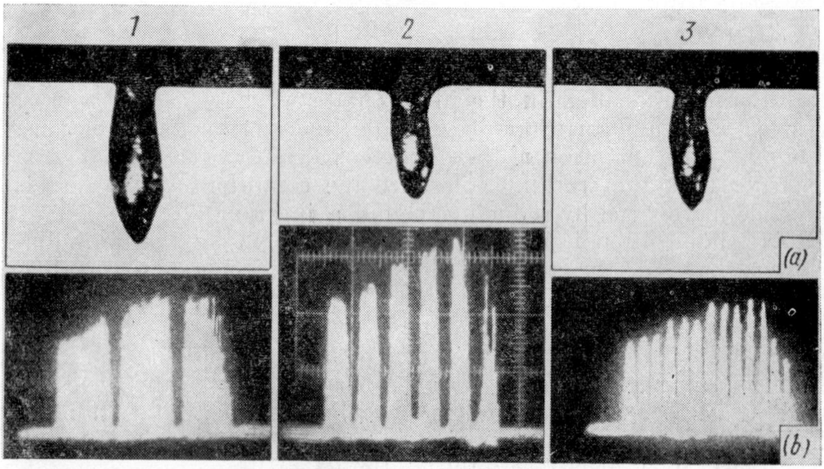

FIG. 74. Profiles of holes (*a*) drilled by three (*1*), six (*2*) and thirteen (*3*) laser pulses produced by splitting the initial laser pulse ($\tau_i = 1.5$ ms); total energies of pulse series (*b*): *1* — 29 J; *2* — 20 J; *3* — 17 J (×13)

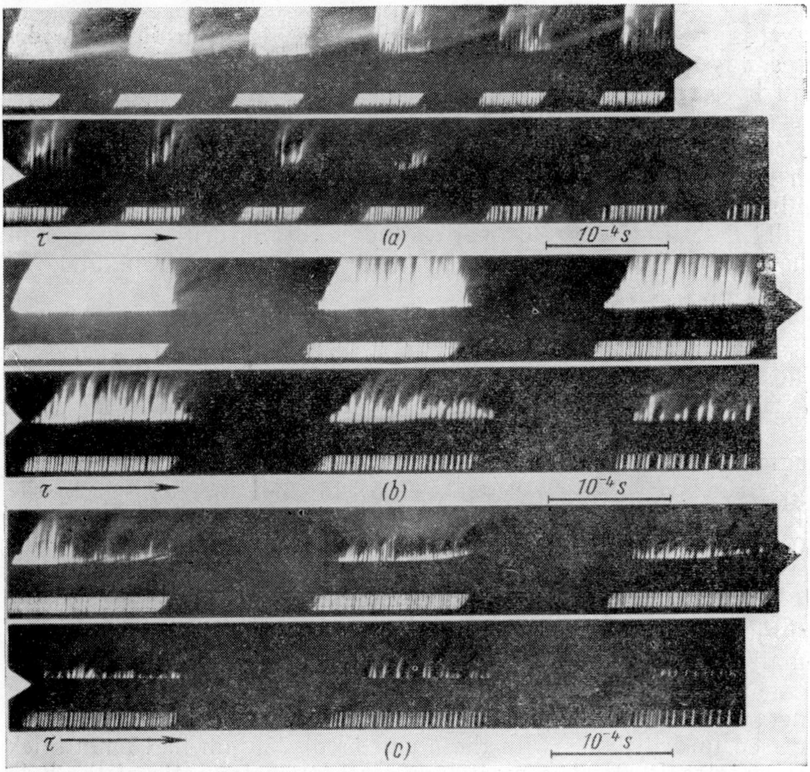

FIG. 75. High-speed photorecordings of ejection of disintegration products of steel Ст. 3 and recordings of laser pulses. Pulse series are produced by splitting the initial laser pulse with the duration about 1.5 ms

(*a*) 13 pulses, 17 J; (*b*) 6 pulses, 16 J; (*c*) repeated drilling of hole (*b*) by six pulses with the total energy of 16 J

pulse. A rotating disk has been used to produce a series of short pulses with the duration of up to 50 μs with variable duty factor. The total number of pulses can be 10 or more so that the rotating-disk modulation of laser radiation simulates pulsed laser radiation with the frequency of about 10 kHz.

Figure 74*a* presents the photographs of the sections of specimens with holes drilled by the series of pulses produced by splitting a laser pulse with the constant intensity. Figure 74*b* presents the oscillograms of these pulses.

Figure 75 presents the high-speed photorecordings of the process of ejection of the disintegration products during the laser pulses and between them. A sharp decrease or almost a total absence of ejection of the disintegration products seen on some recordings at the end of the pulse series is determined by decreasing radiation intensity on the tail of the laser pulse. This can be seen on the recordings of the radiation emitted from the laser in the lower part of the photographic film. The same factor is responsible for the fact that the amount of the molten material on the walls of the hole is only slightly smaller than for single-pulse drilling despite the short duration of the pulses in the drilling series. Hence, to improve the quality of drilling we should shorten not only the pulse duration but also the front and the tail of the pulses.

(c) APPLICATION OF MULTIPULSE TECHNIQUE

The MPT can widen significantly the scope of laser drilling. The MPT drilled holes can have the *h*/*d* ratio of up to 50 and the holes of fine quality grade can be produced with appropriate drilling regimes.

The MPT has been used [166] to drill the holes with the diameter varying from 4 to 100 μm and the *h*/*d* ratio from 20 to 50 in various materials (aluminium, silicon, germanium, zinc, molybdenum, ruby, ceramics, etc.). The holes were drilled with a *Q*-switched ruby laser using a projection machining system (the "conjugated pupil" system) providing for good reproducibility, small entrance cone and good quality of the surface.

The calcium tungstate-neodymium laser with the pulse energy of 1 J and the pulse repetition rate varying from 5 to 10 Hz is used to drill holes with the diameter of 750 μm in 15-mm thick steel plates [200]. The maximum depth/diameter ratio of the hole is 20.

The following industrial applications of the MPT have been reported: (1) drilling of holes with the diameter of 50 μm and spacing of 50 μm in 0.5-mm thick ferrite plates for computer memories; (2) drilling of 50-μm holes in 0.6-mm plates of polycrystalline glass for bilateral printed circuit interconnections; (3) drilling of holes in ceramic insulators for coaxial cables; (4) drilling of holes with the diameters from 5 to 80 μm in high-melting materials up to 0.2 mm thick used for manufacturing diaphragms for electron and ion beams.

Table 13 summarizes the data on MPT laser drilling of some materials.

The MPT applications are developing now due to the fact that in the recent years there have been built reliable pulsed lasers with the pulse repetition rate up to 1-10 kHz and the radiation parameters convenient for the MPT, namely, the lasers with CO_2, nitrogen, yttrium-aluminium garnet and other active mediums. These lasers make possible high-productivity industrial applications of the MPT.

TABLE 13. Parameters of Typical MPT Drilled Holes and Drilling Regimes

Material	Hole parameters				Drilling regime				Technique		Number of holes	
	d, μm	h, mm	h/d	diameter tolerance, μm	$\bar{E}$, J	τ, 10^{-4} s	q_0, W/cm^2	n	near field	far field	one	system of holes
Ceramics	200	3.2	16	—	1.4	5	$4 \cdot 10^6$	40	—	×	×	×
	200	1.0	5	30	1.6	10	$5 \cdot 10^6$	1	—	×	—	×
Ferrite	200	1.0	5	±4	0.3	1	$1.2 \cdot 10^7$	7	×	—	—	×
	50	1.0	20	—	0.05	0.9	$6 \cdot 10^7$	10	×	—	—	×
	100	0.7	7	±15	1.5	10	$2 \cdot 10^7$	1	—	×	—	×
	100	0.5	5	±5	0.2	0.9	$6 \cdot 10^7$	9	×	—	—	×
Stainless steel	50	1.2	24	—	0.2	0.9	$1.2 \cdot 10^8$	12	×	—	—	×
	50	0.1	2	±2	0.2	0.8	$1.5 \cdot 10^8$	1-5	×	—	×	×
Ruby	10	0.4	40	—	0.1	Q-switching	—	—	×	—	×	—
Brass	30	0.1	3.5	±1.5	0.05	1.0	$4 \cdot 10^7$	1-3	×	—	×	×
Polycrystalline glass	50	0.6	12	—	0.3	1.0	$5 \cdot 10^7$	3-5	×	—	—	×

4.10. PROFILE DEVELOPMENT IN LASER-DRILLED HOLES IN MATERIALS WITH VARIOUS THERMAL PROPERTIES

The accuracy of laser drilling of opaque materials is essentially determined by the dominant mechanism of the disintegration of the material by laser radiation with various spatial and temporal structures.

The shape and size of the laser-drilled holes can be reliably reproduced if the hole cross section coincides with the shape of the focussed spot of the laser radiation. The correlation between the hole cross section and the focussed spot shape is especially noticeable for nonround holes. Thus, the analysis of drilling of irregular-shape holes in various materials contributes to our understanding of the factors responsible for accuracy of laser drilling.

Let us analyze the development of the hole cross section in specimens with various thermal properties. The experimental study is done by simultaneous high-speed photorecording of the front and back surfaces of a laser-drilled plate. The photographic film is not sensitive to the laser wavelength ($\lambda = 1.06$ μm).

The focussed spot on the surface of the workpiece of the specified size and shape is the image of the diaphragm through which the laser beam is passed from the emitting end of the laser. The multipulse technique is used for drilling with the pulse duration and pulse rate varying in a wide range. The pulse energy and the radiation power density are also varied. The number of drilling pulses is 6 in all experiments.

Figure 76*a* and *b* shows the formation of the hole drilled by the first (*a*) and second (*b*) laser pulses with the energy of 0.5 J and duration of 450 μs. The time interval between the photographs is 32 μs. At the initial stage of drilling the shape of the affected spot on the front surface of the specimen is seen to correlate with the shape of the focussed spot. The first photographs show the traces of small particles ejected from the affected zone at large angles. At the next stage of drilling the development of the hole continues, and its shape is preserved. The mean velocity of the disintegration boundary can be estimated if we know the moment when the hole is drilled through the plate (Fig. 76*a*, photographs Nos. 11-12). In these experiments the velocity is 1.1 m/s. When the primary through hole is drilled its development begins gradually and is finished by the end of the second pulse (Fig. 76*b*). Machining of the walls of the hole is finished when the correlation between its cross section and the focussed spot shape becomes good enough.

When the pulse energy is decreased by 10% no significant difference is found in the character of development and in the

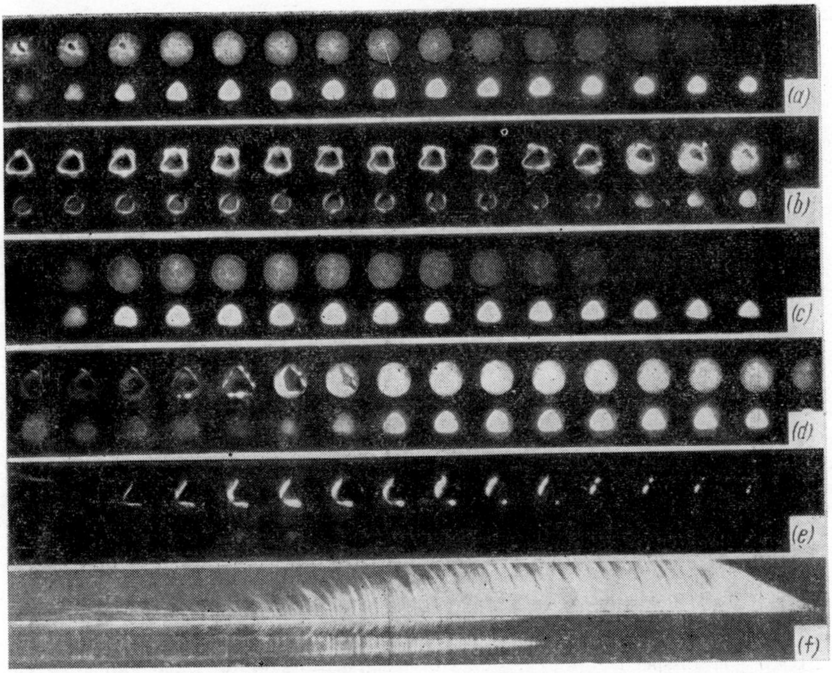

FIG. 76. High-speed photorecordings of development of delta-shaped holes in 0.35-mm ceramic plates (*a*, *b*, *c*, *d*, *e*) and 0.4-mm ferrite plate (*f*)
Below—front surface of plate (×24); above—back surface of plate (×28)

final size of the hole (Fig. 76*c* and *d*). As early as 30 μs after the beginning of the first pulse the traces of a large number of ejected particles can be seen (Fig. 76*c*). The development of the cross section also consists in gradual growth of the initial through channel (Fig. 76*d*, photograph *8*) to the specified size and shape determined by the focussed spot. However, as the disintegration rate becomes lower the through hole is formed only in the middle of the second pulse. The subsequent pulses do not produce a noticeable change of the cross section of the hole even when the pulse energy is increased by 10% (Fig. 76*e*).

The situation is different when tungsten plate 0.3 mm thick is drilled with laser pulses under the same conditions (Fig. 77*a*, *b*, *c* and *d*). The affected zone on the front surface of the specimen has the same shape as the focussed spot for about 100 μs and then it becomes round. No light from the back surface is recorded during the first laser pulse. As shown by Fig. 77*a* the second pulse heats the bottom of a round cavity. The correlation between the shape of the focussed spot and the shape of the affected zone is

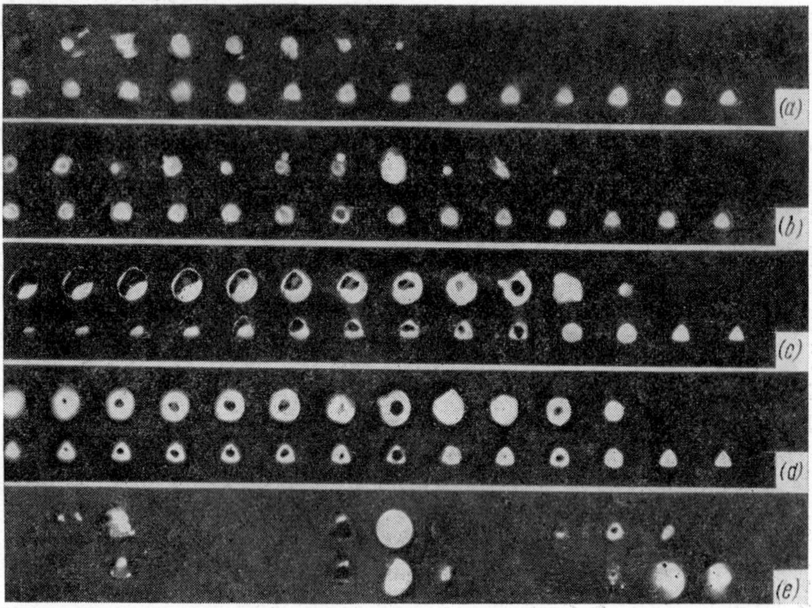

FIG. 77. High-speed photorecordings of laser drilling of tungsten plates 0.3 mm thick (*a*, *b*, *c*. *d*) and 0.05 mm thick (*e*)

maintained for 100 μs, and then the affected zone border becomes blurred and changes with time. The primary through channel is formed by the midpoint of the second laser pulse. The formation of the through hole is accompanied by the ejection of the disintegration products from the exit both in the form of small drops and in the form of large particles. The through channel exists for not more than 50 μs, and then it becomes sealed by the molten material. Subsequently the through hole is formed once more, and after about 60 μs it is again sealed.

The events are repeated with subsequent pulses. For instance, during the fourth pulse (Fig. 77*b*) the exit of the sealed hole starts to glow after about 100-200 μs, and then the disintegration products are ejected and the through hole is formed which exists for not more than 50 μs and after this is sealed. Then the material is again ejected, and the through hole is again formed. In Fig. 77*b*, photographs 9-10, near the edge of the hole there is seen a molten metal drop whose brightness is higher than the brightness of the hole edges and whose size is comparable to the hole diameter. Subsequently the hole is again sealed, again formed and so on.

When the laser pulses of the same parameters are used to drill a tungsten plate 50 μm thick the flow zone has the shape of the focussed spot for 60-70 μs from the beginning of the pulse. During the pulse the shape of the hole changes and by the end it differs from the triangle shape of the focussed spot, the hole diameter being larger than the size of the initial affected zone (Fig. 77*c*). A 10% decrease in the pulse energy results finally in the complete sealing of the through hole (Fig. 77*d*).

The above photographs show that the identical conditions of laser drilling can produce different disintegration effects in different materials depending on their physical and thermal properties.

In laser drilling of ceramics the dominant effects are due to surface vaporization and thermoelastic stresses. Under these circumstances the process is characterized by a small amount of the molten material in the drilling zone and the correlation between the hole shape and the focussed spot shape. This type of disintegration processes is typical of the material with low thermal conductivity. For instance, the high-speed photographs of the laser drilling in ferrite (Fig. 76*f*) show that the hole diameter increases steadily to a given size, and then the material ejection comes to an end. Obviously, the hole size is not affected by variation in the pulse energy in a rather wide range.

The development of holes in tungsten involves melting and washing out of the melt resulting in continuous motion of the melt inside the hole. The development of holes is accompanied also by the ejection of large amounts of vapour and the molten material drops whose size is comparable to the hole diameter. The drilling processes are similar in the specimens with the six-fold difference in thickness. Thus, the laser drilling regimes employed for ceramics and ferrite cannot be successfully applied for drilling tungsten. The better results can be obtained with shorter laser pulses of higher radiation power density (Figs. 77*e* and 78). Figure 77*e* shows the development of a hole in a tungsten plate 0.05 mm thick under the effect of the 4th, 5th and 6th pulses with the duration of 30 μs and the repetition rate about 6 kHz. The satisfactory results are also obtained with $\tau = 50$ μs and $q_0 = 5 \times 10^7$ W/cm^2. The pulse repetition rate also affects the drilling process. The maximum pulse rate can be taken to be the reciprocal of the time of cooling of the material in the drilling zone. For the above drilling regimes the cooling time is somewhat longer than the pulse duration at $h/d < 1$ and is about 50% of the pulse duration at $h/d > 1$. Hence the repetition rate must be less than $1/2\tau$ in the first case and than $2/3\ \tau$ in the second case.

The pulse duration τ must be also chosen to satisfy the condition that no change of the hole shape occurs owing to the motion

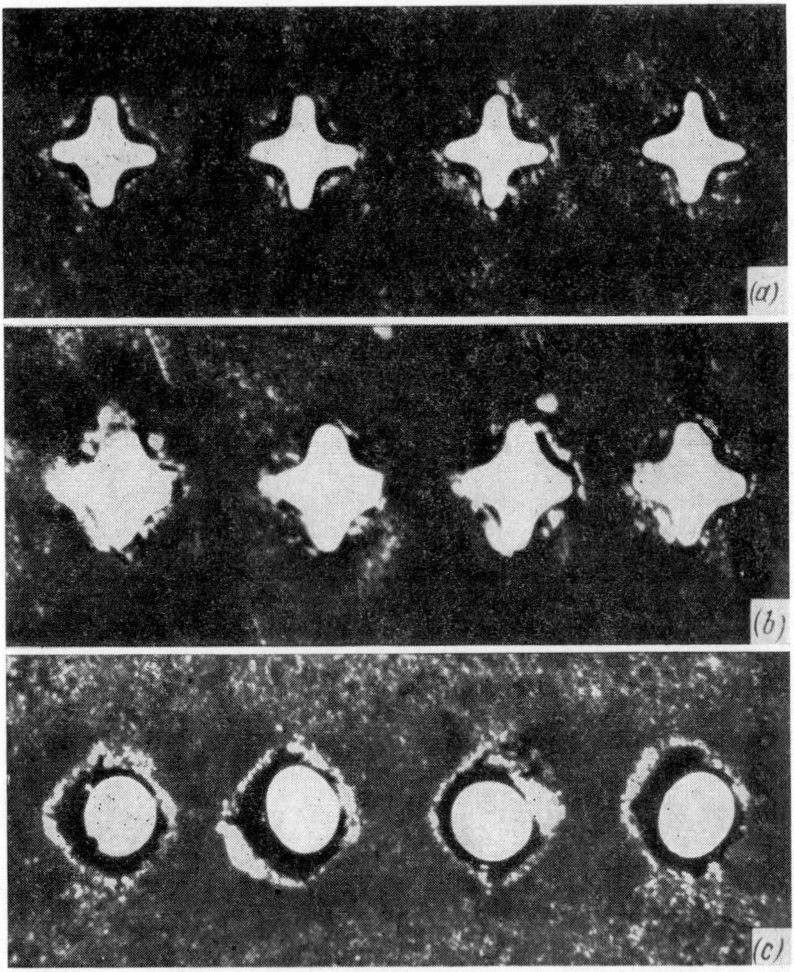

FIG. 78. Development of specified hole shapes in 0.2-mm steel plates according to laser pulse duration and radiation power density
(*a*) 50 μs, 5×10^7 W/cm²; (*b*) 100 μs, 5×10^7 W/cm²; (*c*) 450 μs, 10^7 W/cm²

of the molten material during the time τ. The time τ depends not only on the thermal properties of the material (heat conductivity and heat diffusivity) but also on the properties of the melt (surface tension and viscosity) and, naturally, on the laser pulse parameters (radiation power density, spatial and temporal structure, etc.). Moreover, the pulse duration depends on the parameter determining the distortion of the hole shape. For instance,

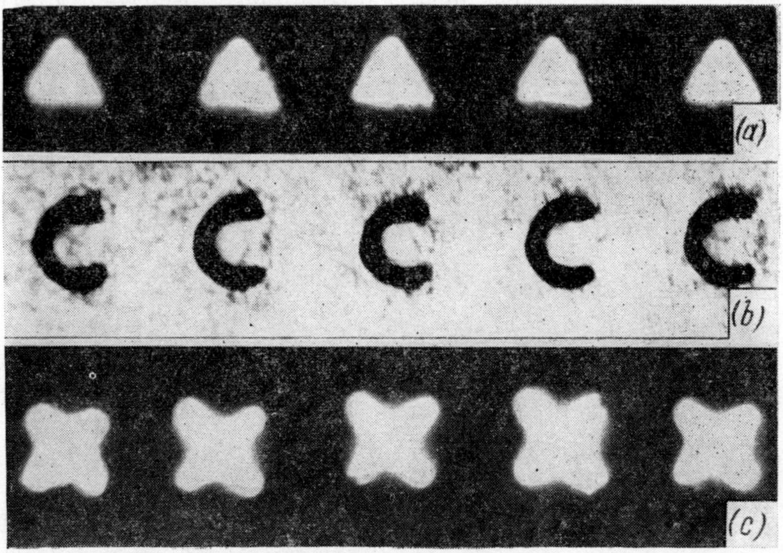

FIG. 79. Laser-drilled holes of various shapes in 0.12-mm stainless steel plate (*a*, *c*) and 0.4-mm ceramics plate (*b*) (×60)
The pulse durations: steel — 50 μs, ceramics — 150 μs

for a triangular hole the distortion of the shape can be described by the parameter

$$\delta = h/l \tag{323}$$

where h is the buckling of the hole wall and l is the side of the triangle in the liquid phase. The distortion of the hole shape can be specified by a given value of δ (for instance, $\delta \geqslant 0.1$).

The time during which the triangular shape can exist in the liquid phase can be estimated by finding the time required for rounding off one of the acute angles. It can be assumed that, the triangle with the side l is transformed into the circle of the same area with the radius r_l. The time τ_d of distortion of the shape can be estimated as

$$\tau_d \approx \frac{0.2 l \mu}{\sigma} \tag{324}$$

where μ is the dynamic viscosity and σ is the surface tension. Assuming $l = 0.02$ cm, $\sigma = 10^3$ dyne/cm and $\mu = 10^{-2}$ Poise we obtain $\tau_d \approx 0.4 \times 10^{-8}$ s. Hence we see that the acute angle shape cannot be preserved under real conditions. In fact, Eq. (324) presents too simple a picture for the surface tension is compensated not only by the viscosity but also by the vapour reactive pressure, which leads to an increase in the time τ_d.

Thus, appropriately selected laser radiation parameters make it possible to drill holes of specified shapes (Fig. 79). The optimum laser drilling regimes for which 10% variations in the pulse energy and duration do not affect the final size and shape of the holes differ significantly for various groups of materials.

5
APPLICATIONS OF CONTINUOUS-WAVE LASERS

As mentioned in Chapter 2 the gas lasers with gas mixtures and the yttrium-aluminium garnet-neodymium lasers are the most widely used continuous-wave lasers whose power is sufficient for technological heating applications.

In a number of countries intense research and development work is underway with the gas-discharge CO_2-N_2 lasers which can produce continuous-wave output of tens of kilowatts. These lasers can be used for the heat treatment of large workpieces of ceramics or metals at a large or small distance from the laser.

5.1. CONTINUOUS-WAVE LASER RADIATION EFFECTS IN IRRADIATED MATERIALS

Let us analyze the effect of the continuous-wave laser radiation with the wavelength $\lambda = 10.6$ μm incident on the surface of the material. If the radiation intensity is not too high the reflectivity for this radiation is rather high for the metallic materials. Table 14 presents the absorption coefficients for some materials [160].

TABLE 14. Absorption Coefficients of Some Metals for Radiation with Wavelength of 10.6 μm

Metal	Absorption coefficient	
	polished	oxidized at 600 C for 2 h
Au	0.010	—
Al	0.034	0.25-0.50
Fe	0.050	0.33-0.74
Zr	0.083	0.45-0.56
Ti	0.094	0.18-0.25

A thick oxide film on the surface of the material significantly increases absorption of radiation. If the radiation power

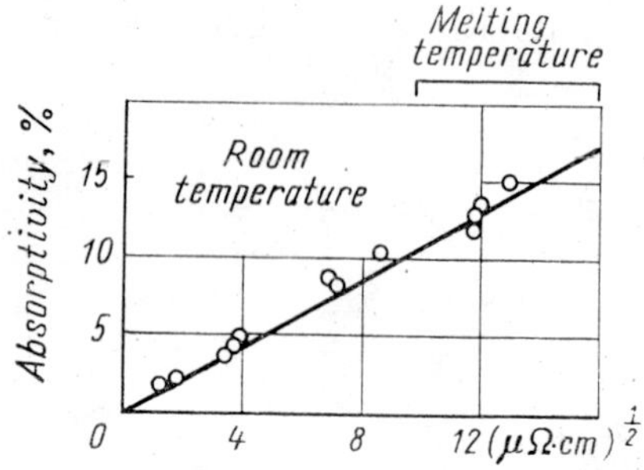

FIG. 80. Absorptivities of some metals at room temperature and at melting temperature as a function of square root of direct-current resistance

density is high then even when the metal surface is polished radiation for a few seconds or tens of seconds results in decreasing the reflectivity R owing both to the temperature dependence of R and to the growth of the oxide film on the surface, the specimen being in an oxidizing medium, for instance, in the air. The absorptivity of metals can be varied by machining the surface [158]. If the radiation power density is high enough to produce a hole in the material the laser beam can be focussed upon the hole bottom owing to multiple reflections from the hole walls. Under such circumstances the absorptivity can increase to 60% and sometimes to almost 100%; it is independent of the optical properties of the metal or the external conditions.

For polished specimens the absorptivity A for the laser radiation with the wavelength of 10.6 μm is described by the well-known Hagen — Rubens equation:

$$A = 112.2\left(\sigma_0^{-1}\right)^{1/2} \tag{325}$$

where σ_0 is the direct-current conductivity, $ohm^{-1}\ m^{-1}$. For instance, for copper $\sigma_0 = 5.8 \times 10^7\ ohm^{-1}\ cm^{-1}$.

The determinations of the absorptivity of some metals for radiation with $\lambda = 10.6$ μm have been described in reference [158]. The absorptivity is measured at the room temperature and at the melting point. Figure 80 presents the absorptivities of the polished metal specimens* *vs* the square root of the direct-current resistance [Eq. (325)]. Among the metals used in the experiments stainless steel has the highest and copper the lowest absorptivity at the room temperature. At the melting point the absorptivity is higher but not more than 15% for steel. To increase the absorptivity and hence the efficiency of laser machining the surface treatment of the workpiece can be used. The following four types of surface treatment have been studied [158]: (a) roughing of the surface; (b) formation of a thin nonmetallic film on the surface; (c) coating the surface with fine metallic or nonmetallic

* Before measurements the polished metal specimens are put into a cleansing solution and then into benzine.

powder; (d) disintegration of a thin surface layer by radiation with a focussed laser beam.

Roughing of the surface is done with emery cloth of varying grade. At the room temperature absorptivity increases with increasing roughness of the surface of iron and stainless steel specimens. At the melting temperature the absorptivity of the rough surface is close to that of the polished surface. A thin nonmetallic film on the surface, for instance Fe_2S_3 on stainless steel and Al_2O_3 on aluminium, increases the absorptivity which is a function of the treatment time required for the formation of the film. For instance, for stainless steel the absorptivity increases with the treatment time up to about 80% and after that remains constant (at the room temperature). Coating of the surface with a thin layer of metallic or nonmetallic fine powder increases the absorptivity almost to 100% at the room temperature. Laser radiation of the surface typically produces an increase in its absorptivity as the oxide film is formed with laser treatment in the air.

It proves to be rather difficult to choose optical materials for the laser focussing system with $\lambda = 10.6$ μm which would possess low absorptivity, good workability and appropriate characteristics and performance. The materials generally used in focussing systems for CO_2 lasers are described in Table 15.

TABLE 15. Optical Properties of Materials Used for Manufacturing Lenses and Windows Transparent for Radiation with $\lambda = 10.6$ μm

Material	Refraction index	Absorptivity, cm^{-1}	Hygroscopicity
NaCl, crystal	1.49	$\ll 0.01$	Yes
KCl, crystal	1.454	$\ll 0.01$	Yes
BaF_2, crystal	1.42	0.077	Slight
KRS-5	2.38	0.02	Slight
Ytran-2	2.19	≈ 0.55	No
Ytran-4	2.4	≈ 0.1	No
Germanium	4.0	≈ 0.07	No

If the irradiated specimen is in an inert gas atmosphere or in vacuum the radiation incident on the surface gives rise to a surface heat source whose intensity $q(r)$ depends on the radiation power density $J(r)$:

$$q(r) = (1 - R_0)\, J(r) = A_0 J(r) \tag{326}$$

Here R_0 is the reflectivity and A_0 is the absorptivity. The subscript 0 denotes the averaged values of the parameters for the time of irradiation.

For a single-mode CO_2 laser the divergence is close to the diffraction value, and the spatial distribution of the intensity can be described by the diffraction function. Then the spatial distribution of the heat source intensity can be described by the gaussian function:

$$q(r) = q_0 \exp(-kr^2) \tag{327}$$

where q_0 is the maximum intensity of the source.

Let us estimate the mean intensity of the heat source on the surface of the material under the radiation of the CO_2 laser. As the laser radiation power is about 10^2-10^4 W, the focussed spot area is about 10^{-2} cm^2 and the absorptivity (for metals) is about 0.1 then the mean heat source intensity (for metals) q_0 is about 10^3-10^5 W/cm^2. As the insulators have higher absorptivities q_0 for them is about 10^4-10^6 W/cm^2 and for high-power gas lasers q_0 is higher by the order of magnitude.

The heating process with these intensities can be analyzed in the framework of the theory of lumped heat sources [117]. Let us at first analyze heating by a stationary heat source. Similar to the analysis of the pulsed sources we can introduce the characteristic times of heating as has been done in Chapter 3 [Eqs. (232), (233) and (236)].

For the CO_2 laser the estimates of the characteristic heating times for steel yield $\tau_1 \approx 10^{-4}$ s, $\tau_2 \approx 10^{-3}$ s and $\tau_3 \approx 10^{-1}$ s. Using the characteristic times of heating we can select the most convenient analytical model according to the radiation intensity and the duration of heating. The above equations are useful for analyzing the heating processes in which the oxide film on the metal surface does not affect significantly the course of the process.

Most metals are covered by a thin film of oxides whose thickness has an important effect on the absorptivity of the metal. The absorptivity for radiation with the wavelength of 10.6 μm increases with increasing thickness of the oxide film. Therefore, if the material is in an oxidizing atmosphere, for instance in the air, and radiation is sufficiently long the heating process consists of two stages. At the initial stage the heating rate is comparatively low while the oxide film is thin. Starting from a moment τ^* when the oxidation process is intensified the heating rate increases sharply.

Let us analyze some features of heating of iron specimens in the oxygen atmosphere or in the air under the effect of the CO_2 laser radiation [126]. The surface oxidation of iron is known to produce a heterogeneous oxide scale of a complex structure consisting of the following successive layers: Fe — FeO — Fe_3O_4 — Fe_2O_3 — O_2 (gas). There are two possible oxidation processes in the iron depending on temperature. When the temperature is below 570 C the oxide scale consists almost entirely of Fe_3O_4, and above it there is a thin layer of the γ- and α-structure (Fe_2O_3).

The oxidation theory analyzes, as a rule, two limiting cases, namely, the thin oxide films (up to 10^3 Å) and the thick oxide scale. The development of thin oxide films (up to 10^3 Å) takes a relatively short time. Wagner [152] has developed a theory of the growth of thick oxide films, which has been repeatedly confirmed by experimental results. According to this theory the rate of growth of the thick oxide film is limited by the diffusion of anions or cations through the film.

The equation describing the initial stage of film growth (thin films) has the following form:

$$\frac{dl}{d\tau} = K_1(T)\exp[-B_1/T(\tau)\,l(\tau)] \tag{328}$$

where $l(\tau)$ is the thickness of the oxide film and $K_1\ (T)$ and B_1 are the kinetic constants depending on the properties of the material. Equation (328) is nonlinear.

The growth of a thick film is described by another equation as the transport processes are different:

$$\frac{dl}{d\tau} = -\frac{D_2}{l}\left[N + K_2(T)\right] \tag{329}$$

where D_2, $K_2(T)$ and N are kinetic constants which depend on the physical properties of the material.

The equation of temperature variation and the boundary condition on the surface of the metal relating the growing oxide film thickness and radiation absorption yield a closed system for describing the heating of the material up to the moment when the contribution of vaporization becomes significant. Generally, this system of nonlinear equations cannot be solved analytically and has to be solved using numerical techniques. However, the system can be solved in terms of simplified models. Typically, one of the following two approaches is used.

The first model assumes that the temperature of the affected zone is stationary so that the growth of the oxide film can be analyzed at a certain temperature depending on the radiation power density. This model has been used for analyzing the growth of an oxide film on chromium irradiated by the CO_2 laser beam [79].

The second model assumes that the growth of the oxide film is stationary since, as shown by the experimental data in [152], the numerical values of kinetic coefficients vary in the wide range of temperatures.

In the second model the oxide film growth is described, instead of Eq. (328), by the what is called inverse logarithmic equation

$$l^{-1} = K - B \log \tau \tag{330}$$

and instead of Eq. (329) by the parabolic equation

$$l^2 = K_p \tau \tag{331}$$

where K_p is a constant.

In this model heating of the material with the growing oxide film can be described by solving the heat conduction equation with the boundary condition in which the radiation power density varies with time. Starting from the time τ^* when the oxidation process is considerably intensified the variation in the absorptivity of the oxide-coated surface of the material is determined only by temperature.

The additional heat sources are the interfaces where exothermic reactions of oxide formation occur. For instance, these exothermic reactions liberate the following amounts of heat: FeO — 64.5 kcal/mole, Fe_2O_3 — 195.2 kcal/mole, and Fe_3O_4 — 266.5 kcal/mole.

The time τ^* can be estimated by the following equation [79]:

$$\tau^* = \frac{\beta^3\delta^2}{8\gamma s}\left(\frac{T_0}{T_D}\right)^5 \exp\left(\frac{T_D}{T_0}\right) \tag{332}$$

Here $\beta = kT_D/q_0 r_1$, $\gamma = x_1\beta T_i/T_D$, T_0 is the stationary temperature of the irradiated surface at the constant absorptivity $A = A_0$, $r_1 = 4.3 r_f (T_0/T_D)^{1/2}$ is the effective radius of the affected zone, $x_1 \approx 60\text{-}100$ Å, T_D is the energy of diffusion activation (in degrees Kelvin), T_i is the initial temperature of the surface, s is the sound velocity in the material, δ is the penetration depth of radiation into the oxide film, and

$$T_0 = T_i + q_0 r_f/\lambda \tag{333}$$

Estimates (332) show that τ^* varies from a few minutes to a few tenths of a second when the ratio T_D/T_0 varies from 25-30 to 15-20.

The above analysis agrees with the results of reference [160] stating that the rate of temperature increase in steel under the effect of the CO_2 laser radiation is different in the temperature ranges above and below the temperature T_τ at which intense oxide growth occurs (the film of α-Fe_2O_3 on steel). Therefore, when a metal in an oxidizing atmosphere is heated by a laser beam [121] there are two possibilities. If the time of radiation is less than τ^* which depends on q_0 according to Eq. (332) the heat source intensity can be described by Eq. (327) with the reflectivity independent of temperature (case 1). If the time of heating is longer than τ^* (case 2) the process consists of two stages; during the first stage absorptivity is independent of temperature and during the second one the temperature increases at a high rate after the temperature T_τ has been reached.

During the first stage of the heating process (cases 1 and 2) the effect of the CO_2 laser radiation is similar to the effect of such heat sources as an electric arc. During the second stage of laser beam heating in an oxidizing atmosphere we should take into account variations in absorptivity owing to the growth of an oxide film. However, there are other possible development processes of the oxide film with heating. If heating results in the disintegration of the oxide film, for instance, owing to the difference between the thermal expansion coefficients of the oxide and the metal then there is no linear increase of absorptivity.

Let us now discuss the experiments with CO_2 laser beam effects * in iron and its alloys [126]. The experiments have been carried out with the continuous-wave single-mode CO_2 laser with the output power about 700 W. The laser beam is focussed on the previously polished surface of the specimen by a NaCl lens with the local distance of 170 mm. The beam divergence is about 10^{-3} radians. If the radiation reflection is not taken into account the radiation power density on the surface of the specimen is about 7×10^4 W/cm^2. The materials used are armco iron, steels of grades 45, У8 and ХВГ, both normalized and hardened, grey cast iron and white pig iron.

The radiation of these samples for about 3-5 s produces the holes with the depth up to 200 μm and the diameter 1.5-1.8 mm depending on the radiation time. The experiments have demonstrated a significant contribution of the surface vaporization to the formation of holes. A thick oxide film is formed on the surface of the irradiated specimens; it is a few tens of micrometres on iron alloys and about 150 μm on armco iron.

The estimates from the above equations show that the oxide film grows practically during all the time of radiation, and its growth is described by the parabolic function [Eq. (331)] as the time of formation of the oxide film with the thickness about 10^3 Å is 10^{-2}-10^{-3} s, that is, it is considerably shorter than the time of radiation in these experiments.

The growth of the oxide film during radiation results in the increase in the absorptivity of the specimen up to 0.3-0.7 depending on the initial state of the surface [160].

* The experiments have been carried out in the laboratory headed by Acad. Prokhorov at the P. Lebedev Physical Institute of the USSR Academy of Sciences using the laser installation built by A. Barchukov.

To estimate the depth of the hole produced by laser radiaton of the material we can make use of the thermal theory of disintegration [34]. To do this we have to estimate the steady-state vaporization rate v_0 and the surface temperature T_0

$$v_0 = \frac{q_0 A_0}{\rho L_b + c\,(T_0 - T_i)} \tag{334}$$

where $A_0 \approx 0.35$, $q_0 = 7 \times 10^4$ W/cm² and

$$T_0 \approx -\frac{L_1}{R} \ln \frac{v_0 \rho L_b}{A_0 q_0} \tag{335}$$

Hence we find $T_0 \approx 3500$ C, $v_0 \approx 0.2$ cm/s and the time of reaching the steady-state vaporization rate $\tau \approx a/v_0^2 \approx 4$ s. These values yield a hole depth which is overestimated by an order of magnitude. One source of errors is that the deformation of the surface of the molten material under the effect of the reactive pressure of the vapour is neglected. The deformation of the liquid phase can be calculated from the equilibrium equation for the radiation power density constant in time [Eq. (263)]:

$$h = \frac{F_s}{\rho g} \frac{\mu^2}{1 + \mu^2} \tag{336}$$

In our case ($F_s \approx 6 \times 10^4$ bars at $q_0 = 4 \times 10^4$ W/cm² and $\mu = 0.02$) we find $h = 5 \times 10^{-3}$ cm. Thus, the deformation of the liquid phase surface occurs; this results in a decrease in the radiation power density owing to defocussing of the laser beam and to an increase in the irradiated area. Another reason for the disagreement between the experimental and predicted results is the fact that the approximated equation (334) does not take into account the heat losses due to heat conduction which make a significant contribution at the radiation power density about 10^5 W/cm². Moreover, the steady-state vaporization takes place only during a fraction of the irradiation period.

The powerful continuous-wave gas laser radiation has an interesting effect on metals [185]. If the laser output power is over the threshold of about 8 kW a cloud of the ionized metal vapour above the irradiated surface shields the surface from the laser beam. This effect prevents deep-penetration melting by the laser beam if the ionized vapour cloud is not blown off with a gas jet.

Let us discuss some features of the CO_2 laser radiation effect on metals in vacuum and in the air. The difference between these two cases lies mainly in the fact that radiation in the air gives rise to an oxide film on the metal surface, which considerably increases the absorptivity of the metal and decreases the radiation density thresholds for melting and vaporization.

Plates of stainless steel [176] are irradiated with the CO_2 laser beam having the power about 200 W in the air and in vacuum. A 2.5-cm disk of stainless steel is put into a massive brass holder contacting with the specimen over its entire perimeter, and the laser beam is focussed on its surface.

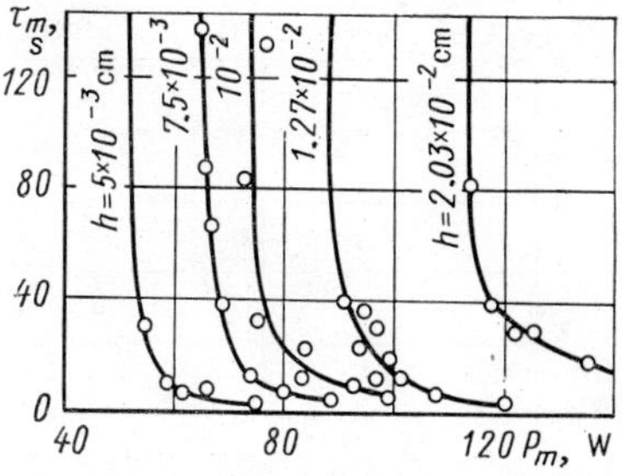

FIG. 81. Critical power P_m as a function of time τ_m for stainless steel in the air ($r_f = 0.04$ cm)

Figure 81 presents the radiation power P_m needed for obtaining the temperature T_m plotted as a function of time τ_m (beginning of melting) for the stainless steel specimens of varying thickness. The shapes of the plots slightly differ with increasing thickness of the specimens. The plots of similar shapes have been obtained for the experiments in vacuum, and for the plates of any thickness it has been found that $P_m(\text{air}) < P_m$ (vacuum) for the same τ_m. The temperature distribution in the specimen is analyzed with the model of a semi-infinite thin disk heated by a heat source with the power AP uniformly distributed over the area πr_f^2 [176]. The solution of the temperature distribution problem for $r \geqslant r_f$ and $\tau \gg r_f^2/a$ is

$$T(r, \tau) = \frac{AP}{4\pi\lambda h}\left[\ln\left(\frac{4a\tau}{Cr^2}\right) + \frac{r_f^2}{2a\tau}\ln\left(\frac{4a\tau}{Cr^2}\right) + \frac{1}{4a\tau}\left(r_f^2 + r^2 - 2r_f^2 \ln r_f/r\right)\right] \quad (337)$$

where $C = 1.781$. For stainless steel and $r_f = 0.1$ cm Eq. (337) is applicable for $\tau > 0.1$ s. Assuming that at the moment τ_m we have $r = r_f$ and $T = T_m$ we obtain

$$T_m = \frac{AP_m}{4\pi\lambda h}\left[\ln\left(\frac{4a\tau_m}{Cr_f^2}\right) + \frac{r_f^2}{2ai_m}\ln\left(\frac{4a\tau_m}{Cr_f^2}\right) + \frac{r_f^2}{2a\tau_m}\right] \quad (338)$$

Hence we see that A is proportional to the slope of the plot of h as a function of P_m (Fig. 82). The segment P_c on the axis P_m is proportional to the power losses which are not taken into account by Eq. (337). These losses due to heat convection transfer do not occur when the metal is heated in vacuum; however, the results do not contradict simplified estimates made with the Newton equation. The value of P_c decreases with decreasing time τ_m and for $\tau_m = 0.1$ s P_c is very small.

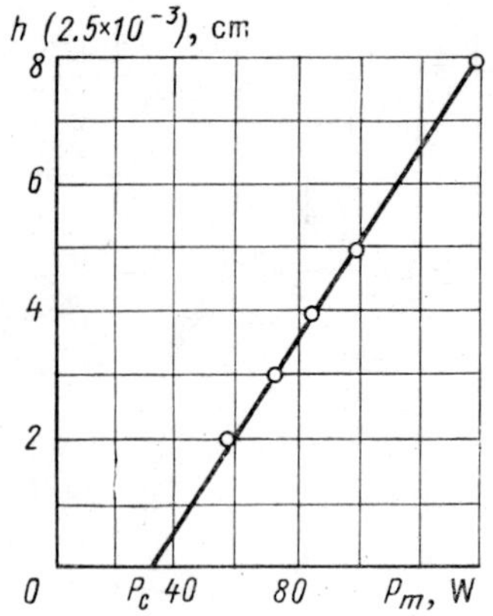

FIG. 82. Relationship between plate thickness h and critical power P_m for $\tau_m = 15$ s

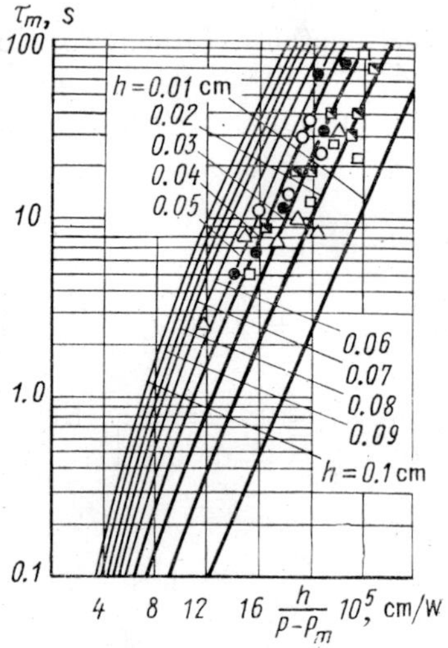

FIG. 83. Relationship between τ_m and $h/(P - P_m)$ for stainless steel irradiated in the air
● — $h=0.0203$ cm; ○ — $h=0.0127$ cm; □ — $h=0.0101$ cm; ■ — $h=0.0076$ cm; △ — $h=0.0058$ cm

The plot in Fig. 82 has been used for determining the value of A [176]. Using the averaged values of a and λ we find $A = 0.16$ that is three times the value of A in vacuum (0.05). Equation (337) shows that the plot of $\ln \tau_m$ as a function of $h/(P\text{-}P_m)$ must be a straight line for $\tau_m \geqslant 10$ s and $T = T_m$. Figure 83 presents the plots for Eq. (337) and the experimental results for various h and $r_f = 0.04$ cm. Some differences between the predicted and experimental results are due to the errors in determination of r_f.

Figure 84 presents the theoretical curves for heating of steel plates with the CO_2 laser radiation in vacuum. Though the measured value of r_f is 0.04 cm most of the predicted points lie on the plot for $h = 0.02$ cm. Equation (338) with the variables $\ln \tau_m$, $4\pi\lambda T_m h/AP_m$ and a/r_f^2 can be used for predicting P_m and τ_m for other metals with the given h and r_f. The respective plots are given in Fig. 85. To make use of these plots we must know the values of λ, a and A for $T = T_m$ (in vacuum and in the air). A good agreement is found between the plots of Fig. 85 for the nickel plate and the experimental results. The plots of Fig. 85 are valid for times of radiation satisfying the inequality $h < (a\tau)^{1/2}$. If $h > (a\tau)^{1/2}$ the critical radiation power for melting can be found from the following relationship:

$$P_m = \pi\lambda T_m r_f / A \qquad (339)$$

Let us now discuss the vaporization of optical materials under the continuous-wave CO_2 laser radiation. The dielectric ma-

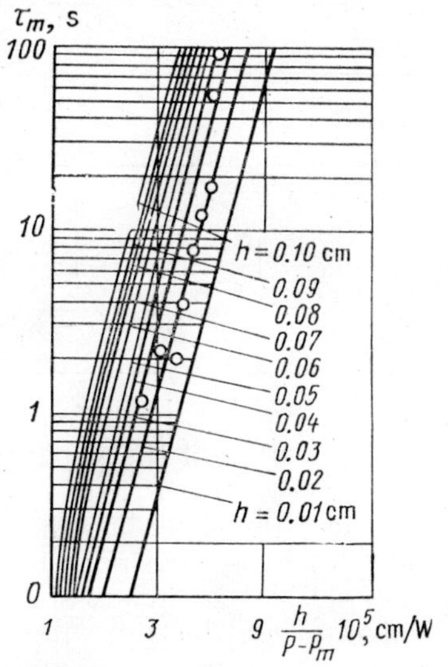

FIG. 84. Predicted plot of $\ln \tau_m$ as a function of $h/(P - P_m)$ for stainless steel irradiated in vacuum

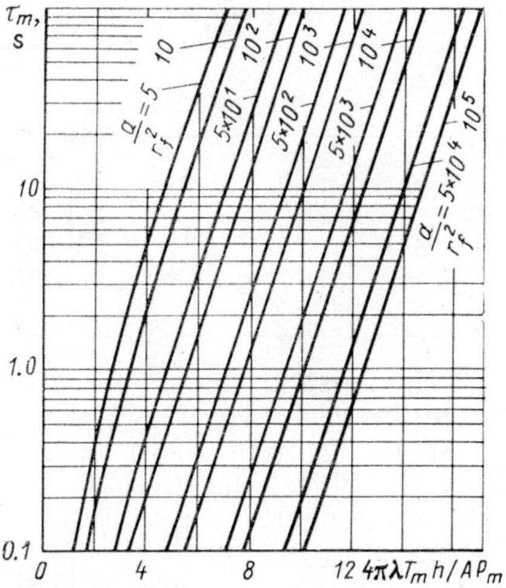

FIG. 85. Plot of τ_m as a function of $4\pi\lambda T_m h/AP_m$ for thin metal plates with the thickness $h \leqslant (a\tau_m)^{1/2}$

terials which are opaque to the radiation with the wavelength of 10.6 μm absorb this radiation via a molecular mechanism and therefore the shielding of radiation by disintegration products starts for these materials at lower power densities than for metals [75].

The polished specimens of quartz glass (grade KB), optical glasses (grades ТФ-5, К-8 and ЛК-5), neodymium glass (grade КГСС-7), polycrystalline glasses, LiF and NaCl crystals, and polymethylmetacrylate (PMMA) opaque for the radiation with $\lambda = 10.6$ μm are irradiated with the CO_2 laser beam of constant intensity of $(0.5\text{-}2) \times 10^4$ W/cm^2 [75]. All the above materials, except for NaCl, have a high absorption coefficient at the wavelength of 10.6 μm (from 15 cm^{-1} for LiF to 500 cm^{-1} for quartz glass). High-speed photorecording is carried out to analyze the development of a hole in the specimen accompanied by ejection of a vapour jet. The initial vaporization rate of the quartz glass is 1.3 cm/s. The predicted vaporization rate according to the equation $\bar{v}_s = (1 - R)q_0/(\rho L_0 + cT_b)$ for SiO_2 at the temperature of vaporization T_b agrees with the experimental results. Vaporization is a markedly pulsating process with the frequency of pulsation of 5-10 Hz. The mean rates of vaporization during pulsations ($\bar{v}_p$) and under the steady-state conditions ($\bar{v}_s$) are given in Table 16 [75].

TABLE 16. Vaporization Rates During Pulsations and under Steady-State Conditions

Material	$\bar{q}$, kW/cm	$\bar{v}_p$, cm/s	$\bar{v}_s \times 10^{-2}$, cm/s	Materia	$\bar{q}$, kW/cm	$\bar{v}_p$, cm/s	$\bar{v}_s \times 10^{-2}$, cm/s
Quartz glass	19	0.64	8.5	Glass TΦ-5	16	1.6	1.2
Same	8.3	0.4	5.7	Glass K-8	16	1.2	1.7
Glass ЛK-5	17	1.0	4.3	Polycrystalline glass	16	1.4	1.5
Same	5.2	0.66	1.3				
				LiF crystal	19	0.07	—

The pulsation of the hole development is caused by the competition between two processes, namely, the ejection of the vapour and shielding of the laser beam. In the glasses ЛK-5, K-8 and others the pulsations die out at a rate a few times higher than that for the quartz glass, and at the hole depth of 10-20 mm the absorption by the vapour becomes constant, and a steady-state regime is established which is complicated by the brittle failure of the material. This self-maintained vaporization regime is qualitatively similar to the vaporization of metals with radiation absorption in the plasma cloud [2]. The optical thickness αl_h calculated from the experimentally measured steady-state vaporization rate is 0.7-1.1, that is, it is close to unity (l_h is the depth of the laser-produced hole and α is the absorption coefficient).

Let us analyze the initial stage of the attenuating pulsations using a one-dimensional model (when the specimen thickness is less than 1 cm this is the main regime of the laser radiation process) [75]. If M is the total mass vaporized during the time τ, m is the mass of the vapour in the hole at the moment τ and $F(\rho, \tau, M)$ is the flow rate of vapour at the entrance of the hole we can write the mass conservation equation as

$$\dot{M} - \dot{m} = F(\rho, \tau, M) \qquad (340)$$

where the point over M and m denotes differentiation with respect to time. If the density of the solid is ρ_0 and the vapour density averaged over the hole volume is ρ we obtain

$$M/\rho_0 = m/\rho \qquad (341)$$

If the distribution of the ejected vapour density is spatially random the absorption outside the hole is low and absorption inside the hole is proportional to the mass m of vapour inside it in the approximation of weak absorption ($\alpha l_h < 1$). Then the radiation power density q_0 (taking into account the Fresnel reflection losses) is partially absorbed by the vapour inside the hole (the mass coefficient of absorption K_m), and the remaining radiation power is absorbed by the solid vaporizing it (the effective enthalpy of vaporization Q):

$$q_0 = q_0 K_m m + \dot{M} Q \tag{342}$$

The system of equations (340)-(342) approximates the vaporization of dielectric materials taking into account the radiation absorption by the vapour. The flow rate function $F(\rho, \tau, M)$ must be found from the solution of the nonstationary gas dynamic problem taking into consideration the vapour density distribution in the hole. However, the problem can be simplified, in particular by averaging the vapour density over the hole volume, and the flow rate function can be obtained in the quasistationary form where the averaged density $\bar{\rho}$ follows the ejection of the material from the hole. Hence Bernoulli's equation for the compressible gas adiabatically expanding with the exponent $\varkappa$ yields [75]

$$F(\rho) = \rho S v_v = \rho S \left\{ \frac{2\varkappa}{\varkappa - 1} [1 - (\rho/\rho_s)^{\varkappa-1}] P_s/\rho_s \right\}^{1/2} \tag{343}$$

where P_s and ρ_s are the pressure and density of the stationary vapour, v_v is the vapour velocity and S is the irradiated area.

The vapours of dielectric materials consist mostly of two- and three-atom molecules, that is, $\varkappa = 1.3\text{-}1.4$, and $F(\rho) = \gamma\rho$ where γ is a constant. Substituting this flow rate function into Eq. (340) and eliminating m from the resulting equation [using Eq. (342)] we obtain the second-order ordinary differential equation for $M(\tau)$

$$\ddot{M} + \dot{M}(\theta + \Phi/M) - g/M = 0 \tag{344}$$

where $\theta = q_0 K_m/Q$, $\Phi = \gamma\rho_0$ and $g = \gamma\rho_0 q_0/Q$.

Equation (344) has no general analytical solution. It is solved [75] with an approximation technique using the fixed values $\theta = 20$, $\Phi = 10$ and $g = 1$ according to the experimental parameters. The power series expansion of the function M in τ in the vicinity of the origin ($M = 0$, $\tau = 0$) has the first extremum for $\tau = 1/6$ s. Figure 86 presents M plotted as a function of time for the time interval from 0 to 0.5 s.

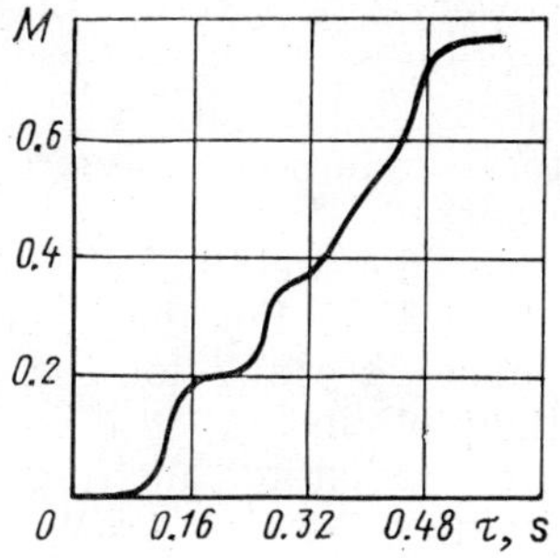

FIG. 86. Vaporized mass M (arbitrary units) as a function of time τ

The pulsation of the heating process during transition to the steady-state self-maintained regime is typical for all the materials whose vapour absorbs the incident laser radiation. Pulsations of the vaporization process are observed when polymethylmetacrylate is irradiated by the CO_2 laser beam [92]. When the incident radiation power is high the polymethylmetacrylate specimens burn in the air producing large soot flakes.

5.2. CO_2 LASER BEAM SEAM WELDING

A higher laser radiation power increases the depth of the molten pool on the one hand and the rate of welding on the other hand. Gas lasers make possible the deep-penetration melting of metals with high h/d ratios (where h is the depth of the molten pool and d is its diameter).

Let us discuss the experimental results on the melting of stainless steel with the CO_2 laser beam having the output power of 20 kW (gas dynamic laser) and 8 kW (gas discharge laser) [185]. With the welding rate of 2.1 cm/s the gas dynamic laser produces the molten pool 1.9 cm deep, the depth/width ratio being about 6 in the steel plate. With the welding rate of 4.2 cm/s the penetration depth decreases to 1.27 cm.

The beam of the 8-kW gas-discharge laser almost melts through an 0.95-cm steel plate with the welding rate of 1.27 cm/s. The welding rate of 1.7 cm/s produces a welded seam with the full penetration of the steel plate 0.635 cm thick. The 8-kW laser can be also used for butt welding of two 0.95-cm plates with the gap of 500 μm between them. Almost full penetration is obtained in this welding experiment.

The above experimental results are shown in Fig. 87. The solid line represents the experimental data on the electron beam welding. When the parameter $P_0/T_m h\lambda$ is larger than 100 (here h is the thickness of the plate) the heat conduction is not significant, and the welding rate is proportional to the radiation power. With lower radiation powers the welding rate is proportional to the squared power. Figure 88 presents the plot of the 30% efficiency power as a function of the plate thickness for various materials (stainless steel, low-carbon steel, aluminium, copper). The efficiency is the ratio of the power required only for melting to the total radiation power (taking into account the heat conduction losses).

The welding efficiency in the 20-kW laser experiments is about 45%. For the laser power of 8 kW the welding efficiency is about 28% and for the power of 3.8 kW it is about 22%. From the plots in Fig. 88 it can be found that the following laser powers are needed for melting 1-cm plates of the following materials: stainless steel — 4 kW, low-carbon steel — 14 kW, aluminium — 16 kW, and copper — 60 kW. The welding efficiency does not depend on the width of the welding zone directly but it is determined by the welding rate multiplied by the welding zone width.

It follows from the above-said that the deep-penetration welding is possible with lower laser powers but with respectively lower efficiencies.

In the above experiments with laser welding of stainless steel plates an argon jet is directed at the welding zone either perpendicular or parallel to the plate surface [185]. The deepest penetration is obtained with a perpendicular argon jet.

When the argon jet is parallel to the plate surface the brightness of the welding zone is higher than in the first case; this type of welding produces a very narrow continuous seam with smaller variations in properties along its length. If no gas is blown in the welding zone a stationary metal vapour cloud appears above the surface; it effectively shields the welding zone from the laser beam. This results in decreasing the depth of the molten pool.

The threshold power of this effect is about 8 kW. The formation of the stationary ionized vapour cloud can be prevented not only by blowing off the cloud by a gas jet but also by supplying the gases which prevent ionization to the welding zone.

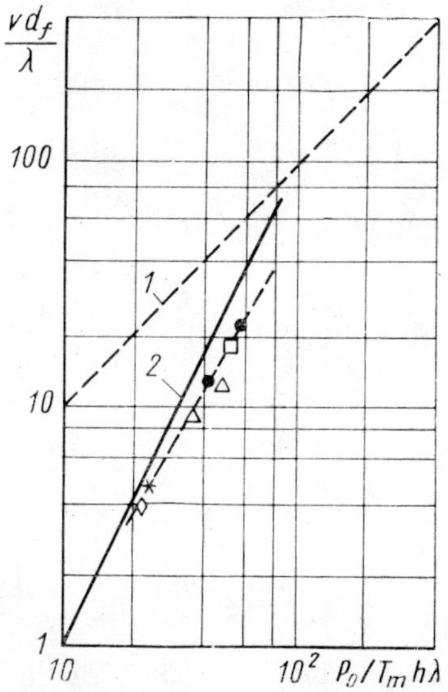

FIG. 87. Normalized welding rate vd_f/λ as a function of normalized laser power $P_0/T_m h\lambda$

1—no heat transfer; 2—electron beam welding; △—8 kW; ●—20 kW; *—3.8 kW; □—10 kW; ◇—4 kW

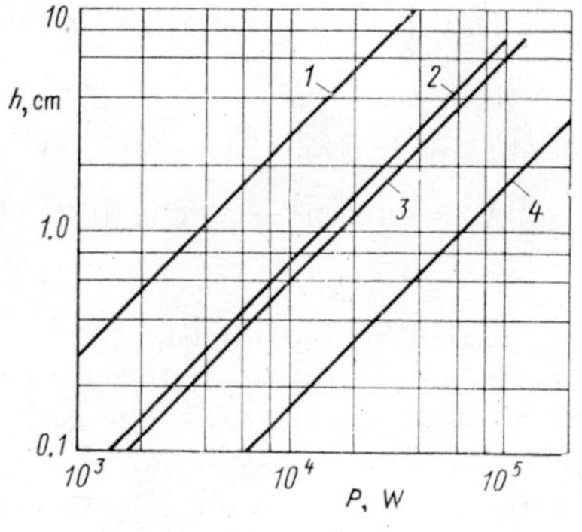

FIG. 88. Melting power P as a function of plate thickness h

1—stainless steel; 2—low-carbon steel; 3—aluminium; 4—copper

If we neglect the process of the formation of a narrow and deep channel in the material the heat propagation in seam welding can be analyzed within the framework of the theory of moving heat sources [117]. For instance, deep-penetration welding of thick materials can be analyzed with the model of a line heat source moving with a constant velocity.

Such a model has been used by the authors of reference [206] who have taken into account also the temperature dependence of thermal parameters. Using Kirchhoff's transformation $s = \int_0^T \lambda(T)\, dT$ for the quasistationary heat conduction equation in the moving coordinate system we obtain, instead of the equation for temperature distribution $\nabla\lambda\nabla T = \rho c v \nabla T$, the following equation for the function s (we neglect the transition processes at the beginning and end of the welding):

$$\nabla^2 s = \frac{v}{a}\nabla s \tag{345}$$

where $a = \bar{K}/c\rho$ and $\bar{K} = T^{-1}\int_0^T \lambda(T)\, dT$.

For a line heat source Eq. (345) has the following solution in the cylindrical coordinate system:

$$s = (q_0/2\pi)\exp(ur\cos\Phi)\, K_0(ur) \tag{346}$$

where K_0 is the modified Bessel function which has a singularity at the origin. If we assume that the total power of the source P_0 is distributed over the depth h then

$$q_0 = P_0/h \tag{347}$$

Defining a normalized power input per unit depth

$$X = P_0/hs \tag{348}$$

we obtain, instead of Eq. (346),

$$X = 2\pi\exp(-\, ur\cos\Phi)/K_0(ur) \tag{349}$$

where $u = v/2a$.

Estimates from Eq. (349) show that for slowly moving heat sources or at large distances from the heat source the isotherms are almost circular as the solution is almost independent of Φ owing to smallness of ur. For fast-moving heat sources the isotherms are egg-shaped according to the theory [117].

For practical applications the most interesting parameters are the width of the molten pool and its maximum depth for a given source power and velocity. The half-width of the molten pool is

$$y = r\sin\Phi \tag{350}$$

The function (350) has a maximum when

$$\frac{dY}{d\Phi} = 0 = \frac{\partial Y}{\partial\Phi} + \frac{\partial Y}{\partial r}\frac{\partial r}{\partial\Phi} \tag{351}$$

As X is constant along the isotherm we have

$$\frac{dX}{d\Phi}=0=\frac{\partial X}{\partial \Phi}+\frac{\partial X}{\partial r}\frac{\partial r}{\partial \Phi} \tag{352}$$

Elimination of $\partial r/\partial\Phi$ from Eqs. (351) and (352) yields the condition of the maximum pool width:

$$\frac{\partial X}{\partial \Phi}\Big/\frac{\partial X}{\partial r}=\frac{\partial Y}{\partial \Phi}\Big/\frac{\partial Y}{\partial r} \tag{353}$$

Using Eqs. (350) and (351) we find

$$\cos\Phi=-K_0(ur)/K_0'(ur) \tag{354}$$

Let us define the normalized width of the molten pool

$$Y=vb/a=2ub \tag{355}$$

where $b=2y$ is the full width of the molten pool. Then

$$Y=4ur\left[1-K_0^2(ur)/K_0'^2(ur)\right]^{1/2} \tag{356}$$

$$X=2\pi\exp\left[urK_0(ur)/K_0'(ur)\right]/K_0(ur) \tag{357}$$

Equations (356) and (357) for the normalized width of the molten pool (and welding rate) and for the normalized power input per unit depth of penetration are both functions of the variable ur which can in principle be eliminated from these equations. In general this can be done numerically but this can be done analytically in both the high and lower power (or velocity) limits.

At high velocities (high ur values) the following asymptotic expressions can replace the modified Bessel functions:

$$K_0(ur)\approx(\pi/2ur)^{1/2}\exp(-ur) \tag{358}$$

$$-K_0(ur)/K_0'(ur)\approx\left[1-(1/2\,ur)\right] \tag{359}$$

then we have

$$X\approx(8\pi\, eur)^{1/2} \tag{360}$$

$$Y\approx4\,(ur)^{1/2} \tag{361}$$

or

$$Y\approx(2/\pi e)^{1/2}\;X\approx0.483X \tag{362}$$

The asymptote (362) is shown in Fig. 89 which presents the normalized velocity Y as a function of the normalized power X similar to Fig. 87. The factor 0.483 is the fraction of the input power spent only for melting the metal. At low velocities of a line

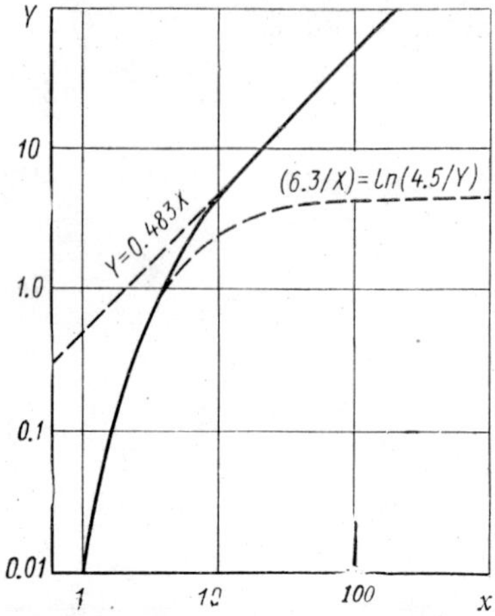

FIG. 89. Normalized velocity Y as a function of normalized power X for deep-penetration welding

heat source (small ur) the asymptotic expressions for the modified Bessel functions are

$$K_0(ur) \approx \ln\left(2e^{-C_E}/ur\right) \quad (363)$$

where $C_E = 0.577 \ldots$ is Euler's constant and

$$-K_0'(ur) \approx 1/ur \quad (364)$$

Hence we find

$$X \approx 2\pi/\ln\left(2e^{-C_E}/ur\right) \quad (365)$$

$$Y \approx \exp(\ln 8 - C_E - 2\pi/X) \quad (366)$$

To obtain a required weld depth and welding rate with lower radiation power we have to narrow the weld width or to decrease the cross section of the laser beam which naturally has a minimum.

For the intermediate range of variations in ur Eqs. (356) and (357) must be solved numerically (the solid line in Fig. 89 [206]).

TABLE 17. Parameters of Deep-Penetration Laser Welding

Metal	P_0, kW	h, mm	b, mm	v, mm/s	P_0/h, kW/m	$P_0/hs = X$	$Y = bv/a$	Refs
Steel	0.125	0.125	0.45	38.0	2×10^3	56	3.5	[212]
	0.25	0.250	0.70	14.8	1×10^3	28	1.1	
	0.25	0.417	0.75	4.7	0.6×10^3	17	0.72	
Same	0.25	0.125	0.50	21.1	2×10^3	56	1.2	
	0.25	0.203	0.50	12.7	1.23×10^3	34	1.3	
	0.25	0.250	1.00	4.2	1×10^3	28	0.86	
Inconel	0.25	0.100	0.25	63.5	2.5×10^3	74	3.2	
	0.25	0.250	0.45	16.9	1×10^3	29	1.52	
Nickel	0.25	0.125	0.45	14.8	2×10^3	20.6	0.53	
Ni-Cu alloy	0.25	0.250	0.62	6.4	1×10^3	21.3	0.57	
Titanium	0.25	0.125	0.37	59.0	2×10^3	50	3.0	
	0.25	0.250	0.55	21.1	1×10^3	25	1.61	
Steel	8	8.9	2.3	12.5	1×10^3	28	5.9	[185]
	8	6.4	2.1	16.7	1.25×10^3	35	7.2	
	20	20.2	3.3	21.2	0.9×10^3	27	14.3	
	20	12.7	2.3	42.4	1.57×10^3	44	19.9	

Note. P_0 is the radiation power, h is the thickness of the specimen, b is the width of the weld, v is the welding rate.

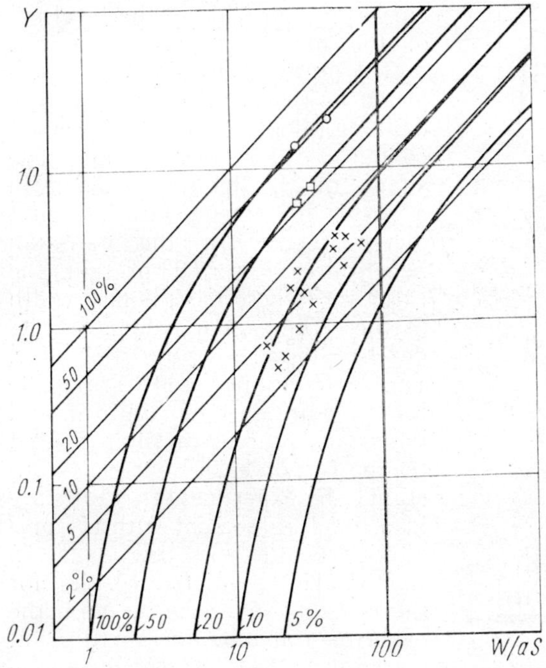

FIG. 90. Comparison of predicted results and experimental data on laser welding
○—20 kW; □—8 kW; ×—250 W (thin plate)

Figure 90 makes a comparison of the predicted and experimental results for deep-penetration laser welding. The experimental data in Fig. 90 are obtained in the above experiments with 8- and 20-kW CO_2 lasers and with a 250-W CO_2 laser welding of thin plates of various materials [212]. Increasing power of the laser results in higher efficiency. This is attributed to the fact that though the metals have high reflectivities the laser-drilled hole "traps" the incident laser radiation, and the energy losses are reduced [160].

Table 17 summarizes the experimental results on high-power laser welding.

Let us analyze the predicted temperature fields in materials heated with CO_2 laser beam and compare them with the experimental results according to reference [159]. The experiments are carried out with a CO_2 laser with a 60-Hz power supply. The output power of the laser oscillates with the frequency of 120 Hz (Fig. 91). The radiation power density distribution over the cross section of the laser beam can be approximated by the gaussian

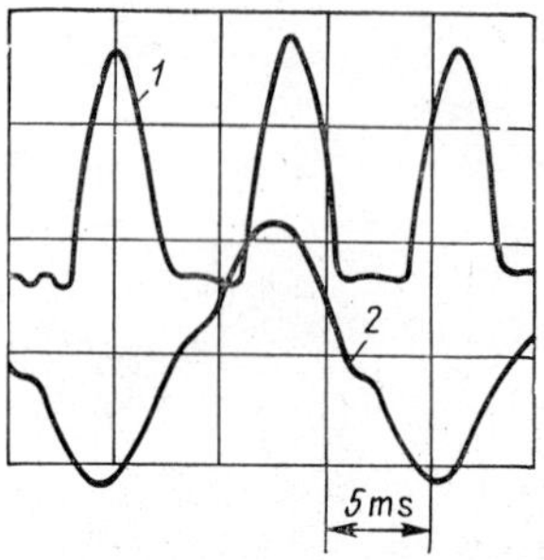

FIG. 91. Output power of CO_2 laser (1) and pumping power (2) as a function of time

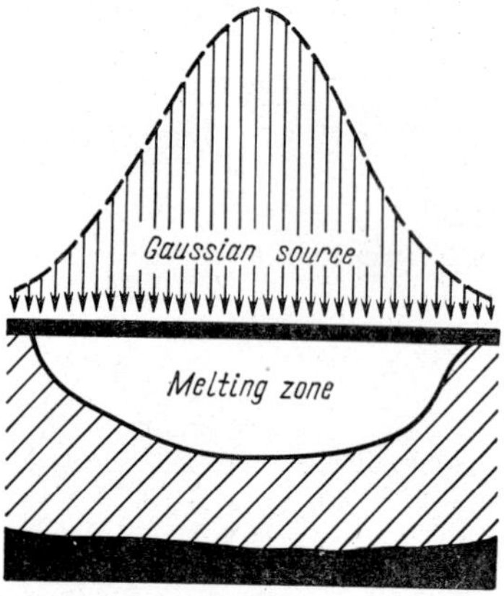

FIG. 92. Typical molten pool produced by CO_2 laser beam on surface of 0.3-mm stainless steel plate coated with Fe_2S_3 film

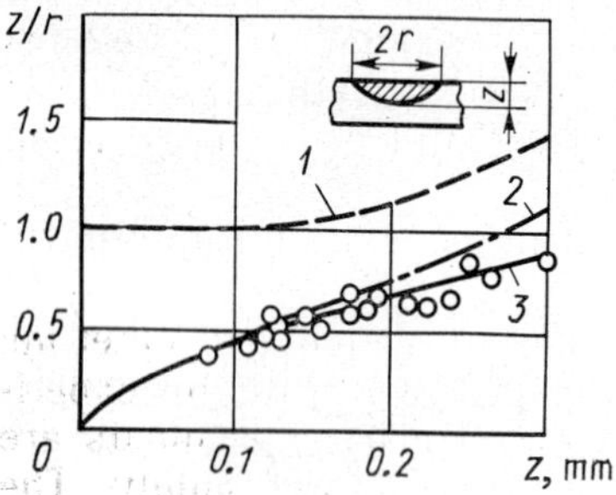

FIG. 93. Ratio z/r as a function of penetration depth z

1—point source; 2—uniform circular source; 3—gaussian source

curve 0.5 mm wide at the level e^{-1} of the maximum power density.

Figure 92 presents the laser-produced molten pool in the 0.3-mm thick stainless steel plate coated with a Fe_2S_3 film. The shape of the molten pool is in a good agreement with the predicted results for the gaussian heat source. Figure 93 presents the ratio z/r (where r is the half-width of the molten pool and z is its depth) plotted as a function of z. The calculations [159] are made for the point, circular uniform and gaussian sources assuming that the power of each source is constant. The experimental results are in a good agreement with the predicted results for the gaussian source. The predicted results for the circular source agree with the experimental data for small z but for high z the agreement disappears. The predicted results for the point source model differ considerably from the experimental data.

Figure 94 presents the penetration depth z as a function of velocity v of a heat source for stainless steel previously treated to produce the Fe_2S_3 film increasing the radiation absorption. There is a satisfactory agreement between the predicted and experimental results at low velocities. At high velocities the experimentally measured molten pool depths are larger than the predicted ones.

Figure 95 presents the predicted temperature distribution (in dimensionless units) over the axis of the seam on the back side of the 0.3-mm thick aluminium plate for the input power varying with the frequency of 120 Hz (see Fig. 91). This plot is compared with the temperature distribution produced by a constant heat source.

For the quasistationary state the temperature in the centre of the focussed spot on the surface of the material irradiated by the continuous-wave laser beam is

$$T(0,\ 0,\ \infty) = \frac{AP}{2\sqrt{\pi}\ \lambda r_f} \qquad (367)$$

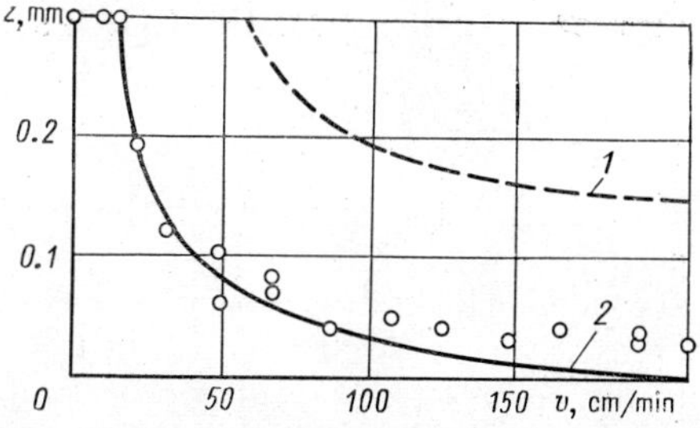

FIG. 94. Penetration depth z in stainless steel as a function of velocity v of heat source

1—point source, 2—gaussian source

Hence the minimum radiation power required for melting the surface of a semi-infinite body is

$$P_{min} = \frac{2\sqrt{\pi}\ \lambda r_f T_m}{A} \qquad (368)$$

The absorptivity of the polished metal specimen at the temperature of melting is

$$A = 4\sqrt{\frac{\pi c \varepsilon_0 \eta_f}{\lambda'}} \qquad (369)$$

where c is the velocity of light, ε_0 is the dielectric constant, η_f is the direct-current resistance at the temperature of melting and λ' is the wavelength of the laser radiation. Hence the minimum

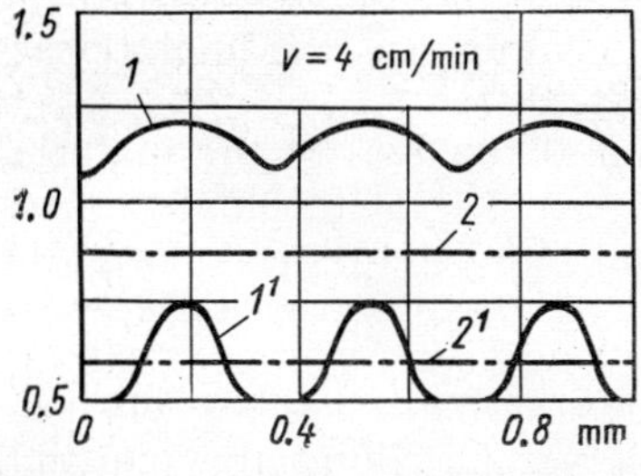

FIG. 95. Variation in temperature (1) on the back side of Al plate ($h = 0.3$ mm) due to pulsating-intensity heat source (1') and constant temperature (2) due to constant-intensity heat source (2')

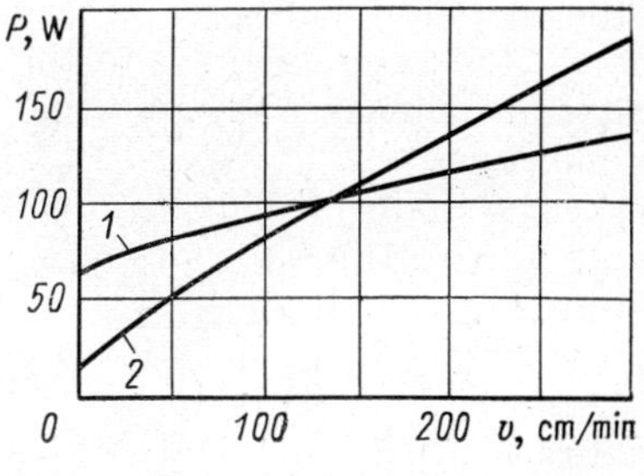

FIG. 96. Power P required for complete-penetration welding of polished plates of Al (1) and stainless steel (2) as a function of welding rate v ($h = 0.3$ mm)

power for melting for the CO_2 laser is

$$P_{min} = 7.9 \times 10^{-4} \frac{\lambda' T_m}{\sqrt{\eta_f}} \tag{370}$$

Estimates show that the typical continuous-wave CO_2 laser power is sufficient for melting metal specimens even with polished surfaces. The minimum powers of melting for the following metals are: titanium — 200 W, stainless steel — 200 W, iron and zirconium — 300-400 W and copper and silver — 9-10 kW.

Figure 96 shows the effective melting power predicted as a function of welding rate in the quasistationary state for aluminium and stainless steel plates 0.3 mm thick. At the welding rates below 150 cm/min the stainless steel plate can be welded at lower radiation powers than the aluminium plate but at the welding rates above 150 cm/min the situation is reversed owing to the fact that the welding rate v is proportional to a^{-1}.

5.3. CUTTING OF MATERIALS BY CONTINUOUS-WAVE LASER RADIATION

One of the most promising applications of high-power continuous-wave lasers is separation of materials (cutting, scribing and thermal cleaving). Cutting is done with the radiation from the CO_2 and garnet lasers. The CO_2 lasers are the most advantageous ones for cutting since they have high efficiencies (about 15-20%). Moreover, the radiation at 10.6 μm is absorbed by numerous materials including most metal oxides, glasses, quartz, natural organic materials (wood, leather, etc.), man-made materials, etc.

There are three possible basic designs of a laser-cutting installation for continuous cutting — the laser moving along the material being cut, the material moving relative to the focussed spot of the laser beam, and the laser beam moving relative to the material. The second and the third designs are the most practicable ones.

The material can be cut by the total removal of the material along the cutting line or by a partial removal of the material for instance, by drilling a system of small-diameter holes along the cutting line in a plate and subsequent breaking of the plate along this line [191]. This technique of cutting is known as scribing. Plates of brittle materials can be cut by thermal cleaving when no material is removed. The heat source moving along the material produces thermal stresses in it and small cracks; subsequently the material is broken along the heat source effect line.

During laser cutting the typical procedure is to blow a gas through the cutting zone; the gas flow removes the disintegration products for inflammable materials or initiates chemical reactions in the cutting zone for metallic materials. For the materials of the first type such inert gases, as nitrogen are used, and for metals oxygen is used.

The amount of energy required for cutting materials with continuous-wave laser radiation is determined by the optical and thermal properties of the material. As discussed above, the polished metals have high reflectivities for radiation with $\lambda = 10.6\ \mu m$; however, the efficiency of the process is significantly increased by increasing absorptivity due to high temperature and oxidizing atmosphere.

Sullivan and Houldcroft [205] have developed the laser-cutting technique with an oxygen blow (gas-jet laser cutting). Figure 97 illustrates this cutting technique. The laser beam is focussed on the surface of the material by a special optical system. A jet of oxygen is directed parallel to the laser beam on the material. The oxygen jet has two effects: it increases the absorptivity of the material contributing to the formation of the oxide film on the surface and removes the film and the melt from the cutting zone until the material is completely cut. The laser beam produces a heat source with a high energy concentration, therefore, it provides for a smaller cutting zone width, smaller size of the heat-affected zone and higher cutting rate than any other thermal cutting technique. The width of the cutting zone is close to the diameter of the laser focussed spot or somewhat smaller while the size of the heat-affected zone is 0.05-0.2 mm.

Let us now discuss some features of the oxygen-jet laser cutting technique. The cutting rate depends on the thermal properties of the material and its thickness and grows with increasing laser power. Figure 98 [179] presents the plots of the cutting rate as a function of the laser power for various thicknesses of the material and gas flow rates. The laser beams of 300-500 W cut steel and titanium at the cutting rates of a few metres per minute but such powers are not sufficient for cutting high-heat-conductivity metals, for instance, copper and aluminium. High-power laser beams (up to 20 kW) cut these metals at high rates even without an oxygen blow [185].

Table 18 presents the cutting rates for various materials [165, 179, 185, 186, 191].

In cutting noninflammable materials the gas jet mostly carries out the cleaning of the cutting zone blowing off the vapour and molten drops and cooling the material adjacent to the cutting zone. Cooling is especially important for cutting dielectric materials without glazing and charring. The gas jet also protects the optical system from the disintegration products of the material.

Continuous-wave laser radiation cutting with inert gas jet is especially convenient for cutting brittle and inflammable ma-

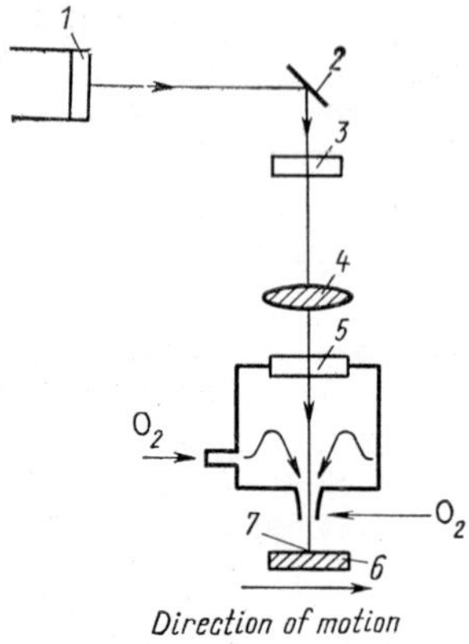

FIG. 97. Diagram of laser-cutting installation with air blow

1—laser; 2—mirror; 3—shutter; 4—lens; 5—window; 6—workpiece; 7—focussed spot

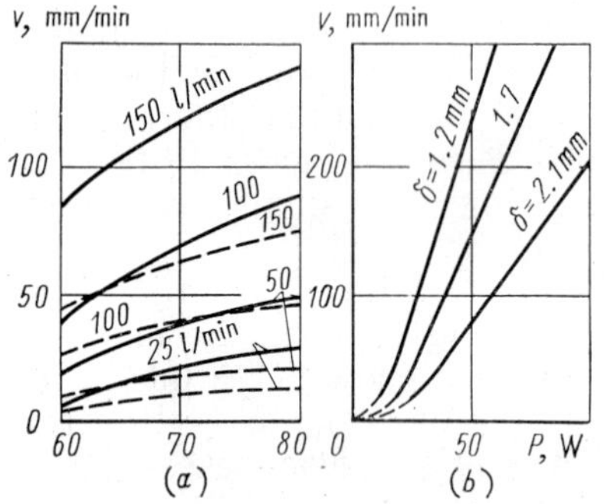

FIG. 98. Cutting rate v as a function of laser power P

(a) low-carbon (—) and stainless (– – –) steel with various oxygen flow rates; $h = 1$ mm; (b) quartz glass of various thickness

terials (ceramics, glass, plastics, wood, etc.). These materials have a high absorptivity for the radiation with $\lambda = 10.6\ \mu m$.

Babenko and Tychinsky [3] have described experiments in cutting metals and dielectric materials with the continuous-wave CO_2 laser beam using the multimode regime at the mean power of 500 W. The laser beam is focussed by a 150-mm lens into a focussed spot with the diameter of 0.5 mm. The gas is supplied through a conical nozzle with the outlet diameter of 1.5 mm. The laser beam also passes through the nozzle. The gas pressure in the laser beam cutter is 3 kgf/cm^2 and the gas flow rate about 0.5 l/s.

Oxygen is typically employed in laser cutting of metals. The cutting rate depends significantly on the oxygen pressure in the cutter at low pressures and slightly at pressures over 2.5 kgf/cm^2. The quality of cutting is better at high gas flow rates and smaller distances between the nozzle and the material surface.

Figure 99 presents the threshold cutting rate plotted as a function of the laser power for steel plates of various thickness [36]. It can be seen that the cutting rate depends significantly on the thickness of the specimen and the laser radiation power.

The authors of reference [3] have reported the experimental results on cutting dielectric materials with the CO_2 laser beam.

Laser cutting of dielectric materials has the following significant advantages compared to the conventional techniques: high productivity with good quality of cutting, easy automation of the process, small deformations of the workpiece and the lack of contamination of the workpiece by emulsion and the products of dis-

TABLE 18. Laser Cutting Rates for Various Materials

Material	Power									
	100 W		200 W		250 W		850 W		⩾8 kW (no gas jet)	
	δ, mm	v, m/min	δ, mm	v, m/min	δ, mm	v, m/min	δ, mm	v, m/min	δ, mm	v, m/min
Low-carbon steel	1.0	1.6	4.75	0.635	0.5	0.635	2.2	1.8	6.3	2.3 *
Stainless steel	1.0	0.94	0.46	0.635	0.5	2.6	9	0.36	4.7	1.27 **
		—	0.15	0.762						
Titanium	—	—	0.51	0.203 [1]	0.6	0.2 [1]	0.5	3.24	—	—
Niobium	—	—	0.13	0.127 [1]	—	—	—	—	—	—
Tantalum	—	—	0.13	0.127 [1]	—	—	—	—	—	—
Zirconium	—	—	0.25	0.914 [1]	—	—	—	—	—	—
Plastic-coated steel	—	—	0.89	0.635	—	—	—	—	—	—
Silicon carbide	—	—	1.57	0.762	—	—	—	—	—	—
Wood	—	—	17.5	0.203	18-50	0.1-0.2	5	4.5 [2]	—	—
Asbestos cement	—	—	6.35	0.025	6.3	0.025 [1]	—	—	—	—
Glass	—	—	0.381	0.101	4.0	0.1 [1]	—	—	9.5	1.52 **
	—	—	1.57	0.381	—	—	—	—	—	—
	—	—	0.20	0.101	—	—	—	—	—	—
Nylon	—	—	0.76	6.10	—	—	—	—	—	—
Plexiglas	—	—	—	—	—	—	—	—	25.4	1.52 ***
Aluminium	—	—	—	—	—	—	—	—	12.7	2.3 *
Quartz	1.2	0.5	—	—	—	—	—	—	—	—
Rubber	2.0	1.9	—	—	—	—	—	—	—	—
Ceramics	—	—	—	—	—	—	6.5	0.6 [2]	—	—

Notes: [1] — air blow, [2] — argon blow; in the rest of the experiments oxygen is used.

* — power is 15 kW. ** — 20 kW. *** — 8 kW.

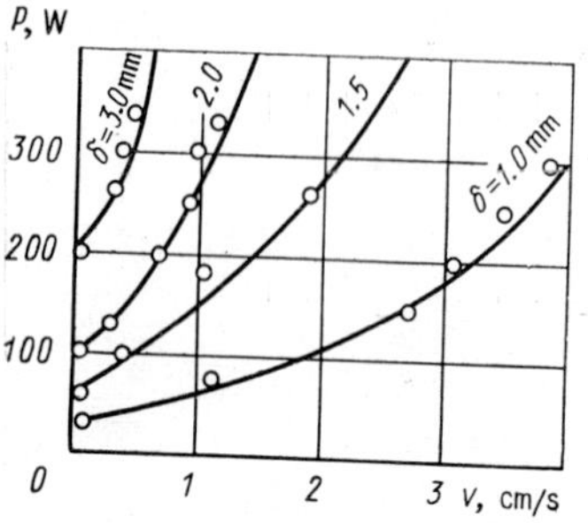

FIG. 99. Relationship between threshold cutting rate v of laser cutting and laser power P for steel 12X18H10T

integration of the material and the cutting tool. The cut surface is practically always clean, smooth and has regular-shaped edges. The thickness of the damaged layer is less than 0.1 mm. At the edges of organic materials there are traces of some charring. In cutting thick workpieces with laser beams self-channeling of radiation is found to occur, that is, the width of the cut from the back surface is significantly smaller than it could be expected from the geometric divergence of the laser beam.

Plastics, wood, ceramics, glass and other materials can be cut by a laser beam when the workpiece thickness is up to 50 mm. The cut width is not more than 1 mm though the laser beam diameter is about 7 mm at the distance of 20 mm from the lens focus. The reason for this difference is that cutting proceeds in the following way. The laser beam focussed on the surface vaporizes the material, and the vapour cloud screens radiation. The gas blown from the nozzle is mixed with the vaporization products making up a hot gas jet which cuts the material.

Laser cutting is a highly promising technique for the textiles industries. The multilayer stacks of fabrics with the maximum thickness of 20-30 mm can be cut by a laser beam without charring (a high-quality cut is produced). For laser cutting of such fabrics as nylon or glass fabric the layers of fabric are separated with paper tissue or the fabric is moistened to prevent their welding.

Let us now discuss briefly the cutting of ceramics by scribing. As ceramics have poor heat conductivity and high Young's modulus the conventional gas-jet laser cutting does not often produce good results. This results in the superheating of the material near the cut and its disintegration owing to high thermal stresses. To reduce the total power input the pulsed regime is typically used. A series of laser pulses with high radiation power density and low mean power is used to drill a series of small-diameter holes in ceramics. Finally, the material is cleaved along the line of holes.

Let us analyze the cutting of materials with the continuous-wave laser beam using the following simple model. A heat source of a constant power P moves with the constant velocity v along the surface of an infinite plate with the thickness δ. At the steady-state stage of the process the heat source power P equals the laser radiation power P_l absorbed by the material if there is no gas blow. If there is a gas blow through the cutting zone which initiates chemical reactions after the temperature T_0 has been reached there occurs also absorption or liberation of power $P_x = Qm$ where Q is the specific heat of the chemical reactions and phase transitions in the cutting zone, $m = 2y_0\rho\delta v$ is the rate of disintegration of the material, and $2y_0$ is the width of the cut (equal to the width of the isotherm with T_0).

As the gas jet removes some power ΔP the power of the heat source can be written as [3]

$$P = P_l + P_x - \Delta P \tag{371}$$

The heat source power is a complicated function of the coordinates as it is nonuniformly distributed over the surface and bulk of the material. To simplify the problem we can use the model of a point heat source assuming that the chemical heat source coincides with the laser focussed spot, that the cut width is equal to the width of the zone with the temperature $T \geqslant T_0$ and that the thermal parameters are temperature-independent and the kinetics of chemical reactions does not limit the cutting rate [3]. With these assumptions the temperature distribution in the infinite plate with the thickness δ at the steady-state stage of heating [117] is given by

$$T(r, x) = \frac{P}{2\pi\lambda\delta} \exp\left(-\frac{vx}{2a}\right) K_0\left(\frac{vr}{2a}\right) \tag{372}$$

where $r = \sqrt{x^2 + y^2}$ is the radius-vector and $K_0(u)$ is the modified Bessel function.

Let us introduce the dimensionless parameters

$$\xi = \frac{vx}{2a}; \quad \eta = \frac{vy}{2a}; \quad \rho = \frac{vr}{2a}; \quad W = \frac{P}{2\pi\lambda\delta T_0}$$
$$W_l = \frac{P_l - \Delta P}{2\pi\lambda\delta T_0}; \qquad W_x = (2/\pi)\psi\eta \tag{373}$$

Here $\psi = Q/cT_0$ is a dimensionless parameter describing the material and the blown gas, which does not depend on the laser-cutting conditions. Substituting T_0 into Eq. (372) gives

$$W_l = \exp(\xi) K_0^{-1}(\rho) - \frac{2}{\pi}\psi\eta \tag{374}$$

To determine the cut width we have to determine the maximum width of the isotherm with T_0 from the condition $\partial T/\partial x = 0$ for $r = r_0$. Transformations yield the following system of equations:

$$W_l = K_0^{-1}(\rho_0)\exp\left[-\rho_0\frac{K_0(\rho_0)}{K_1(\rho_0)}\right] - \frac{2}{\pi}\psi\eta \tag{375}$$

$$\eta = \pm\rho_0\sqrt{1-\left[\frac{K_0(\rho_0)}{K_1(\rho_0)}\right]^2}$$

where $K_1(\rho_0)$ is the modified Bessel function of second kind and of order 1.

From the system of equations (375) we can find the relationship $W_l(\eta, \psi)$ plotted in Fig. 100 [3]. At large values of the argument we can expand $K_0(\rho_0)$ and $K_1(\rho_0)$ and obtain

$$W_l = \frac{2}{\pi}\eta(1.9-\psi)+0.3 \tag{376}$$

It can be seen that at $\psi < 1.9$ we have $\partial W_l/\partial\eta > 0$ that is, an increase in η results in an increase in W_l and in the stability of the cutting regime. When $\psi > 1.9$ then $\partial W_l/\partial\eta < 0$ and the cutting regime is unstable. The parameter W_l can then be negative, that is, laser cutting can proceed only at the expense of the chemical reaction.

Let us evaluate the applicability of the above model. It cannot be applied at low cutting rates when the heat transfer from the plate cannot be neglected. The model does not yield correct results at high cutting rates as the cut width can prove to be smaller than the diameter of the actual heat source. The model is valid at the heat source velocities (cutting rates) varying in the range $(ab)^{1/2} < v < 2a/r_f$ where b is the heat loss coefficient. For metals this range is $0.1 < v < 1$ cm/s.

The quasistationary temperature in the centre of a gaussian heat source is

$$T(0, 0) = \frac{P}{4\pi\lambda\delta}[-\mathrm{Ei}(-U^2/4)] \tag{377}$$

where $U = vr_f/2a$ is the dimensionless parameter proportional to v and $\mathrm{Ei}(-U)$ is the exponential integral function. Similar to Eqs. (373) we can write $W_x = (2/\pi)\varphi U$ where $\varphi = (y_0/r_f)\psi$. The parameter φ depends on the cut width. Taking into account the heats of chemical reactions Eq. (377) can be transformed into

$$W_l = \frac{2}{-\mathrm{Ei}(-U^2/4)} - \frac{2}{\pi}\varphi U \tag{378}$$

Figure 100 presents W_l plotted as a function of φ and U. In this model we also have the region of stable regime (at $\varphi < 1.9$) and of unstable regime (at $\varphi > 1.9$). For small values of U the results obtained from the point source and the distributed source models are similar. For the U values exceeding 0.8 these models yield different results. Figure 100 shows that at high values of U the cutting regime is stable even at $\varphi > 1.9$. For instance, when $\varphi = 4$ the regime is stable at $U > 1.4$.

When the velocity v of the circular gaussian heat source is high the temperature field in the moving coordinate system tends to a simple distribution function [117]. This heat source moving along the surface of an infinite plate with the thickness δ is equivalent to the gaussian strip heat source of the same intensity P_l which has a gaussian distribution along the width of the plate and uniform distribution along the thickness of the plate. If the velocity of such a heat source is high the heat will be transferred only perpendicular to the direction of the source. Then the temperature is a function of y and can be estimated from the following equation [117]:

$$T(y) = \frac{P_l}{v\delta\sqrt{4\pi c\rho t_0\lambda}} \exp(-y^2/4at_0) \tag{379}$$

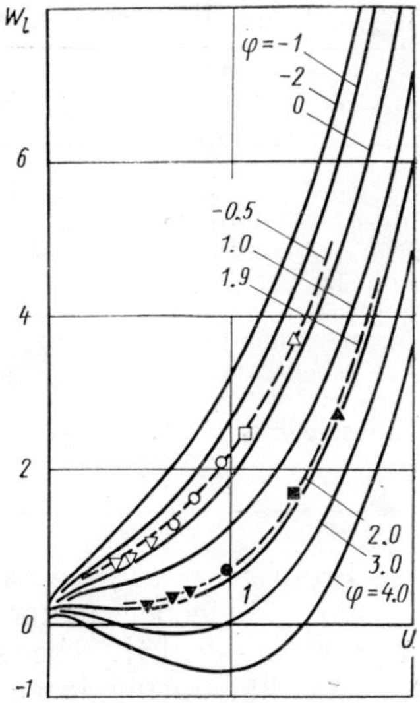

FIG. 100. Predicted plots of W_l as a function of U in laser cutting for model of gaussian heat source

Experimental results for steel 12X18H10T are represented by open symbols and for steel Ст. 3 by shaded symbols; thickness 0.5 mm (△, ▲), 1.0 mm (□, ■), 1.5 mm (○, ●), 3.0 mm (▽, ▼)

where $t_0 = \frac{1}{4}ak$, and k is the concentration coefficient of the heat source. The minimum power P_l^* needed for obtaining the disintegration temperature on the surface T_0 can be found from Eq. (379) by substituting $T = T_0$ and $y = 0$:

$$P_l^* = T_0 v\delta c\rho \sqrt{\frac{\pi}{k}} \tag{380}$$

Equation (380) indicates that P_l^* grows with increasing T_0 and v.

The power requirements for cutting can be described by the specific energy of cutting $S_0 = cT_0$ which is a parameter of the material and independent of the cutting conditions. This parameter is found to be a constant for a number of materials in a wide range of cutting parameters (rate, plate thickness and cutting power) [3]). Table 19 presents the specific energies S_0 of cutting for some materials.

TABLE 19. Specific Energy S_0 of Cutting for a Number of Materials

Material	S_0, kJ/g	Material	S_0, kJ/g
Textolite	50	Pine wood	0.9
Fibre-glass laminate	47	Oak wood	5.4
Laminated plastics	2	Plywood	5.4
Cardboard	0.8	Vinyl plastics	1.8
Plexiglas	2.0	Oil- and petrol-resistant rubber	2.5
Asbestos cement	28	Vacuum rubber	2.1
Sheet asbestos	20	Polycrystalline glass	25
Ceramics	30	Simple glass	31
Composite materials	80	Quartz glass	45

From $S_0 = cT_0$ we can determine ψ for dielectric materials. For most dielectric materials when cutting with the air blow $-1 \geqslant \psi \geqslant -2$ [3]. The small range of negative values of ψ suggests that during laser cutting of dielectrics only a relatively small proportion of the material is vaporized, and the major contribution to the cutting process is done by the removal of fine particles and molten drops by the gas blow. The situation is different for phenol-formaldehyde-base materials, such as textolite and fibre-glass laminates. Assuming that for these materials T_0 is about 1000 C we find $-5 > \psi > -10$. The laser beam converts these materials into a caked sticky mass which cannot be ejected by the gas jet from the cutting zone, thus resulting in high energy consumption for their cutting.

5.4. LASER INITIATION OF THERMOCHEMICAL PROCESSS

The chemical effects of laser radiation can be determined by the following factors: local heating increasing the number of particles with the energy above the activation energy for a chemical reaction in the irradiated zone; photon absorption by the molecule leading either to dissociation or to excitation to the level at which the activation energy for a chemical reaction is small while the temperature of the material can remain practically constant; activation of radiation and chemical transformations in the material similar to the processes generated by fast particles passing through the material which give rise to ions, radicals and excited states; and macroscopic heating of the irradiated material.

Let us analyze some features of laser activation of chemical reactions in gases [141, 156]. The laser-induced chemical transformations can be produced via various mechanisms depending on the laser radiation power density and the state of the irradiated material, for instance, the photochemical mechanism can be replaced by the thermal one. It is typically difficult to distinguish between these two processes, and they can be reliably identified only at relatively low radiation power densities.

The experimental evidence of a nonthermal mechanism of laser-induced effects has fundamental importance as it enables us to demonstrate the selectivity of the laser chemical effects. In a laser-irradiated material there can be a competition between the thermal and photochemical mechanisms. It is important to develop a criterion for identifying the ranges of dominance of these mechanisms. The authors of reference [6] have analyzed the model of the laser-induced first- and second-order reactions and estimated the critical laser pulse duration τ_{cr} so that the thermal reaction mechanism dominates at $\tau_i > \tau_{cr}$ and the photochemical mechanism dominates at $\tau_i < \tau_{cr}$.

The chemical reactions can be either endothermic or exothermic. The endothermic reactions always need a certain activation energy. However, the exothermic reactions have to be activated, too, since the molecules must be drawn together for reaction, and the activation energy is spent for overcoming their repulsion.

Bond breaking and initiation of reaction consume not only the energy from the translational motion of the molecule but also the energy of rotational and vibrational degrees of freedom. The vibrational energy of molecules makes a major contribution to chemical transformations. For instance, the probability of dissociation of a nonexcited molecule is very low even if its energy of the translational motion is higher than the binding energy. It is mostly the molecules whose vibrational energy is higher than the binding energy that can dissociate.

To enter a chemical reaction the molecules increase their vibrational energy to the activation energy E^* in the interactions with each other so that the characteristic time of the process is the time during which the molecule increases its energy to E^*.

A fundamentally new approach to the initiation of chemical reactions is to excite not the molecule as a whole but rather individual bonds of the molecule by laser radiation up to the temperatures of tens of thousands of degrees; such excited molecules can enter selective reactions.

There are two possible pathways for selective initiation of reactions by laser radiation. The first is to excite selectively certain molecular energy states resulting in the required reaction. The second is selective excitation of one of a mixture of compounds making it to enter a reaction. Lasers provide a unique opportunity for exciting the vibrational levels since no other light source can give the required radiation intensity in the infrared range. The high monochromaticity of laser radiation makes possible selective excitation of one of a few close energy levels. The laser-induced chemical reactions provide for high selectivity of reactions and for a considerable difference between the temperature for vibrational degrees of freedom and the temperature for translational motion of molecules.

The laser-induced chemical reactions can be distinguished from the reactions with the "equilibrium" molecules and classified into the group of reactions of selectively excited molecules [137]. This provides a mechanism for controlling chemical reactions by selectively exciting the required levels in molecules and maintaining this excitation until the reaction starts.

The laser-induced photochemical reactions can differ significantly from the similar reactions induced by conventional light sources. The main difference consists in the development of the photochemical waves and waves of shielding of the material from the incident high-intensity laser radiation.

The steady-state laser radiation incident on gases can stimulate in them supersonic perturbations in the form of a propagating reaction zone known as a photochemical wave, for instance, a photodissociation wave [103-106]. The velocity u of the photochemical wave is typically close to the velocity of the photodissociation wave. For instance, for the H_2-Cl_2 mixture (the absorption cross section of Cl_2 at $\lambda = 3300$ Å is $\sigma \approx 2.6 \times 10^{-19}$ cm^2) we have $u = 1.2 \times 10^6$ cm/s at $q_0 = 10^{26}$ photons/cm^2 s. The experimental observations of the photochemical waves in gases have been reported in references [103-106].

Most experiments with the laser-induced photochemical reactions have been done with pulsed ruby lasers, for instance, bromination of olefins in the gaseous phase [209], two-photon dissociation of AgCl [197], two-photon photolysis of iodoform [203], etc. Continuous-wave CO_2 laser radiation can be used to study the photochemical reactions due to the stepwise excitation of the vibrational levels. The reactions in laser-irradiated mixtures of N_2F_4 and SF_6 with various gases (NO, CF_4, C_2F_6, N_2 and N_2O) have been studied by Basov *et al* [137], and the reactions in the mixtures of the same gases with the compounds BCl_3, PF_5, and SiH_4 which absorb the laser radiation in reference [136]. All the above laser-induced reactions have the explosion-type rates and are accompanied by luminescence. The composition of reaction products differs greatly from that obtained by conventional heating of the mixture.

Karlov *et al* [23] have reported the initiation of the chain reaction in the gas mixture of BCl_3 and H_2 owing to multistage dissociation of the BCl_3 molecules stimulated by CO_2 laser radiation. Dissociation of BCl_3 molecules gives rise to atomic chlorine which initiates the chain reaction of dissociation.

The CO_2 laser radiation pulses are observed to produce two-stage dissociation of the NH_3 molecules [33]. First, the laser radiation excites the vibrational levels of about 50% of the NH_3 molecules. Then the excited NH_3 molecules are irradiated by the light of a flash lamp whose photon energy ($\lambda \gtrsim 2200$ Å) is not sufficient for photodissociation from the ground vibrational state but enough for dissociation of the excited molecules.

Resonance absorption by the vibrational levels of molecules of laser radiation increases the gas temperature up to about 10^3 K in a fraction of a second at the radiation power densities as low as 10^2 W/cm^2. This technique of gas heating differs from the conventional heating techniques by the fact that at first laser radiation increases the temperature of the vibrational degrees of freedom, and then the energy is transferred to other degrees of freedom.

The surface thermochemical reactions stimulated by the continuous-wave laser radiation are of particular interest for technological applications, for instance, such reactions as thermal decomposition of some inorganic and organic compounds, surface oxidation, reduction of metals, recrystallization of thin films, etc.

The applications of these laser experiments involve variation in the surface properties of bulk or thin-film materials owing to a decrease or increase in electric conductivity [52, 84], mechanical strength, chemical resistance, etc.

The thermochemical surface reactions have the following general mechanism. The laser radiation increases the surface temperature and the temperature of the material layers close to the surface. As the surface is in contact with a chemically active medium, or the surface has a layer of absorbed gases, the material enters a chemical reaction with them; the rate of the reaction is proportional to $\exp(-\varepsilon^*/kT)$ where ε^* is the activation energy. During the subsequent stage the process depends on the rate at which the reactants are supplied through the layer of reaction products rather than on the reaction rate.

Much experimental work has been done with laser pyrolysis of coal [81, 182, 207-209] attempting to raise the reaction temperature for obtaining a higher content of acetylene in the reaction products for the equilibrium of the reaction $2CH_4 \rightleftarrows C_2H_2 + 3H_2$ is shifted towards acetylene.

Some studies have been made of the interaction between the laser-vaporized material of the target and the ambient gas in the reaction chamber. Of the products obtained by laser heating of a graphite target in the atmosphere of H_2 [207, 208], C_2H_2, CH_4, C_6H_6, cyclohexane and olefines $\mathbf{C}_4$ [113] have been studied.

The formation of CCl_4, SiH_4 and GeH_4 under laser radiation of graphite, silicon and germanium targets in the chlorine and hydrogen atmospheres has been reported in [210].

The vapours of the compounds of group VI elements with bismuth, arsenic and antimony are irradiated with the light from the neodymium and ruby laser at the radiation power densities of the order of 10^7 W/cm^2, and the products are analyzed by mass-spectrometry [161, 162]. The composition of the vapours is found to differ significantly from that of the conventionally heated vapours. For instance, laser heating produces the particles of compounds Bi_3X which are not found with conventional heating. It is suggested that the Bi_3X particles are not produced by the gas phase reactions but ejected directly from the condensed phase.

The continuous-wave laser radiation can be used for the vacuum deposition of thin films. For instance, a 25-W CO_2 laser radiation is used for vacuum deposition [177]. The laser radiation is focussed by a 270-mm lens on the target in a vacuum chamber (about 5×10^{-6} mm Hg) which is slowly rotated to reduce absorption by the vapour. The deposition rate is about 10^3 Å/min with vaporization of the tabletted material. The thin films of MgF_2, Si_3N_4, SiO, ZnS, TiO_2 and Al_2O_3 are deposited on the substrates of germanium, silicon and glass.

The propagation of a high-intensity laser beam in liquids is strongly affected by various impurities and inhomogeneities in the liquid. The heat effects in chemical reactions are determined by the persistent local heating centres; the breakdown in liquid [13, 147, 163, 167] gives rise to strong shock waves ($P \approx \approx 10^4$-10^5 kgf/cm^2).

The free-generation laser pulse (1-8 J) produces the pyrolysis of liquid benzene the only products of which are hydrogen and carbon black [97, 100]. The laser-induced pyrolysis of benzene occurs only when the laser pulse energy is higher than the threshold energy and is accompanied by the development of shock waves in the liquid. The threshold pulse energy depends on the content of fine powder suspended in the liquid benzene (irrespective of its chemical nature). The laser technology can be applied for initiating the surface thermochemical reactions in manufacture of microminiaturized solid-state electronic devices where laser treatment is followed by chemical treatment.

Let us now discuss briefly some thermochemical reactions on the surface of solids. A simple illustration of such reactions is given by thermal decomposition. A suspension of the starting material is applied on a substrate and heated according to a given pattern by the laser beam with such an intensity q which is sufficient for decomposition of the material throughout the layer. The radiation intensity for the reaction must satisfy the condition $q_1 < q < q_2$ where q_2 is the radiation intensity at which the material is vaporized and q_1 is the intensity at which the bottom layer of the material is heated to the decomposition temperature.

Veiko and Libenson [20] have carried out the experiments on laser-induced decomposition of copper formate $(HCOO)_2Cu$ which is dissolved in ethylene glycol, mixed with glass frit and applied on the polycrystalline substrate. At the temperature of 295-325 C formate decomposes in the reaction $(HCOO)_2Cu \rightarrow Cu + H_2O + CO + CO_2$, the volatile components being removed; as a result, the substrate surface is coated by a copper film, and the unreacted material is washed off by water. The free-generation neodymium-laser radiation is used in these experiments. The thickness of the starting material layer is 50-75 μm. The copper layer produced by conventional thermal decomposition has the thickness of about 1 μm while the laser technique produces copper layers of smaller thickness. A simple one-dimensional heat conduction model yields the following estimates: $q_1 \approx 1.3 \times 10^3$ W/cm^2 and $q_2 \approx 6 \times 10^4$ W/cm^2. The experimentally measured threshold radiation intensity is about 10^3 W/cm^2.

When a chromium film is irradiated by the laser through a mask [142] the chromic oxide Cr_2O_3 is formed on the irradiated parts of the surface producing an image of the mask which can be "developed" by etching with concentrated HCl using an aluminium catalyst. Hydrochloric acid dissolves chromium and practically does not affect Cr_2O_3 so that the plating remains only on the irradiated parts making a positive image of the mask. The laser-induced oxidation can be used for manufacturing passive multilayer thin-film microcircuits.

Laser heating of the material surface (for instance, of ferrite surface) in a reducing atmosphere (hydrogen or carbon monoxide) can activate the reduction of metals from oxides (comprising ferrite) according to the reaction $MeO + H_2 \rightarrow Me + H_2O$.

It is of interest to study the thermochemical reactions in photosensitive materials stimulated by laser radiation, for instance, in photoresist materials and various materials sensitized to radiation of various wavelengths. The laser radiation produces thermochemical, rather than photochemical, reactions in them.

The thermochemical effect of laser radiation on photosensitive materials can be illustrated by reduction of palladium from its photosensitive salts, which makes possible selective application of palladium on a substrate. The reduced palladium is subsequently used as a catalyst for chemical deposition of metallic (copper) layers. The suggested reaction sequence for palladium reaction in this case is $3K_2PdCl_4+6FeCrO_4 \rightarrow 3Pd+2Fe_2(CrO_4)_3+$ $+ 2FeCl_3 + 6KCl$. The palladium reduced on the irradiated parts of the surface acts as a catalyst for the deposition of metals from salt solutions.

The neodymium laser pulses with the duration of 1 ms have been reported to stimulate the synthesis of a new allotropic modification of carbon [132].

6

CHANGES IN STRUCTURE AND PROPERTIES OF MATERIALS IN LASER-AFFECTED ZONES

Application of laser technology to various materials requires the knowledge of variations in the structure, composition and properties of laser-treated materials. Laser heating can be used to study the phase transformations in heating of materials, for instance, of iron-carbon alloys, at superhigh rates. Finally, the studies of the structure and properties of the laser-treated materials produce the results of fundamental scientific interest.

6.1. STRUCTURE AND PROPERTIES OF MATERIALS IN LASER-AFFECTED ZONES

A wide range of materials including iron-carbon alloys (steel and cast iron), pure nonferrous metals and their alloys are irradiated by pulsed and continuous-wave laser light, and the structure and properties of the laser-affected zone are studied using such techniques as metallographic analysis, electron microscopy and spectroscopic analysis [28, 39, 50, 66, 73, 93, 138, 213].

Discussed below are the typical changes in the structure and properties found in iron and iron-carbon alloys (containing from 0.02 to 4% of carbon and other alloying elements) under laser treatment [73]. These materials are selected for these experiments because their properties are sufficiently well-known. Many of these steels are used in rapid-heating experiments with high-frequency current [61]. A comparison of these results with the results for laser radiation can yield new information on structural transformations in these materials.

As can be seen from Figs. 101-103 and 106-107 in the laser-irradiated specimens of armco-iron, steels 20, 45 and У8, white pig iron and grey cast iron the laser-affected zones consist of two layers separated by a distinct boundary owing to the fact that the metal of the first layer is molten.

For the most alloys studied (apart from steels ХВГ and ШХ15) the microstructure of the first layer is distinguished by the columnar-dendritic structure (Figs. 104 and 105). Owing to

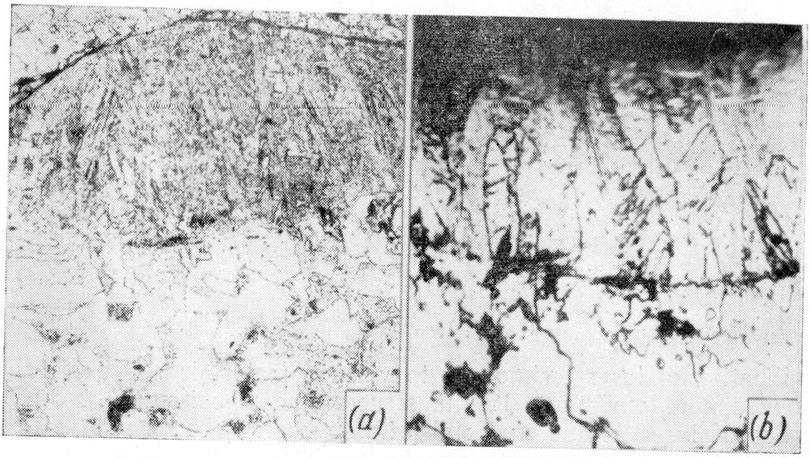

FIG. 101. Microphotographs of laser-irradiated zone in armco-iron
(*a*) cross section (×70); (*b*) microstructure (×450)

high crystallization rates there is observed the predominant growth of the major axes of dendrites which are perpendicular to the boundary between the layers. The transcrystalline structure of the first layer is observed only with repeated polishing and etching of specimens.

Electron-microscopic studies (magnification about 5000 and higher) indicate that the columnar crystals of the first layer have dendritic structure and that owing to high crystallization rates microscopic liquation is widespread. In the cross section the columnar crystals look like equiaxed grains consisting of individual oriented blocks with very small dendritic branches inside the blocks (steel 45 specimens).

The first layer of the affected zone in armco-iron specimens exhibits a much higher microhardness (H_{50} = 280-320 kgf/mm^2 compared to the microhardness of 100 kgf/mm^2 of the initial structure), which can be attributed to hardening producing low-carbon martensite and to strain hardening due to thermal stresses. The metallographic and X-ray diffraction results (see Fig. 102*b*) show that the first layer of the irradiated zone in steel 45 has the structure of acicular martensite with some residual austenite; the microhardness of this layer is H_{50} = 820-830 kgf/mm^2. The first layer of the affected zone in steel У8 (see Fig. 103) consists of acicular martensite and considerable amount of residual austenite. The microhardness of this layer is H_{50} = 880-920 kgf/mm^2 despite the high content of residual austenite (about 70%); this fact can be attributed to the specific features of laser treatment (phase strain

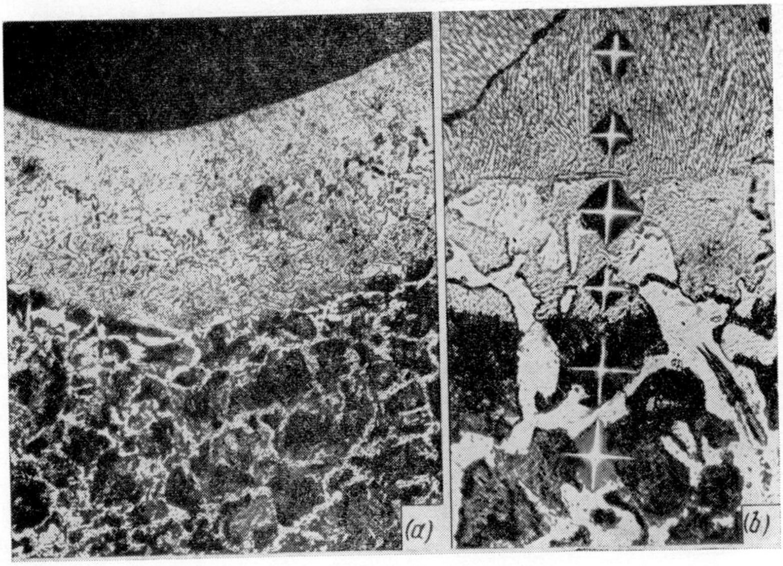

FIG. 102. Microphotographs of laser-irradiated zone in steel 45
(a) ×340; (b) ×600

hardening, high rates of cooling and heating, etc.). In the low-alloy steels ХВГ and ШХ15 (see Figs. 104 and 105) no marked dendritic structure is observed in the first layer which consists of relatively equiaxed grains and has the martensite-austenite structure wherein large martensitic needles are found and carbides are absent owing to the fact that the temperature of the layer is considerably higher than the critical point A_{c_m}, and carbides are completely dissolved.

In grey pig iron (see Fig. 106) the first layer consists mainly of individual drop-like ledeburite regions and eutectic regions oriented along the direction of heat transfer owing to longitudinal growth. There is an almost total lack of graphite inclusions in the first layer. There is a considerable porosity which can be attributed to liberation of gases adsorbed by the graphite inclusions with primary crystallization of the pig iron. The smooth spherical inner surface of the pores also suggests their gaseous origin.

The poor wettability of the grey pig iron by the molten metal can be also attributed to liberation of gas from numerous graphite inclusions disintegrated by the moving melting front produced by laser heating. The microhardness of the first layer in the grey pig iron is $H_{50} = 950$ kgf/mm^2 indicating the dissolving of graphite in the liquid metal and subsequent hardening

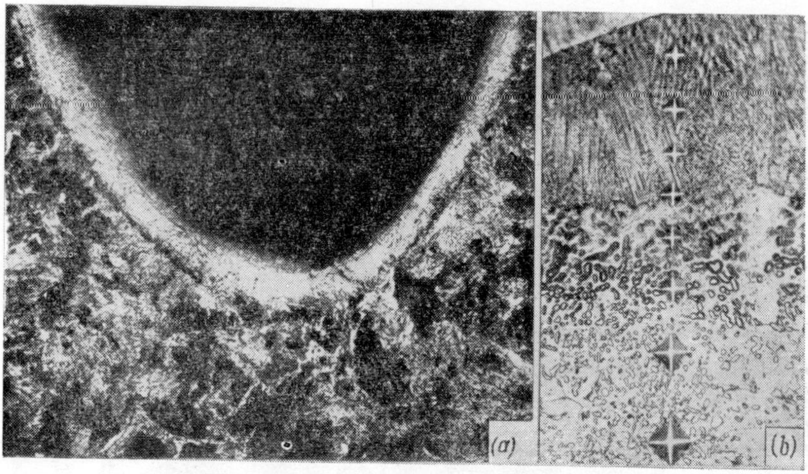

FIG. 103. Microphotographs of laser-irradiated zone in steel У8
(a) ×200 (initial structure of lamellar pearlite); (b) ×600 (granular pearlite)

owing to rapid heat transfer to the bulk of the material. However, the microhardness decreases even in the lower regions of the first layer owing to incomplete dissolving of graphite inclusions.

The structure of the first layer in the hypoeutectic white cast iron is distinguished by columnar crystallization of primary austenitic dendrites developed from the melting boundary (see Fig. 107) where dendritic crystals with marked branches are found in contrast to steel and grey pig iron. Considerable residual stresses are developed in this layer resulting in cracking. The structure corresponds to a lower carbon content than in the first zone of grey iron and is typical of the highly supercooled hypoeutectic white cast iron. Compared with the ledeburite of the second layer (the heat-affected zone) and of the starting structure the first layer has a considerably finer structure. The diameter of the austenite columns in the cementite of the first layer is less than 0.5 μm (in the starting structure it amounts to 2 μm). The lack of large precipitates of the primary austenite substantiates the suggestion that the crystallization process is essentially nonequilibrium. The microhardness of the surface of the first layer is H_{50}=900-970 kgf/mm². In the deeper regions of the first layer the microhardness varies depending on the initial structure. For the initial pearlite structure the microhardness is H_{50} = 900-925 kgf/mm². For the ledeburite structure with the initial microhardness of H_{50} = 750 kgf/mm² it increases up to H_{50} = 1200 kgf/mm² after laser heating. This difference suggests that no diffusion of carbon between the grains takes place as the lifetime of the molten state is very short.

FIG. 104. Microstructure of laser-irradiated zone in steel ХВГ (×340)

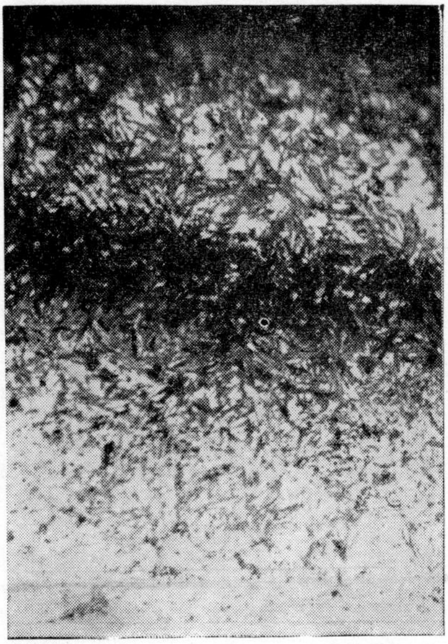

FIG. 105. Microstructure of laser-irradiated zone in hardened steel ШХ15 (×450)

The second layers of the heat-affected zones in alloys have different structures depending on the carbon content, the content of alloying elements, initial structures and high-temperature gradients. The second layer in the armco-iron (see Fig. 101*b*) has a fine-grained structure with the grain diameter varying from 3 to 20 μm (the grain diameter in the initial structure is 80-100 μm). The decrease in the grain diameter is attributed to phase recrystallization $\alpha \rightleftarrows \gamma$ and plastic deformation due to thermal stresses and subsequent recrystallization. The electron-microscopic results show that the second layer at the contact with the first layer has the lamellar pearlite structure (with the interlamellar distance of about 1 μm). Since the carbon content in the armco-iron is not sufficient for the formation of equilibrium pearlite the resulting structure should be regarded as a pseudopearlite produced by rapid cooling of superheated austenite with lower carbon content. The microhardness decreases with depth from $H_{50} = 280$ kgf/mm² to $H_{50} = 80\text{-}100$ kgf/mm².

Owing to a relatively wide temperature range of α-γ equilibrium the second layer in steel 20 has a two-phase structure in which the α-phase is a low-carbon martensite near the molten zone and ferrite far from it.

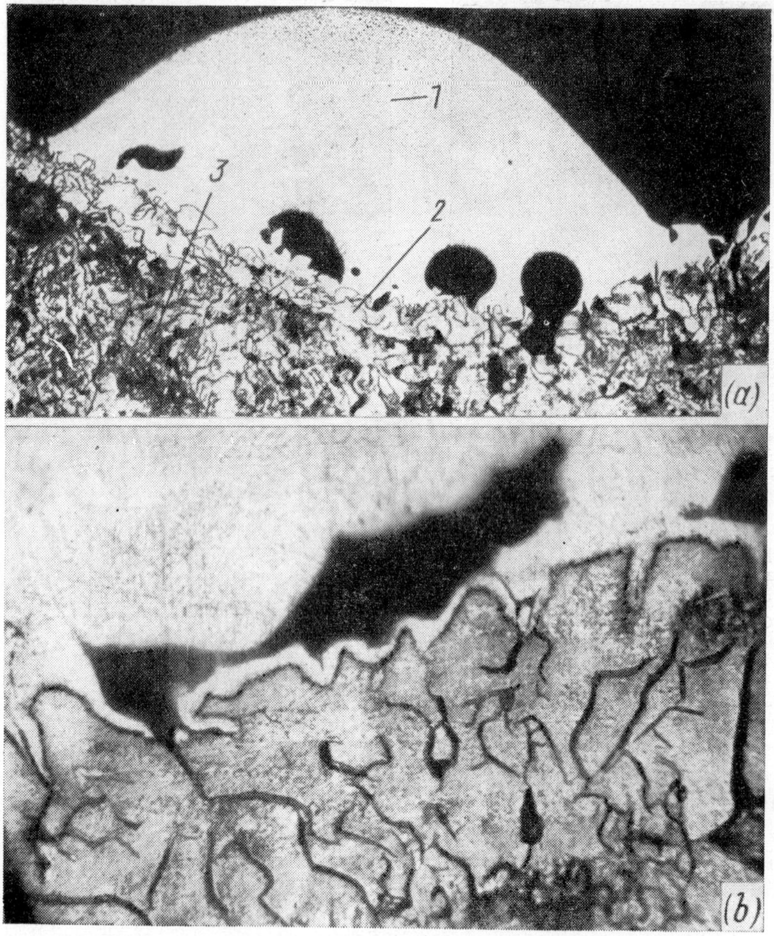

FIG. 106. Laser-affected zones in grey cast iron
(a) general view (×125); 1—molten and fast-solidified region; 2—heat-affected region; 3—initial structure; (b) microstructure of heat-affected region (×600)

In steel 45 at the sites of the second layer where the pearlite grains were previously located there is found a structure which correlates with the periodicity of the carbon distribution according to the position of lamellar cementite in former pearlite (see Fig. 102b). The ferrite network is preserved at the boundaries of the grains owing to the fact that the period when the temperature is equal or above the hardening temperature for steel is not long enough for homogenization of austenite and sufficient carbonization of ferrite. The microhardness of former pearlite grains

FIG. 107. Microstructure of laser-irradiated white pig iron at boundary of molten zone and heat-affected zone (×450)

is $H_{50} = 820$-850 kgf/mm^2 and that of former ferrite grains is 320-350 kgf/mm^2 (for the initial structure $H_{50} = 100$ kgf/mm^2). This increase in microhardness of ferrite is explained by strain hardening with the $\alpha \rightleftarrows \gamma$ phase recrystallization and deformation of ferrite grains owing to the volume changes in the neighbouring grains following martensite transformation. Some carbon diffusion from former pearlite grains to ferrite grains could take place.

In steel У8 specimens the inhomogeneity of the second layer is attributed to large temperature gradient with depth (from the melting temperature to A_{c_1} across the layer of tens or hundreds of micrometres). As the laser heating rates amount to 10^5-10^6 K/s the complete austenitization of pearlite is shifted towards high temperatures occurring in a wider temperature range. The diffusion equilibration of the carbon concentration in the material can be uncompleted even at the temperatures much higher than the critical point A_{c_3} resulting in the formation of inhomogeneous austenite and hence inhomogeneous martensite after hardening. In steel У8 specimens the lamellar pearlite in the deep part of the second layer bordering the initial structure is completely transformed but austenite is not homogenized owing to high heating rates and small times of exposure to temperatures above the critical point A_{c_m} so that the layer contains what is called pseudo-pearlite (see Fig. 103*a*). The stratified structure of the regions of former lamellar pearlite can be due to the microinhomogeneity of austenite with respect to the distribution of manganese and silicon, which correlates with the distribution of carbon. In specimens with globular pearlite structure (see Fig. 103*b*) cementite is not fully dissolved. In specimens with the initial structure of both lamellar and globular pearlite the second layer of the laser-affected zone contains no carbides. Acicular martensite against the background of residual austenite is found in the second layer after prolonged etching of specimens.

In specimens of steels ХВГ and ШХ15 (see Figs. 104, 105) the structure of the second layer is also inhomogeneous with depth. The upper part of the second layer bordering the first layer consists of austenitic regions with microhardness $H_{50} = 350$-450 kgf/mm². There are also regions with martensite-austenite structure with microhardness $H_{50} = 660$-750 kgf/mm². The temperature in the upper part of the second layer is above the critical point A_{c_m} providing for complete dissolving of carbides. This increases the chromium content and the amount of the residual austenite. The lower part of the second layer adjacent to the initial material consists of acicular martensite with residual austenite and carbide particles uniformly distributed over the layer, which provides for high microhardness ($H_{50} = 870$-950 kgf/mm²).

Let us now briefly discuss the effects of free-generation laser pulses on the hardened specimens of steels 45, У8, ХВГ and ШХ15. Under these circumstances the first layer in steel 45 has the structure which is similar to that described above for steel 45 with the initial pearlite-ferrite structure while the first layer in steel У8 has a more finely dispersed structure. A tempered region whose microhardness is below the initial level is typically found in the second layer (the heat-affected zone) at the border with the unchanged material.

In the second layer of grey pig iron specimens (see Fig. 106*b*) the pearlite basic structure is transformed into acicular martensite owing to heating above the upper critical point and rapid cooling. The electron-microscopic results indicate that the layers adjacent to the unchanged material have some residual cementite and ferrite lamellar structures suggesting that the metal was in the austenitic state only for a short time and that distribution of carbon in austenite and hence martensite is not uniform. In the second layer at the melting front near graphite inclusions the wedgelike protrusions of 10-20 μm are formed.

The heat-affected zone in white cast iron consists of ledeburite regions and primary austenite grains with changed structure (see Fig. 107). In the layer about 10-20 μm thick adjacent to the first layer acicular martensite with comparatively large needles is found with the needle size decreasing with the depth of the layer. The inclusions of hardening sorbite are found at larger depths. These regions contain small (up to 0.5 μm) carbide particles which did not have time to dissolve. Typically, the second layer has retained needles of secondary cementite which are found in the unchanged structure.

The metallographic and electron-microscopic data for the ledeburite cementite regions in the first and second layers indicate that cementite dissolves a certain amount of austenite and is made to melt without decomposition (graphitization). With solidification the excess austenite forms small dendrites.

In the second layer (the heat-affected zone) there are two types of cementite inclusions. The cementite of the first type has small inclusions of austenite (martensite) whose particle size is smaller than for the austenite dendrites in the molten zone (the first layer) by one-two orders of magnitude. The microhardness of this two-phase structure is higher than in the unchanged structure ($H_{20} = 1270\text{-}1484$ kgf/mm^2) [73]. The cementite of the second type in the heat-affected zone does not show the above evidence of decomposition but its microhardness is higher than that of the unchanged cementite ($H_{20} = 960\text{-}1270$ kgf/mm^2 compared to $H_{20} = 890\text{-}960$ kgf/mm^2 in the unchanged material).

Noteworthy also is the lack of a smooth melting front as the melting is delayed near the large grains of primary austenite whose melting point is higher than that of ledeburite. The melting is deeper near the needles of the secondary cementite in the austenite (pearlite) grains.

The presence of hardening zones of the molten material and thermal hardening zones of the solid material is common for all iron-carbon alloys studied.

The X-ray diffraction results for iron-carbon alloys show that the laser-induced structural changes are largely similar to the changes produced by hardening or plastic deformation. Laser pulses (free-generation mode) incident on iron alloys produce a considerable decrease in the block size, a decrease in the density of defects and microdeformations of the lattice in the heat-affected zone. For instance, in steel 45 the sizes of blocks in solidified drops of the molten metal ejected by laser radiation to the surface and in the molten zone are 0.33×10^{-5} cm and 0.57×10^{-5} cm, respectively, compared to 3.0×10^{-5} cm in the unchanged metal; the microdistortions $\Delta a/a$ increase to 0.26; and the defect density is increased up to 30×10^{10} cm^{-2} and 9×10^{10} cm^{-2} compared to 0.3×10^{10} cm^{-2} in the unchanged material [73].

The transmission electron-microscopic results for the 0.1-mm foil of carbonyl iron produced by double vacuum remelting (car-

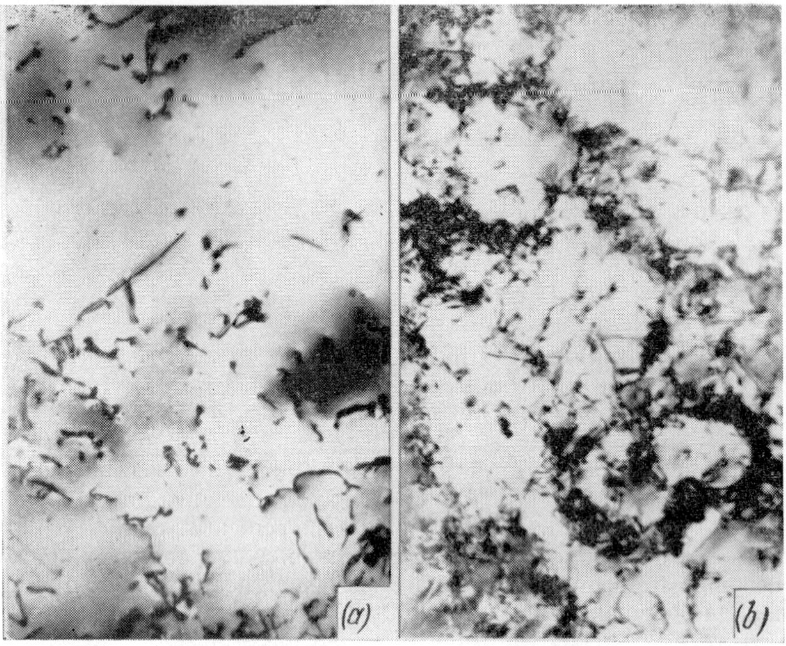

FIG. 108. Dislocations in annealed carbonyl-iron foil before laser radiation (*a*), (×3400) and in zone irradiated by laser beam (*b*), (×3100) with radiation power density of 8×10^4 W/cm^2

bon content being about 0.02%) after laser radiation show that the dislocation density is much higher than before radiation (Fig. 108). For instance, the laser radiation with the power density of about 8×10^4 W/cm^2 (at such power densities the molten material is ejected from the irradiated zone, and a through hole is drilled) produces in the foil a dislocation network similar to that found in highly strained metals. The dislocation density along the network lines is 10^{11}-10^{12} cm^{-2}, that is, higher by approximately three orders of magnitude than the initial density [73]. This significant increase in the dislocation density is caused by high rates of heating and cooling; by high temperature gradients in the laser-affected zones which, in addition to $\alpha \rightleftarrows \gamma$ recrystallization, result in the development of thermal stresses (owing to the difference in the specific volumes and expansion coefficients of the coexisting phases), the latter relaxing by formation, motion and interaction of dislocations; by reactive forces due to the ejection of the molten material and vapour from the irradiated zone and so on.

Thus, the hardening effects observed in laser-irradiated alloys are due to the increase in the density of the structural defects (dislocations, vacancies and their aggregates) produced under radiation, cooling and martensite transformation.

Let us now discuss briefly some metallographic results for the specimens of iron alloys irradiated by powerful continuous-wave CO_2 laser beams [126]. The effects produced by the continuous-wave laser radiation vary according to the radiation power density, the time of radiation, the heat-transfer conditions and the composition of the ambient atmosphere, apart from other factors.

Laser radiation of iron-carbon alloys for about 5 s produces craters 1.5-1.8 mm in diameter and up to 0.2 mm deep. The craters are surrounded by an irregular rim of the solidified metal not more than 0.3 mm wide and not more than 0.1 mm high. Solidified metal drops of not more than 0.3 mm are sometimes found on the specimen surface outside the rim; these drops leave a trace of structural changes on the surface (which is especially distinct on the previously hardened specimens). The crater is surrounded by concentric zones of various colours, the darkest outer zone has the diameter of 6-8 mm. The crater volume is considerably larger than the sum of volumes of the rim and large metal drops ejected on the surface; this fact suggests that the bulk of the metal removed from the crater is ejected in the form of vapour and small liquid particles.

Let us analyze the typical basic structural changes in iron alloys in the zones irradiated by powerful continuous-wave laser beams for about 5 s.

In armco-iron (Fig. 109*a*) in the laser-affected zone with the thickness of about 0.55 mm the grain size is reduced to 10-15 μm (the initial ferrite grains have the size of about 50-60 μm), and this reduction is especially marked in the annular layer between 0.35 and 0.55 mm. Large grains are found in the regions adjacent to the crater. The reduction in the grain size can be attributed to the austenitization processes occurring under different heating conditions and to subsequent decomposition of austenite with cooling. Under these circumstances the rates of origination and growth of ferrite grains are different owing to the difference in the starting sizes of austenite grains formed near the melting front and far from it. The melting front is extended. The solidified layer has a considerable porosity owing to the lack of a sufficient concentration of reducing agents.

In all iron alloys studied in the upper layer of the laser-affected zone or sometimes below the porous zone there is found what is called a slag layer which consists mainly of iron oxides and other products of thermochemical reactions of the alloy with the air and gases dissolved in the metal. Generally, the slag layer

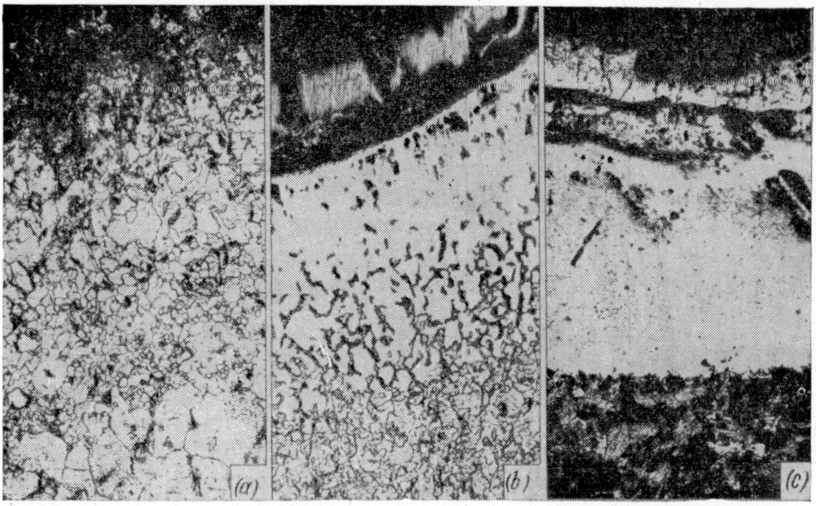

FIG. 109. Microstructures of zones irradiated with continuous-wave laser radiation
(a) armco-iron (×150); (b) steel 45 (×150); (c) normalized steel У8 (×150)

contains many pores and cracks, and its dendritic cellular structure shows signs of oriented columnar crystallization.

In low-carbon steel (0.2-0.3% C) the laser-affected zone has the Widmanstatten pattern, its microhardness being similar to that of the unchanged material.

In steel 45 (normalized specimens) whose initial structure consists of pearlite grains and lamellar ferrite the laser radiation produces a layer of the molten porous metal about 75 μm thick, in addition to the slag layer (Fig. 109b). Significant decarburization is found in this layer and the neighbouring regions. At a distance of about 100 μm from the crater martensite is found whose content increases with increasing distance from the crater. A completely hardened layer lies at the depth of 300-400 μm and below it there is an incompletely hardened layer. In the previously hardened specimen of steel 45 radiation produces a layer (about 400 μm) which consists of a superficially decarburized zone and the lower zone of secondary hardening with the smaller needle size of martensite than in the unchanged material. Below this layer there is a tempered zone (about 400 μm) with decomposition products of martensite.

In specimens of normalized and previously hardened steel У8 structural changes produced by laser radiation are largely similar to those in steel 45 (Fig. 109c). In the hardened layer adjacent to the molten layer in steel У8 specimens the significant etching of the grain borders is seen owing to segregation of impurities.

In steel ХВГ specimens structural changes are similar to those in steel У8, the only difference being that the carbide particles of rounded form can be found both in the laser-hardened layer and in the hardened unchanged metal. The columnar structure can be seen in the molten layer (Fig. 110).

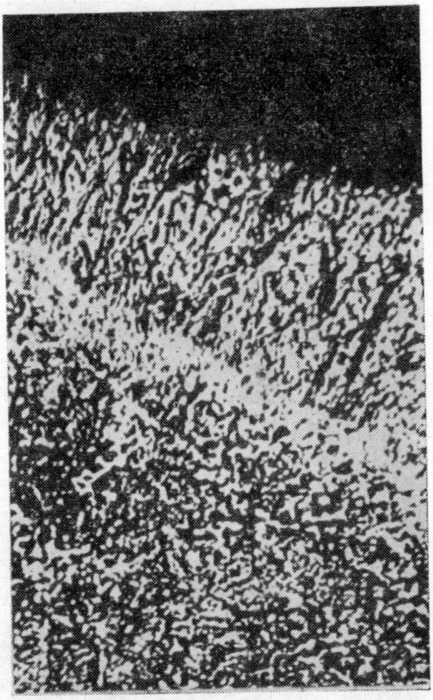

FIG. 110. Microstructure of zone irradiated by continuous-wave laser radiation in hardened steel ХВГ (×600)

In grey pig iron specimens (with the pearlite basic structure) the molten zone (about 110 μm wide) is significantly decarburized; it lacks graphite inclusions, and its porosity is lower than that of the respective zone in steels. The hardened zone below has the size about 275 μm and retains graphite inclusions (Fig. 111*a*).

In white eutectic cast iron structural changes are found in a large layer (about 900 μm) where pearlite grains have a much smaller size (about 4 μm) than in the unchanged material (about 30-40 μm) (Fig. 111*b*). In the heat-affected zone (about 500 μm wide) with the initial structure of rough conglomerate the cementite is molten completely, austenite is partially molten, and small-size ledeburite is found between the retained structures after solidification. In steel У8 with the initial pearlite grain structure the longer laser treatment (about 50 s) produces a considerably larger heat-affected zone (about 6-7 mm). The inner surface of the laser-produced crater is highly oxidized. The layer adjacent to the crater (about 1.3-1.6 mm wide) consists of large-size acicular martensite with the needle size decreasing with increasing distance from the crater. At a distance of about 3 mm from the crater the material has the ferrite-carbide structure.

6.2. CRYSTALLIZATION OF LASER-PRODUCED MOLTEN POOL

Let us analyze the variation in the solidification rate of a small volume of the molten metal with time. The problem can be approximated by a one-dimensional model since the depth of the molten

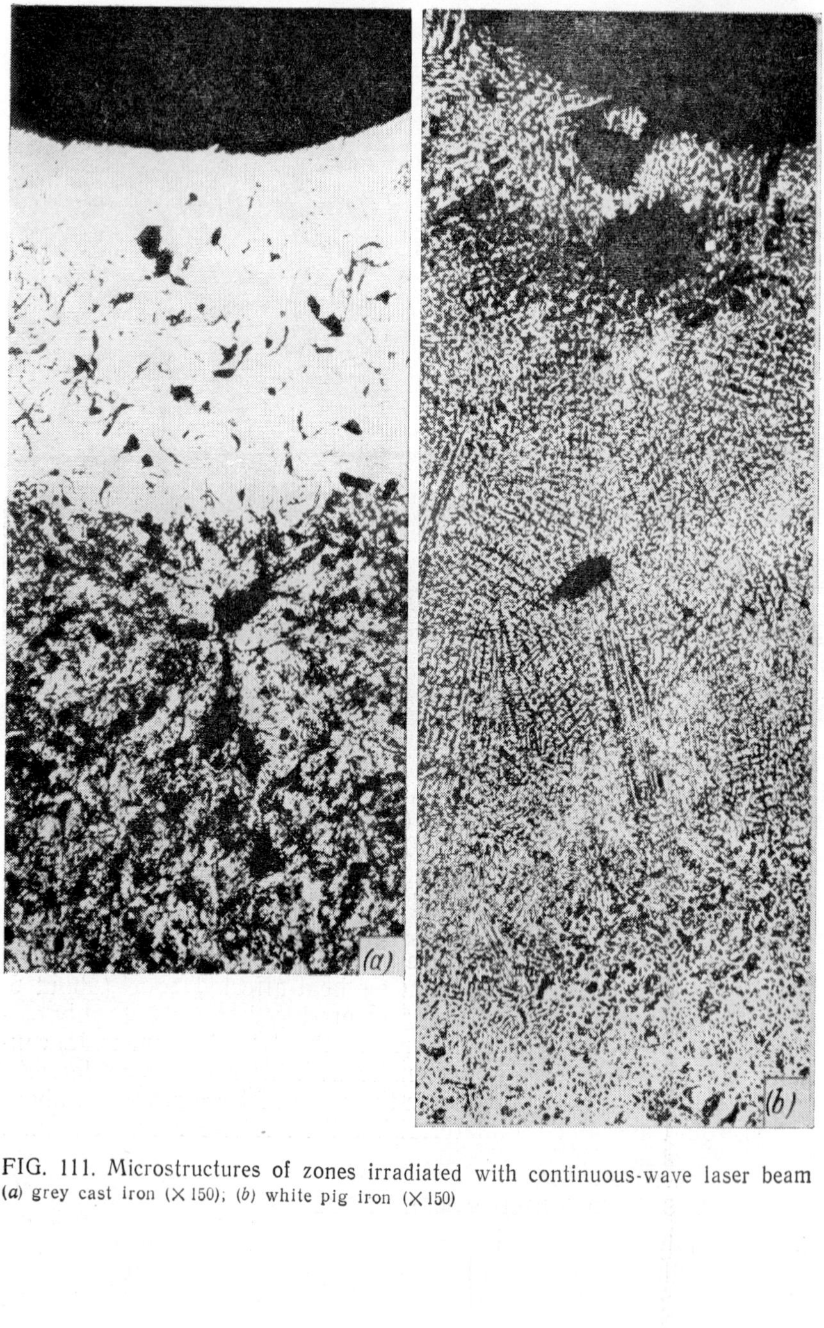

FIG. 111. Microstructures of zones irradiated with continuous-wave laser beam
(*a*) grey cast iron (×150); (*b*) white pig iron (×150)

pool is considerably smaller than its diameter with the laser pulse duration $\tau_i \approx 10^{-3}$ s. Assume that by the end of the laser pulse the superheating of the molten metal is not high and the temperature distribution in the molten pool is uniform owing to convective heat transfer in it. Then the solidification rate v_m can be written as

$$v_m = b\sqrt{a/\tau} \tag{381}$$

where b is a constant found as a root of the following equation:

$$\frac{c\,(T_m - T_0)}{L_m\sqrt{\pi}} = b\exp(b^2) \tag{382}$$

and T_0 is the initial temperature of the metal.

The position of the solidification front is given by

$$z(\tau) = 2b\sqrt{a\tau} \tag{383}$$

It can be estimated from Eqs. (381)-(383) that for iron ($b \approx 0.75$) the molten pool is solidified within a few milliseconds.

The solidification of metal is accompanied by a number of effects which give rise to the deformation of the melt surface and to the waves of the liquid metal travelling towards the edge of the molten pool. These perturbations are caused by the reactive pressure of the ejected vapour, by the convective flows in the molten pool and by other factors.

One of the effects produced by the above factors is the formation of the system of rings concentric to the molten zone on its surface (Fig. 112). No detailed analysis has been made to understand this effect found in molten pools produced by local

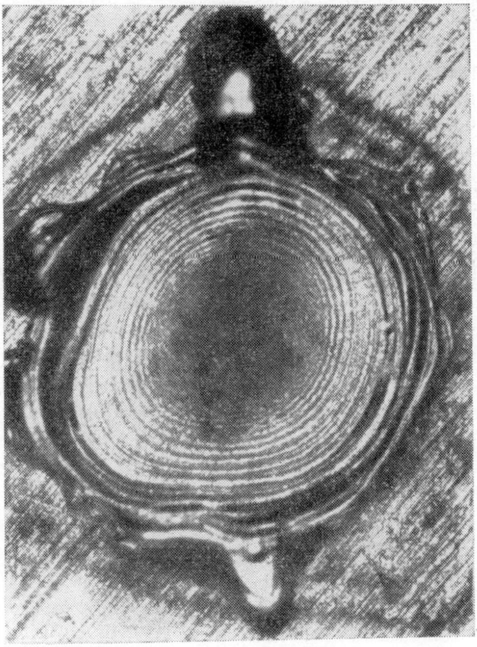

FIG. 112. Concentric rings of solidified metal on surface of laser-irradiated zone in steel ХВГ (×70)

heating of metals, and with the laser heating the effect has not been studied at all. Only the authors of reference [41] have noted that the concentric waves on the surface of the solidified niobium are produced by the reactive pressure of the ejected vapour as the effect is enhanced with increasing radiation power density.

Let us analyze the development of the concentric rings of the solidified metal. The laser-produced molten zone and the rings of the molten metal are small so that we must analyze the capillary waves produced by the vapour reactive pressure acting on the central part of the molten pool. Indeed, let us estimate the capillary constant

$$a_k = (2\sigma/\rho g)^{1/2} \tag{384}$$

to determine whether the waves are gravitational or capillary. Here σ is the surface tension, ρ is the density and g is the gravity acceleration. For the molten iron at $T \approx 1550\text{-}1650$ C we have $\sigma = 1820$ dyne/cm [21] and for typical metals where $\sigma \approx \approx 10^3$ dyne/cm and $\rho \approx 10$ g/cm^3 we obtain $a_k \approx 0.4$ cm; clearly, this value is larger than the wavelengths of the molten metal waves.

Let us now analyze the propagation of annular (that is, axially symmetrical) capillary waves on the surface of the molten pool. The problem is not a steady-state one as the solidification process changes the size of the molten pool. If the problem is precisely formulated it becomes too complex to yield an analytical solution. Both the equations and the boundary conditions of the problem are nonlinear for the arbitrary amplitude of the wave. Moreover, this formulation does not take into account some effects due to rapid solidification and the contribution of these effects to the motion of the capillary waves.

Let us assume that the wave amplitude is small compared to the wavelength. The experimental results show that this assumption corresponds to reality. Then we have to find the profile of the molten pool surface $\xi(r, \tau)$ and the velocity potential $\Phi(r, z, \tau)$. The equation for the velocity potential can be written as [77]

$$\Delta\Phi(r, z, \tau) = 0 \tag{385}$$

For the waves which are strictly periodic in time we have

$$\Phi(r, z, \tau) = \mathrm{Re}\,[\varphi(r, z)\exp(-i\omega\tau)] \tag{386}$$

where Re is the real part of the function $\Phi(r, z, \tau)$ and $\varphi(r, z)$ is a complex function of coordinates which satisfies Eq. (385)

$$\frac{\partial^2\varphi}{\partial r^2} + \frac{1}{r}\frac{\partial\varphi}{\partial r} + \frac{\partial^2\varphi}{\partial z^2} = 0 \tag{387}$$

Assume that the molten pool is radially infinite and has a finite depth h. The function $\varphi(r, z)$ has the following boundary conditions [77]. On the free surface $z = 0$

$$\rho g \frac{\partial \varphi}{\partial z} + \rho \omega^2 g - \sigma \frac{\partial}{\partial z}\left(\frac{\partial^2 \varphi}{\partial r^2} + \frac{1}{r}\frac{\partial \varphi}{\partial r}\right) = 0 \tag{388}$$

On the bottom of the pool $z = h$

$$\frac{\partial \varphi}{\partial z} = 0 \tag{389}$$

The wave amplitude tends to zero:

$$r \to \infty; \ \varphi \to 0 \tag{390}$$

The condition (389) is satisfied only at the first moment. With solidification of the molten metal the magnitude $(h - z) \to 0$ according to $z \approx \tau^{1/2}$ as follows from Eq. (383). Moreover, the momentum of the surface takes some time to vanish.

Equations (387)-(390) have the following solution

$$\Phi(r, z, \tau) = \operatorname{Re} A J_0(kr) \times \cosh k(h - z) \exp(-i\omega\tau) \tag{391}$$

where k is the wave number

$$\lambda = \frac{2\pi}{k} \tag{392}$$

Equation (388) yields the relationship between k and ω:

$$\omega^2 = \left(gk + \frac{k^3\sigma}{\rho}\right)\tanh(kh) \tag{393}$$

If the wavelength is much smaller than the capillary constant, and the depth of the molten pool is much larger than the wavelength Eq. (393) can be transformed into

$$\omega^2 = k^3\sigma\rho^{-1} \tag{394}$$

The profile $\xi(r, \tau)$ of the capillary wave can be found [77] from the following condition:

$$z = 0, \ \frac{\partial \xi}{\partial \tau} = \frac{\partial \Phi}{\partial z} \tag{395}$$

Hence we obtain

$$\xi(r, \tau) = \frac{kA}{\omega} J_0(kr) \sinh(kh) \sin \omega\tau \tag{396}$$

Equation (396) shows that the distance between two neighbouring crests increases with increasing r, and the amplitude of the wave decreases; this result agrees with the experimental findings. With the large molten pool depth (the starting stage of solidification) the group velocity of the waves is

$$v_g = \frac{\partial \omega}{\partial k}\bigg|_{k=k_0} = \frac{3}{2}\left(\sigma k_0 \rho^{-1}\right)^{1/2} \tag{397}$$

According to [77] the attenuation coefficient which characterizes the dissipation of energy in the wave is

$$\gamma = \frac{2\omega^{4/3}\mu}{\rho^{1/3}\sigma^{2/3}} \tag{398}$$

where μ is the dynamic viscosity of the molten material.

Let us now make some estimates. Rapid solidification can result in "fixing" the wave profile $\xi(r, \tau)$ on the molten pool. Let us compare the characteristic time for "levelling" the wave

$$\tau_1 \approx \gamma^{-1} \tag{399}$$

and the time of solidification of the molten material in the region with the linear size of the order of the wavelength

$$\tau_2 \approx \frac{\xi(r, \tau)}{v_m(\tau)} \tag{400}$$

If the condition $\tau_2 \ll \tau_1$ is satisfied the waves of the molten metal are "fixed". For instance, for iron $\mu = 7.6 \times 10^3$ Poise, $\rho = 7.5$ g/cm^3, $\lambda = 0.01$ cm, $k = 628$ cm^{-1} and $\sigma = 1820$ erg/cm^2 so that $\omega = 7.7 \times 10^6$ s^{-1} and $\gamma = 0.8 \times 10^5$ s and therefore we find $\tau_1 \approx 10^{-5}$ s. Using Eq. (381) we can estimate $\tau_2 \approx 10^{-5}$ s. Thus, for these solidification rates the times τ_1 and τ_2 are similar (in iron). Hence, the waves of the molten metal can be "fixed" at the onset of solidification, that is, at a sufficient distance from the centre. The solidification rate decreases with time to about $\tau^{1/2}$ but τ_1 also decreases if we take into account the temperature dependence of the viscosity of the molten material. Thus, the region where the molten material waves can be "fixed" is extended. However, no solidified waves of the molten material can be found in the central region since in this region $\tau_1 \ll \tau_2$ because the rate of decrease of the solidification rate is higher than the rate of viscosity increase. This result agrees with the experimental data.

6.3. ZONES IN OPAQUE MATERIALS IRRADIATED WITH Q-SWITCHED LASER BEAMS

The Q-switched lasers operate mainly in the single-pulse regime and have high peak powers and the pulse durations of a few tens of nanoseconds. The resulting radiation power densities on the focussed spot are 10^9-10^{12} W/cm^2 or more and the pulse front duration is about 10^{-9}-10^{-8} s. Such laser pulses produce the effects in the irradiated metals that significantly differ from the effects produced by free-generation laser radiation discussed above. The effects produced by Q-switched laser pulses depend primarily on the parameters of the pulses.

A number of experiments (see for instance [188]) have been carried out to study some effects produced by high-power light fluxes in irradiated materials. Among such effects are the local hardening of steels, the local strain hardening produced by pressure pulses in the laser-affected zone, etching of the specimen surface by gas and ion jets expelled from the affected zone, etc.

The giant pulse effects on the solid surface can be analyzed to estimate the changes in the properties of materials subjected to high-rate heating (the rates about 10^{10} C/s) with the constant density prior to the vaporization of the material. The analysis of the metal structures in the zones affected by giant laser pulses provides more profound knowledge of the disintegration processes in metals making it possible to develop more accurate theoretical models.

Let us discuss some features of the disintegration processes in opaque materials under the effect of giant laser pulses with the radiation power densities about 10^9 W/cm^2 or more.

The craters produced by laser giant pulses (the power is 30 MW, the focussed spot area is 3×10^{-3} cm^2) have the depth depending on the pulse shape but it is always smaller than the depth of the crater produced by free-generation laser pulses of the same energy (the radiation energy density is 5×10^3 J/cm^2, the pulse duration is 600 μs) [195, 196]. The results are given in Table 20.

TABLE 20 Depths of Craters Produced by Free-Generation Laser Pulses and Giant Laser Pulses in Various Materials with the Same Pulse Energy

Materia	Crater depth, um	
	free-generation pulse	giant pulse
Aluminium	780	3.6
Steel	610	1.1
Nickel	580	1.2
Copper	900	2.2

The disintegration of the material can be significantly affected by the recoil momentums I due to the vaporization of the material by the laser pulse. The recoil momentum of 0.18 dyne s is generated by the vaporization of copper by the giant pulse with the energy of 0.3 J and $\tau_i = 50$ ns in vacuum (about 3×10^{-5} mm Hg) while for the total reflection of the laser radiation the momentum of the laser light pressure is 2×10^{-4} dyne s [194].

The recoil momentum produced by the giant pulse is higher than that produced by the free-generation laser pulse even though the mass ejected by the latter pulse is larger. This fact is explained by the higher velocity of the material ejected by the giant pulse.

The metallographic techniques have been used [211] to study the affected zones in Al, Be, Cu, Pb and low-carbon steel irradiated by giant laser pulses at the radiation power density about 10^8-10^9 W/cm^2. The affected zone consists of two zones, namely, the molten zone whose size is about the diameter of the focussed spot and the external zone which consists of clusters of microscopic craters. Similar craters are found also under a razor blade in the affected zone at a distance of 100 μm from the centre of the focussed spot which is not produced by direct effect of the laser radiation [211]. The most probable cause of the development of craters is suggested to be the liberation on the surface defects of the energy of elastic vibrations which develop in the specimen with local instantaneous heating owing to thermal stresses and the effect of the recoil momentum. In stainless steel 12X18H9T the giant laser pulses (the pulse duration is 8×10^{-8} s, the pulse energy is 0.4 J, the ruby laser) produce shallow (about 5 μm deep) craters with a large diameter (about 600 μm) [66]. No increase in microhardness is found on the surface of the crater and just below it.

The results of disintegration of high-purity single crystals and polycrystalline specimens of nickel and copper and low-purity copper, nickel and steel (about 1% C) specimens caused by giant pulses have been reported in reference [164]. When the surface of a high-purity Ni specimen is placed at a distance of 7 mm from the focal plane of a 5-cm lens the radiation-affected zone consists of a zone of complete disintegration and the outer ring of microcraters 10 μm in diameter. When the specimen is positioned in the focal plane of the lens the affected zone consists only of the zone of complete disintegration while a reduction in the radiation power density produces only the zone of separate microcraters. Determinations of the volumes of the microcrater and the rim surrounding it indicate that the bulk of the metal (about 10^{-10} g) is removed from the crater in the form of vapour and liquid.

The microcraters produced by laser pulses are found to be similar to the craters produced by electric sparks [164]. The authors of reference [164] indicate the small difference in the numbers of the laser-produced craters on high-purity single crystals of nickel and copper and on polycrystalline specimens of the same roughness; this means that the grain borders do not make a significant contribution to the development of craters; however, in high-purity metal specimens the number of craters is smaller than in lower-purity specimens.

The development of microcraters can be caused by the discrete structure of the *Q*-switched laser radiation pulse and electric erosion of the materials at the sites of the lattice defects on the surface owing to discharges between the metal and the plasma cloud.

The development of structural defects due to the effects of giant laser pulses is described in references [89, 188]. For instance, in the annealled copper crystals 0.1 mm thick the defects are found in the vicinity of laser-produced craters with the depth of 10 μm and the diameter of up to 1 mm at a distance of up to 10 μm from the crater. The development of craters is attributed to high hardening rates (about 10^6 C/s), high temperature gradients (about 10^5 C/cm), high reactive pressures of the metal vapour (about 10 kbars) and ion bombardment.

Laser radiation of low-carbon steel having the initial ferrite-pearlite structure by giant single pulses with the energy of 35 J and the duration of 1.8×10^{-8} s (in vacuum of 10^{-2} mm Hg) produces craters 5 μm deep. The molten layer has the acicular martensite structure, the heat-affected zone contains transformed pearlite grains and below this zone there is a mechanically affected zone containing ferrite grains with twinning [89]. No deformation twins are found in the heat-affected zone.

In iron alloys (carbonyl iron and steels 45, У8 and ХВГ) irradiated by laser pulses of 4×10^{-8} s the laser-affected zones have a diameter of 1.5-2.5 mm and the depth in the centre of up to 5 μm. The giant laser pulse produces a molten region in the centre of the laser-affected zone in the air and in vacuum (Fig. 113). When specimens are irradiated in the air the molten region is surrounded by a scale ring which is, apparently, the oxides produced by burning of the ejected particles in the air (Fig. 113*a*). Small drops of the ejected metal are found in the scale ring and beyond it.

When the laser radiation experiments are carried in vacuum the central affected region is surrounded by a concentric band up to 1 mm wide which contains microcraters with the size varying from 10 to 50 μm (Fig. 113*b*). The density of craters and their size decrease with increasing distance from the centre of the focussed spot. Most microcraters are fairly round and surrounded with rims formed by the expelled molten metal.

Figure 114*a* shows the central region of the affected zone produced by a giant pulse in a specimen of the ХВГ steel irra-

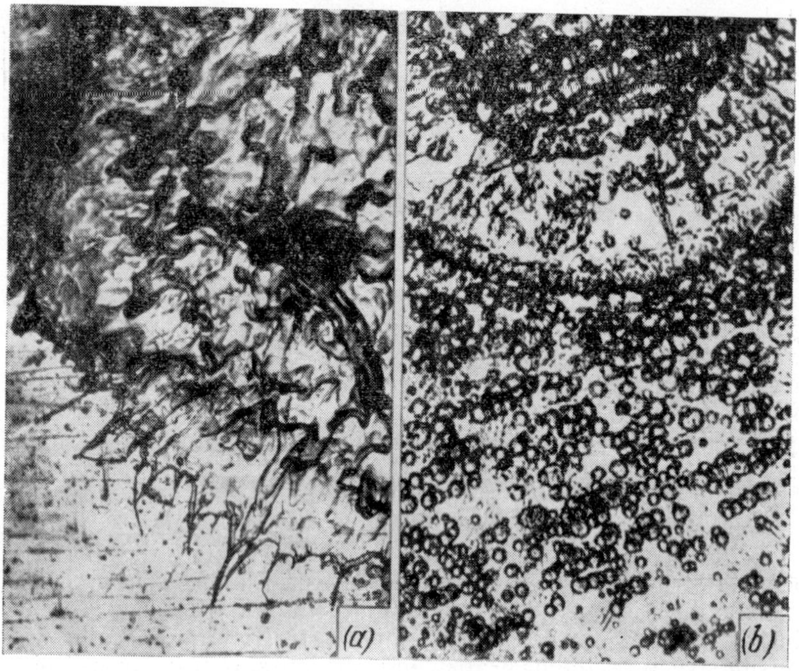

FIG. 113. Microphotographs of zones irradiated by Q-switched laser pulses in steel ХВГ (×70)
(*a*) in the air; (*b*) in vacuum

diated in vacuum (the melting zone). It can be seen that the specimen surface is etched by gas and ion jets. Outside the molten zone the surface is also etched (Fig. 114*b*); this photograph shows also the impact site of the particle ejected from the molten zone. Noteworthy are the characteristic traces directed from the centre of the crater.

The microhardness of steel in the zone affected by the giant laser pulse is increased considerably (from $H_{50} \approx 250$-270 kgf/mm^2 to $H_{50} \approx 600$-750 kgf/mm^2) but only in the molten zone. The hardening zone found on the section of the ХВГ steel specimen etched in the ethanol solution of nitric acid has a depth of about 10 μm (Fig. 115) and a diameter of about 1.9 mm which is almost equal to the diameter of the molten zone.

In the 0.11-mm carbonyl iron foil (double vacuum remelting) the giant laser pulse with the radiation power density of about 5×10^{11} W/cm^2 in vacuum produces mechanical disintegration of the specimen along the grain boundaries (see Fig. 119). The

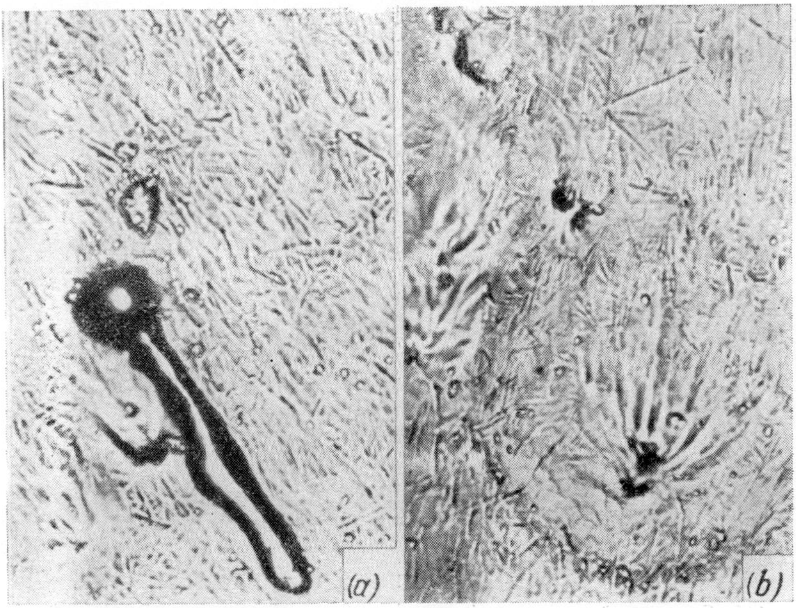

FIG. 114. Laser-molten zone in steel XBГ (×1000)
(*a*) central region; (*b*) outside region adjacent to molten zone

maximum disintegration is observed on the back side of the foil. When the radiation power density is increased to 10^{12} W/cm^2 the through holes are produced in the foil (Fig. 116). It can be seen that the foil surface is molten and the foil is buckled in the direction of the laser beam.

Figure 116 shows the boundary of the molten zone in the foil irradiated by a giant pulse with the radiation power density of 10^{10} W/cm^2. Different structural features can be seen in the molten zone and outside it. Figure 117 presents a microphotograph of a region inside the molten zone adjacent to its boundary. The laser pulse produces the deformation of individual grains, and a substructure can be seen on the surface of the zone which is oriented along the traces of the ejected disintegration products. There is practically no change in the microhardness in the laser-affected zone of iron foil (see Fig. 119) and in the regions around through holes (see Fig. 118*b*) (Figures 118*b* and 119 show the "pinpricks" of the diamond cone of the indenter).

The craters produced by the giant laser pulses in stainless steel 12X18H9T are practically similar to the laser-produced craters in steels of other grades that have been studied [66]. However, no variation in the phase composition and microhardness

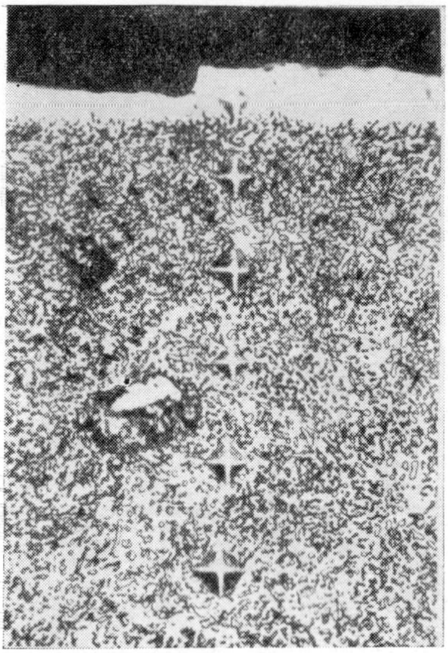

FIG. 115. Cross section of zone irradiated in vacuum by giant pulse in steel XBΓ (×600)

has been found in this austenitic steel after laser radiation in contrast to other steels [66].

To analyze the crater development in thick steel specimens under laser radiation we have to estimate the mass expelled by the laser pulse effect. The quasistationary vaporization theory overestimates the thickness of the vaporized layer. The gas dynamic theory of vaporization [2] also overestimates the depth of a crater produced by a giant laser pulse. A qualitative analysis of crater formation can be made using the approximate spherically symmetrical model of the plasma cloud and assuming that the velocity of the gas particles is proportional to the radius [91]. As shown in reference [183] 99% of the incident radiation is absorbed in the plasma cloud produced by a giant laser pulse irradiating an aluminium foil.

It has been shown [91] that the vaporized mass M is given by the following equation:

$$M = \frac{4\tau_i^2 E}{(0.13p^{0.15})^2} \quad (401)$$

where E is the pulse energy, p is the mean radiation power and τ_i is the pulse duration.

The vaporized mass given by Eq. (401) at $\tau_i = 100$ ns, $E = 5$ J and $p = 38$ MW agrees with the experimental results reported in reference [24].

The small craters found around the molten zone due to irradiation of steel specimens by giant laser pulses in vacuum can be produced, among other factors, by condensation of the expanding vapour into drops which fall on the surface of the specimen. However, the small drops (about 10^{-4}-10^{-5} cm) cannot produce craters with the diameters of tens of micrometres when their velocity is less than 10 km/s, that is, less than the critical velocity for vaporization by their impact on the surface of steel [46]. We

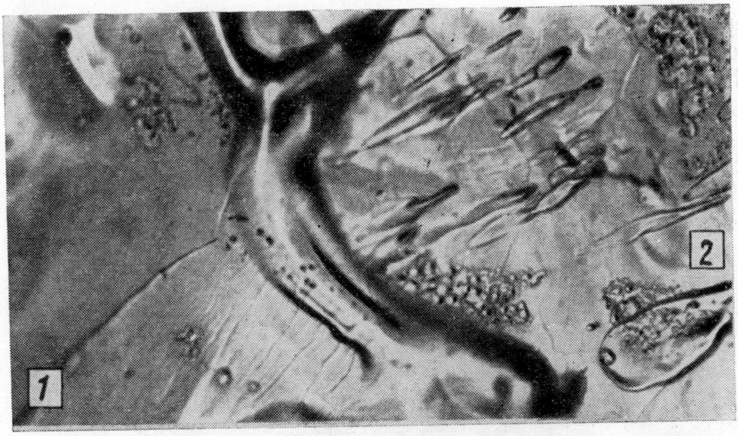

FIG. 116. Boundary of molten zone of armco-iron foil irradiated in vacuum (×600)
1—molten zone; *2*—solid zone

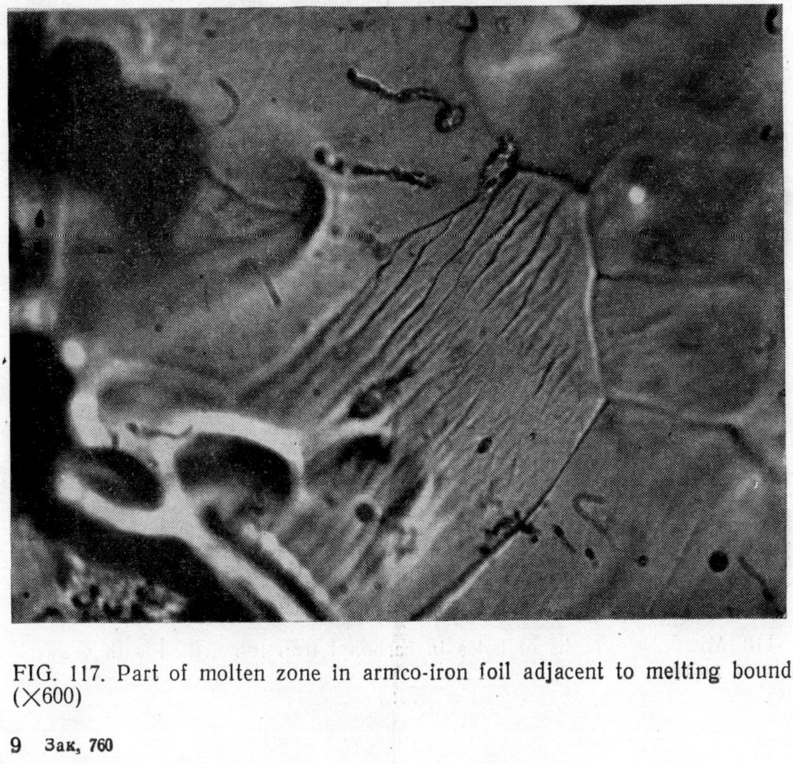

FIG. 117. Part of molten zone in armco-iron foil adjacent to melting boundary (×600)

have repeated the experiment [211] with the razor blade but found no craters under the blade.

The discrete structure of the giant laser pulse seems to be the most probable factor that causes the formation of microcraters with radiation in vacuum. With radiation of metals in the air the Q-switched laser pulse at a high radiation power density produces

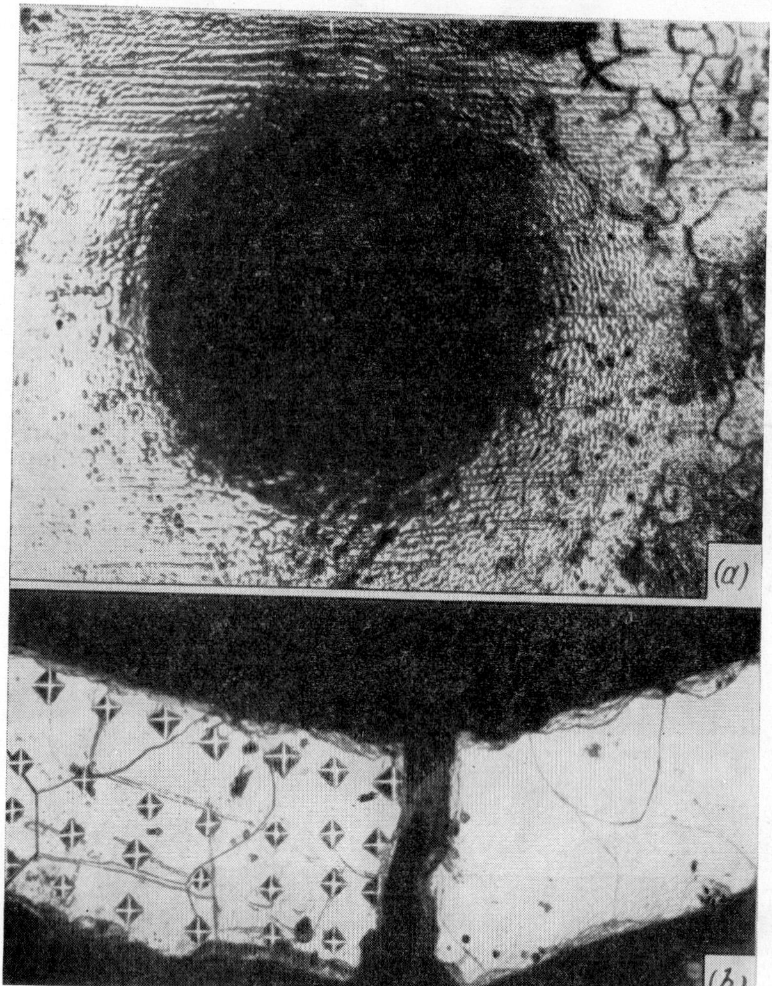

FIG. 118. Microphotographs of holes in carbonyl iron foil drilled with Q-switched laser pulse

(a) view from above (×70); (b) cross section (×340)

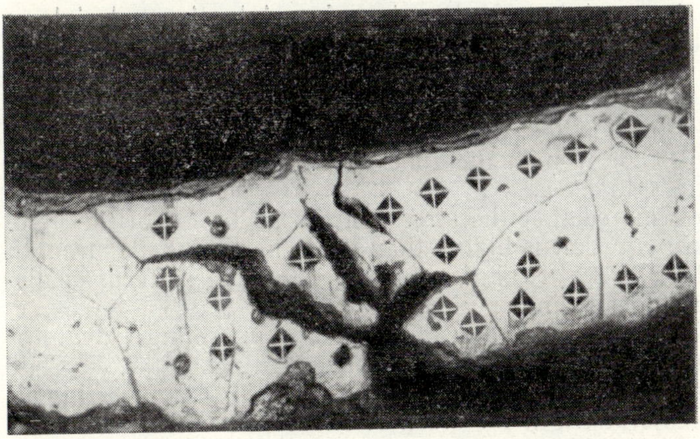

FIG. 119. Cross section of zone irradiated by giant pulse in carbonyl-iron foil ($\times 340$)

the disintegration of the metal surface through the electric spark-over. Apparently, craters are formed by the ejection of a part of the liquid metal in the molten zone as the pulse duration is sufficient for the heat transfer processes to be significant (this can be seen from the fact that the heat diffusion zone for $\tau_i \approx 100$ ns has the size of $\sqrt{a\tau_i} \approx 10^{-4}$ cm while the penetration depth for the radiation with $\lambda = 1.06$ μm is about 10^{-5} cm, that is, it is smaller by an order of magnitude).

Figure 119 shows that a laser pulse with the energy of 5 J and duration about 100 ns produces the disintegration from the back side of an 0.1-mm carbonyl iron plate, and pulses of somewhat higher energy produce through holes in it (see Fig. 118).

Let us estimate the momentum produced by the laser pulse acting on the surface of the material. Under the conditions discussed here the pulse energy is absorbed in the thin surface layer (about 10^{-5} cm). With the above pulse parameters the energy absorbed per gram of the material in the absorption layer is 2×10^5 cal/g, that is, higher than the vaporization heat by two orders of magnitude (1.5×10^3 cal/g).

Assume that the absorbed energy is spent, apart from vaporization, for the ejection of the material and estimate the "cleavage" momentum I using the equation [59]

$$I = 0.8\sqrt{2Em_0} \tag{402}$$

where E is the pulse energy, m_0 is the mass of the absorption layer and the factor 0.8 takes into account the redistribution of energy among the particles during ejection.

If we know the momentum transferred to the material we can estimate the peak stress due to the propagation of the shock wave in the material. The shock wave is reflected by the free surface from the back side of the plate so that the amplitude of the tensile stress is doubled. This prediction agrees with the experimental results for iron at the shock wave pressures up to 1.6×10^6 kgf/cm^2 [46].

It is estimated that the stress on the back side of the plate is 3×10^5 kgf/cm^2 that is higher than the dynamic tensile strength for steel which is 3×10^4 kgf/cm^2 [46]. Thus, the plate must disintegrate and does so in experiments.

If we take into account the decrease in Young's modulus with increasing temperature the predicted value of the dynamic stress becomes less by a factor of about two * but still it is higher than the tensile strength for steel. It should be noted that if the stress were much higher than the tensile strength of the material there would occur not only single but also multiple cleavage of the material and its breaking. As no such breaking is found in experiments, and disintegration occurs mainly along the grain boundaries the above estimates are in a satisfactory agreement with the experimental data.

6.4. MASS TRANSFER IN SOLIDIFICATION AND HEAT-AFFECTED ZONES

Laser radiation (free-generation mode) has been noted to produce significant changes in the concentrations of alloying elements in the laser-irradiated zones of alloys [67, 93, 145]. Despite the short durations of high-power laser radiation of iron alloys mass-transfer processes have been found to occur in the affected zones so that the distribution of the alloying elements after the laser pulse differs from that before the pulse. The redistribution of alloy components has a significant effect on the physical and mechanical properties of the alloy and its performance.

The distribution of components in the laser-affected zone is studied mostly using the local microspectral analysis. The local X-ray diffraction analysis also yields results demonstrating the redistribution of components but this evidence is not sufficiently convincing.

The carbon distribution profiles are measured by using the local spectral analysis in the irradiated zones of specimens of steels 45, Cт. 3, and X12M and armco-iron, and the results prove to be identical (Figs. 120-122). A typical feature is the carbon

* In metals Young's modulus at the melting point is half of that at the absolute zero temperature.

concentration peak found at the depth of 200-350 μm from the bottom of the laser-produced crater depending on the grade of the steel. At the depth of 0.6-1.2 mm the carbon concentration returns to the original level. A similar distribution profile of chromium is found in the laser-radiated zone of steel X12M, the maximum depth of changed concentration being approximately the same for chromium and carbon (about 0.5 mm). Since the diffusion coefficient of carbon in iron is larger than the diffusion coefficient for chromium by 5-6 orders of magnitude the above finding is the evidence against the diffusion mechanism of mass transfer [73]. The concentration profiles of alloying elements such as Mn, Ni, Ti, Cr and C in the laser-affected zone (ruby or neodymium laser) of stainless steel 12X18H9T are determined by means of local spectral analysis using a point source of rectified high-frequency current [67, 73, 115]; the results show significant redistribution of the concentrations of Mn, Ni, Ti, Cr and C in this zone. In the layer adjacent to the laser-produced crater the carbon content remains at the initial level while the concentrations of other alloying elements can either be higher or lower than the respective mean concentrations in the specimen.

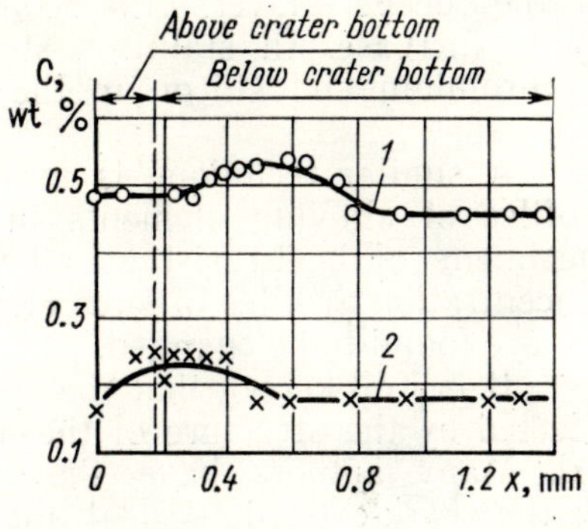

FIG. 120. Distribution of carbon in laser-affected zone
(*a*) steel 45; (*b*) steel Ст. 3

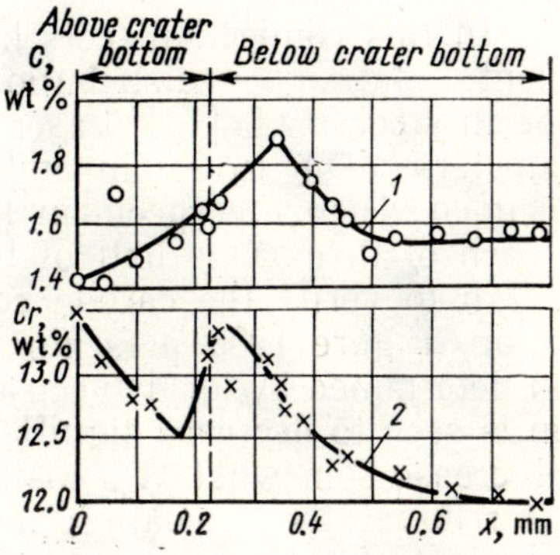

FIG. 121. Distribution of carbon (*1*) and chromium (*2*) in laser-affected zone of steel X12M

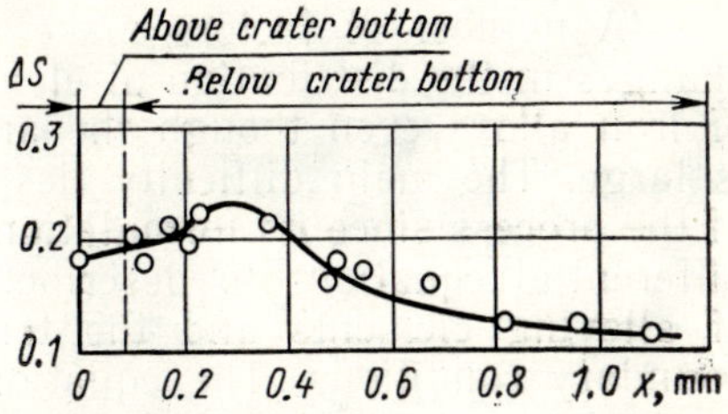

FIG. 122. Distribution of carbon with depth in laser-affected zone of armco-iron

In the deeper layers the concentrations of the above elements increase. In stainless steel specimens the concentrations of chromium and silicon in the laser-affected zones remain unchanged.

A similar technique is used to determine the concentration profiles of alloying elements in the laser-affected zones in the specimens of heat-resistant alloys and cast iron [67, 73]. The concentration profiles of such elements as Mn, Ni, Ti, Nb, Mg, Si, etc. are found to be changed.

It is well-known that the spectral lines of some elements are affected by the structure of the alloy in which they are dissolved [67]. Many authors have shown that there is no such effect for the spark-excited spectral lines even for such significant structural changes as hardening [42]. The specific structural changes produced by laser radiation may affect the spectra of alloying elements producing a wrong impression of significant changes in their concentrations.

In this connection tests have been made to verify the experimental data on redistribution of carbon in the laser-affected zone in steel 35ХГ [73]. Laser pulses produce two holes with the diameters of up to 3 mm and the depth of up to 5 mm in the specimen. Then the specimen is cut into two halves with a hole in each of them. One half of the specimen is annealed at 650 C for 2 h to verify the carbon concentration profile after laser radiation. Figure 123*a* presents the carbon concentration distribution determined by local spectral analysis *. The carbon concentration is seen to decrease significantly at the edge of the crater and to increase somewhat at a distance of 100-120 μm from the edge. In ferrite at the annealing temperature of 650 C the diffusion coefficient of carbon is 3.6×10^{-7} cm^2/s, that is, in two hours the diffusion distance is about 1 mm; this conclusion agrees with the microspectral analysis results (Fig. 123*b*) which show no change in the carbon concentration in the laser-radiated zone of the annealed specimen.

A number of factors make it very difficult to analyze the changes in the distribution of elements in the laser-affected zones of iron alloys even though the amount of experimental evidence is large. The main difficulty lies in the mathematical treatment of the process since its formulation involves a system of nonlinear differential equations to describe the concentration distributions of alloying elements and the temperature field with nonlinear boundary conditions. The equations for concentrations are non-

* After each spark discharge a layer of metal is removed by polishing to the depth which is larger than the depth of carbon burning to eliminate the possible carbon burning as a result of the preceding spark discharge.

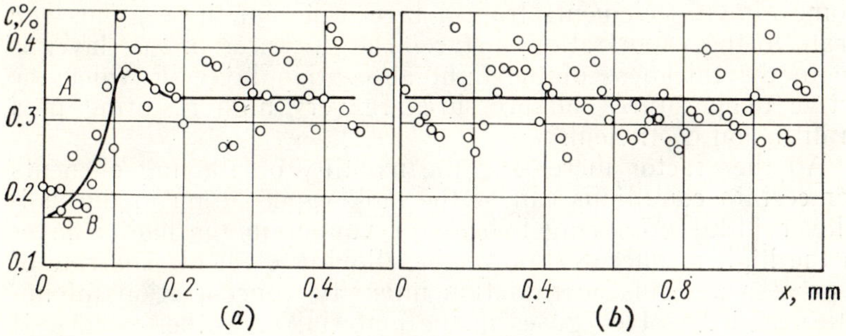

FIG. 123. Distribution of carbon near laser-drilled hole bottom in steel 35ХГ (the origin on the x-axis is the depth of the hole bottom)

(a) unannealled specimen: A — initial concentration; B — carbon concentration in molten zone after solidification; (b) specimen annealled for 2 h at 650 C

linear because of the strong temperature dependence of diffusion coefficients.

Reference [73] discusses the contribution of the diffusion processes (thermal diffusion and the diffusion under pressure and plastic deformation) to the redistribution of alloying elements produced by laser radiation. It has been shown that no possible diffusion processes (even with contribution of plastic deformation) can produce such changes in the concentrations of alloying elements in the laser-affected zone as are found in the experiments. Thus, the concentration changes are produced by the mass-transfer mechanisms other than diffusion. A thin layer of the molten metal is produced by the laser radiation on the surface of the metal at the beginning of radiation; the thickness of this layer decreases with increasing radiation power density. The predicted thickness of the molten layer is plotted as a function of the radiation power density in Fig. 124. In steel the thickness of the molten layer is smaller by a factor of about two than in iron due to a difference in thermal properties.

The molten layer on the surface of the specimen can be enriched or depleted in alloying elements compared with the basic material because of the difference in the vapour tensions of alloying elements and the base metal. For most alloying elements their binding energy in the alloy per atom is lower than the binding energy per atom of the base metal. Therefore, with the moving vaporization front there can be vaporized not only the atoms at the vaporization surface but also the atoms

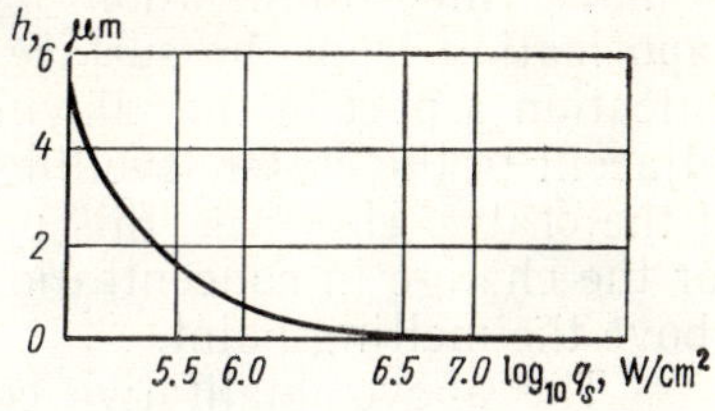

FIG. 124. Thickness h of molten layer as a function of laser radiation power density

of some alloying elements from a layer wherein these atoms can migrate to the vaporization surface. The thickness of this layer is close to the thickness of the light absorption layer in which the effective temperature and pressure are very high resulting in a high diffusion coefficient.

Another factor increasing the mobility of alloying elements under certain conditions can be the bulk vaporization in the molten layer [120] producing bubbles or vapour in the molten layer with the higher concentration of the alloying element. The rate of this process depends on radiation intensity, concentrations of impurities, and dissolved gases in the material and other factors. If the above arguments are valid and the binding energy of alloying elements is lower than the binding energy of the base metal then the thin layer adjacent to the crater must have a higher or lower concentration of alloying elements depending on their vapour tension. Finally, there is one more factor which can change the concentrations of alloying elements in the layer adjacent to the crater. The interface between the molten layer and the solid metal moves during the laser pulse and "withdraws" from the solid phase to the liquid phase those alloying elements whose solubility in the liquid phase is higher than in the solid phase. Hence, we can define the nonequilibrium distribution coefficient k_{ne} by analogy with the equilibrium distribution coefficient as

$$k_{ne} = \frac{C_s}{C_l} \tag{403}$$

where C_s is the concentration of the alloying element in the solid phase and C_l is its concentration in the liquid phase.

The rate of dissolving is determined by the temperature of the liquid phase, the time of contact between the solid and the liquid phase, the nature of alloying elements and their concentrations. For many elements the solubility in the liquid phase is higher than in the solid phase so that the layer at the boundary of the liquid phase has a lower concentration of alloying elements. After the end of the laser pulse the liquid-solid interface ceases to move, and solidification begins. Under these circumstances the vaporization from the surface becomes insignificant. During solidification a part of the alloying element atoms return to the layer adjacent to the crater levelling off the concentration in the vicinity of the crater. However, this process cannot completely compensate for the change in concentration as the temperature is only slightly above the melting point.

The above qualitative considerations suggest that the final distribution of alloying elements in the zone around the crater must differ from the initial distribution. Note that the effective depth of concentration change can never be larger than the depth of thermal diffusion, that is $\sqrt{a\tau}$ where τ is the total time of heating and cooling of the material. When $\tau \approx 10^{-2}$ s and $a = 0.1$ cm²/s the thermal diffusion depth is about 600 μm. The experimental results agree with this value. The diffusion depth depends significantly on the temperature field in the crater zones as the diffusion coefficient D typically varies with temperature as

$$D = D_0 \exp(-\varepsilon^*/kT) \tag{404}$$

where D_0 is the limiting diffusion coefficient and ε^* is the activation energy.

Proceeding from the above discussion we can identify several stages of redistribution of alloying elements in the laser-radiated zones of metals.

The first stage is a nonstationary process resulting in the development of a quasistationary temperature distribution in the crater zone. At $q_0 \geqslant 10^6$ W/cm² the duration of this stage is small and has no noticeable effect on the redistribution of alloying elements. The next stage is the quasistationary process of material disintegration with a quasistationary distribution of temperature. At this stage there occurs the diffusion process of accumulation of alloying elements in the liquid layer produced by the radiation, or its depletion.

At the last stage of the process when the molten layer solidifies after the end of the laser pulse the alloying elements are transferred back to the solid phase, that is, the surface layer is enriched with alloying elements; the alloying elements are also transferred to the periphery of the solidifying layer by the process similar to zone melting. As the duration of this stage is small we can neglect here the diffusion of the alloying elements from the solid phase.

Obviously, the above model of mass transfer does not take into account all the possible mechanisms and therefore does not give a good agreement with the experimental data.

REFERENCES

1. ANISIMOV, S. I., Vaporization of Metal Absorbing Laser Radiation, *Zhurn. eksper. i teor. fiziki*, 1968, 54, No. 1, 339-342.
2. AFANASYEV, Yu. V., KROKHIN, O. N., Theory of Interaction of Laser Radiation with Matter, *Trudy FIAN*, *Kvantovaya Radiofizika*, Nauka, Moscow, 1970, 52, 118-170.
3. BABENKO, V. P., TYCHINSKY, V. P., Gas-Jet Laser Cutting, in *Kvantovaya elektronika*, ed. by Basov, N. G., Sov. Radio, Moscow, 1972, 3-21.
4. BARANOV, M. S., VOSHCHINSKY, M. L., GEINRIKHS, I. N., *Lazernaya svarka metallov* (Laser Welding of Metals), Mashinostroenie, Moscow, 1971.
5. BARANOV, M. S., KONDRATYEV, V. A., UGLOV, A. A., Kinetics of Wire Joining in Pulsed Laser Beam Welding, *Fizika i khimiya obrabotki materialov*, 1972, No. 5, 11-14.
6. BARASHEV, P. P., TAL'ROZE, V. L., Competition of Thermal and Photochemical Mechanisms of Transformation of Materials Irradiated by Laser Pulses, *Khimiya vysokikh energiy*, 1971, 5, No. 1, 30-36.
7. BARKHUDAROVA, T. M., Variation in Intensity Distribution of the Giant Laser Pulse, *Fizika i khimiya obrabotki materialov*, 1969, No. 4, 10-15.
8. KITTEL, C., *Quantum Theory of Solids*, N. Y., 1963.
9. BONCH-BRUEVICH, A. M., IMAS, Ya. A., Laser Effects in Metals, *Fizika i khimiya obrabotki materialov*, 1967, No. 5, 3-10.
10. BRAGINSKY, V. B., MINAKOVA, I. I., STEPUNIN, P. M., Absolute Measurements of Energy and Power from the Radiation Pressure in the Optical Range, *Pribory i tekhnika eksperimenta*, 1965, No. 3, 183-187.
11. BREKHOVSKIKH, V. F., KOKORA, A. N., UGLOV, A. A., Spatial Distribution of Heat Source Intensity in Steel Produced by Laser Beam, *Fizika i khimiya obrabotki materialov*, 1967, No. 6, 3-9.
12. BREKHOVSKIKH, V. F., RYKALIN, N. N., UGLOV, A. A. Possible Effect of Gases in Metals on the Laser-Affected Zone, *Doklady AN SSSR*, 1970, 190, No. 5, 1059-1062.
13. BUZUKOV, A. A., POPOV, Yu. A., TESLENKO, V. S., Experimental Study of the Explosive Process due to Focussing of Single Laser Pulse into Water, *Zhurn. prikl. mekhaniki i tekhn. fiziki*, 1969, No. 5, 17-24.
14. BURAVLEV, Yu. M., *Vliyanie struktury na rezul'taty spektral'nogo analiza splavov* (Effects of Alloy Structure on the Spectral Analysis Results), Metallurgizdat, Moscow, 1963.
15. BUTKOVSKY, A. G., *Teoriya optimal'nogo upravleniya sistemami s raspredelennymi parametrami* (Theory of Optimum Control of Systems with Distributed Parameters), Nauka, Moscow, 1965.
16. VALITOV, R. A., KALININ, Yu. A., KUZ'MICHEV, V. M., Determination of Laser Energy and Power, *Izmeritel'naya tekhnika*, 1965, No. 5, 37-41.
17. VEIKO, V. P., IMAS, Ya. A., Disintegration of Ferrite by Laser Radiation, *Fizika i khimiya obrabotki materialov*, 1967, No. 2, 120-123.

18. VEIKO, V. P., IMAS, Ya. A., LIBENSON, M. N., Heating of Materials by Laser Radiation, *Fizika i khimiya obrabotki materialov*, 1967, No. 1, 27-32.
19. VEIKO, V. P., KOKORA, A. N., LIBENSON, M. N., Experimental Study of Temperature Distribution in the Laser-Affected Zone in Metals, *Dokl. AN SSSR*, 1967, 179, No. 1, 68-71.
20. VEIKO, V. P., LIBENSON, M. N., *Lazernaya obrabotka* (Laser Machining), Leningrad, 1973.
21. VERTMAN, A. A., SAMARIN, A. M., *Svoystva rasplavov zheleza* (Properties of Molten Iron), Nauka, Moscow, 1969.
22. ANISIMOV, S. I., DMITRIENKO, B. I., LESKOV, L. V., *et al*, Effect of Reflectivity of Surface on Vaporization of Metals by High-Intensity Light Flux, *Fizika i khimiya obrabotki materialov*, 1972, No. 4, 10-14.
23. KARLOV, N. V., KARPOV, N. A., PETROV, Yu. N., *et al*, Development of the Shock Wave with Chain Reaction Initiated in Gas Mixtures by the CO_2 Laser Radiation, *ZhETF Pis'ma*, 1971, 14, is. 4, 214-217.
24. KAPORSKY, L. N., KOKORA, A. N., ROMANENKOVA, G. A., *et al*, Effects Produced by the *Q*-Switched Laser Pulse on Steel, *Fizika i khimiya obrabotki materialov*, 1968, No. 1, 3-10.
25. GAGLIANO, F. P., LUMLEY, R. H., WATKINS, L. S. *Laser in Industry*, Proceedings of the IEEE, 1969, v. 57. No. 2, 114-147.
26. GARASHCHUK, V. P., Pulsed Laser Welding (Review), *Avtomatickeskaya svarka*, 1966, No. 9, 16-23.
27. GENERALOV, N. A., KOZLOV, G. I., MASYUKOV, V. A. Clarification of Molecular Iodine and Breakdown in It Due to Laser Pulses, *Zhurn. eksper. i teoret. fiziki*, 1970, 58, is. 2, 438-440.
28. GLIKMAN, L. A., KRYLOV, K. I., RUBASHKINA, Z. M., Laser Effects in Steel, *Trudy LITMO, Voprosy kvantovoy elektroniki*, 1967, is. 61, 58-68.
29. GOLOVEIKO, A. G., Ejection of Molten Material of Electrodes in Electric Discharge, *Izv. VUZOV Energetika*, 1966, No. 6, 83-89.
30. GOL'DFARB, E. M. *Teplotekhnika metallurgicheskikh protsessov* (Heat Technology of Metallurgy), Metallurgiya, Moscow, 1967.
31. GREVTSEV, N. V., KARABUTOV, V. G., KRASULIN, Yu. L., Energy Parameters of Ruby Laser, *Fizika i khimiya obrabotki materialov*, 1968, No. 6, 13-20.
32. GRIDNEV, V. N., MESHKOV, Yu. Ya., CHERNENKO, N. F., Critical Points of Ti-Cr Alloy with Rapid Heating, in *Voprosy fiziki metallov i metallovedeniya* (Problems of Physical Metallurgy), Kiev, 1963, No. 17, 143-145.
33. AMBARTSUMYAN, R. V., LETOKHOV, V. S., MAKAROV, G. N., *et al*, Two-Stage Photodissociation of Ammonia Molecules Due to Laser Radiation, *ZhETF Pis'ma*, 1972, is. 12, 709-711.
34. ANISIMOV, S. I., IMAS, Ya. A., ROMANOV, G. S., *et al*, *Deystvie izlucheniya bol'shoy moshchnosti na metally* (Effect of High-Power Radiation on Metals), Nauka, Moscow, 1970.
35. DITER, G. E., Hardening Effects Produced with Shock Waves in Strengtheining Mechanisms in Solids, *Amer. Soc. Metals*, 1962.
36. DUL'NEV, G. N., ISPIRYAN, R. A., YARYSHEV, N. N., Heat Conduction with Constant and Pulsed Local Heating, *Trudy LITMO*, Leningrad, 1967, is. 31, 5-19.
37. DUSHMAN, S., *Scientific Foundations of Vacuum Technique*, N. Y., 1949.
38. EMLYUTINA, L. D., Measurement of Power and Energy of Laser Radiation, *Obzory po elektronnoy tekhnike*, is 18 (87), Elektronika, Moscow, 1969, 24, 25.
39. ZHIRYAKOV, B. M., FANNIBO, A. K., YURYSHEV, N. N., Some Deformation Effects Produced by Laser Radiation of Metals, *Zhurn. priklad. mekh. i tekh. fiz.*, 1967, No. 4, 145-146.

40. ZHIRYAKOV, B. M., FANNIBO, A. K., YURYSHEV, N. N., Interaction of Pulsed Laser Radiation with Metals for Quasistationary Laser Operation, in *Ispol'zovanie opticheskikh kvantovykh generatorov v sovremennoy tekhnike* (Application of Lasers in Modern Technology), P. 1, LDNTP, Leningrad, 1969, 44-49.
41. ZHIRYAKOV, B. M., FANNIBO, A. K., YURYSHEV, N. N., Technological Application of Quasistationary Ruby Laser, *Fizika i khimiya obrabotki materialov*, 1970, No. 3, 14-24.
42. ZHUKHOVITSKY, A. A., SHVARTSMAN, L. A., *Fizicheskaya khimiya* (Physical Chemistry), Metallurgiya, Moscow, 1968.
43. ZIMAN, J., *Principles of the Theory of Solids.*, Cambridge, 1964.
44. ZAKHAROV, V. P., TSVIRKO, Yu. A., CHUGAEV, V. N., Recrystallization of Thin Semiconductor Films under the Effect of Laser Radiation, *Dokl. AN SSSR*, 1966, 170, No. 5, 1056-1058.
45. ZAKHAROV, V. P., POL'SKY, Yu. M., CHUGAEV, V. N., Kinetics of Structural Changes in Thin Germanium and Carbon Films under Laser Radiation, *Fizika i khimiya obrabotki materialov*, 1970, No. 5, 55-60.
46. ZEL'DOVICH, Ya. B., RAIZER, Yu. P., *Fizika udarnykh voln i vysokotemperaturnykh gidrodinamicheskikh yavleniy* (Physics of Shock Waves and High-Temperature Hydrodynamic Phenomena), Nauka, Moscow, 1966.
47. ZUBAREV, T. N., SOKOLOV, A. K., Variation with Time of Radiation of Ruby Laser with Spherical Mirrors, *Dokl. AN SSSR*, 1964, 159, No. 3., 539-540.
48. ZUEV, I. V., RYKALIN, N. N., UGLOV, A. A., Deep Penetration Melting of Metals by Electron Beam, *Fizika i khimiya obrabotki materialov*, 1968, No. 5, 7-15.
49. IVANOVSKAYA, S. I., SADOVSKY, V. D., Effect of Heating Rate on Structural Transformations with Electrohardening of Steel, *Trudy UIFM*, Sverdlovsk, 1951, is. 13, 10-31.
50. IERUSALIMSKAYA, A. N., SAMOILOV, V. N., ULYAKOV, P. I., Structural Changes of Material Irradiated by Laser Pulses, *Fizika i khimiya obrabotki materialov*, 1968, No. 4, 26-34.
51. BONCH-BRUEVICH, A. M., IMAS, Ya. A., ROMANOV, G. S., *et al*, Variation in Reflectivity During the Time of Effect of Laser Pulse, *Zhurn. tekhn. fiz.*, 1968, 38, No. 5, 851-855.
52. AVOTIN, S. S., KRIVCHIKOVA, E. P., PAPIROV, I. I., *et al*, Variation in Electric Resistance of Laser-Irradiated Beryllium, *Zhurn. eksper. i teoret. fiz.*, 1972, 62, is. 1, 288-293.
53. VALITOV, R. A., KOKODY, N. G., KUBAREV, A. V., *et al*, *Izmerenie kharakteristik opticheskikh kvantovykh generatorov* (Measurement of Laser Parameters), Moscow, 1969.
54. IOFFE, A. F., *Fizika poluprovodnikov* (Physics of Semiconductors), AN SSSR, Moscow, 1957.
55. BATANOV, V. A., BUNKIN, F. V., PROKHOROV, A. M., *et al*, Vaporization of Metal Targets by High-Power Laser Radiation, *Zhurn. eksper. i teoret. fiz.*, 1972, 63, is. 2, 586-608.
56. LOLADZE, T. N., BETANELI, A. I., SEMILETOVA, A. I., *et al*, Possible Local Alloying of Steel Surface, in *Ispol'zovanie opticheskikh kvantovykh generatorov v sovremennoy tekhnike* (Applications of Lasers in Modern Technology, P. 1, LDNTP, Leningrad, 1971, 11-14.
57. ISHCHENKO, E. M., KLIMKOV, Yu. M., *Opticheskie kvantovye generatory* (Lasers), Sov. Radio, Moscow, 1968.
58. KAGANOV, M. I., LIFSHITS, I. M., TANATAROV, L. V. Electron-Lattice Relaxation, *Zhurn. eksper. i teoret. fiz.*, 1956, 31, No. 2, 232-237.
59. KALMYKOV, A. A., KONDRAT'EV, V. N., NEMCHINOV, I. V., Ejection of Instantaneously Heated Material and Derivation of the Equation of

State for It from the Pressure and Momentum Values, *Zhurn. priklad. mekhaniki i tekhn. fiz.*, 1966, No. 5, 3-16.
60. CARSLAW, H. S., JALGER, J. C., *Conduction of Heat in Solids*, Second edition, Oxford.
61. KIDIN, I. N., *Fizicheskie osnovy elektrotermicheskoy obrabotki metallov i splavov* (Physical Fundamentals of Electrothermal Treatment of Metals and Alloys), Metallurgiya, Moscow, 1969.
62. KARASEV, I. G., KIRILLOV, V. M., NORSKY, V. E., *et al*, Kinetics of Disintegration of Metals by Free-Generation Laser Radiation, *Zhurn. tekhn. fiz.*, 1970, 40, No. 9, 1954-1959.
63. KIRILLOV, V. M., ULYAKOV, P. I., Proportions of Liquid and Gaseous Products Produced in Metals by Focussed Laser Radiation, *Fizika i khimiya obrabotki materialov*, 1971, No. 1, 8-12.
64. KOVALENKO, V. S., Study of Surfaces of Laser-Drilled Microholes, in *Elektrofizicheskie i elektrokhimicheskie metody obrabotki* (Electrophysical and Electrochemical Processing Techniques), 1970, is. 5, 18-22.
65. KOZDOBA, L. A., Techniques for Solving the Nonlinear Heat-Conduction Problems, *Fizika i khimiya obrabotki materialov*, 1968, No. 4, 3-9.
66. KOKORA, A. N., ZHUKOV, A. A., Some Effects Produced by Laser Radiation in Metals, *Fiziko-khimicheskaya mekhanika materialov*, 1967, 2, No. 3., 359-360.
67. KOKORA, A. N., RIKMAN, E. P., Distribution of Alloying Elements in the Laser-Affected Zone, *Fizika i khimiya obrabotki materialov*, 1968, No. 1., 11-15.
68. KOKORA, A. N., UGLOV, A. A., Some Effects Produced by the *Q*-Switched Laser Pulse Incident on Steel, *Fizika i khimiya obrabotki materialov*, 1968, No. 5, 16-21.
69. KOLMOGOROV, A. N., Statistical Theory of Solidification of Metals, *Izv. AN SSSR*, 1937, No. 3., 355-359.
70. KONDRAT'EV, V. A., UGLOV, A. A., Laser Welding of Thin Wires, *Fizika i khimiya obrabotki materialov*, 1969, No. 6, 17-24.
71. KOSHLYAKOV, N. S., GLINER, E. B., SMIRNOV, M. M., *Osnovnye differentsial'nye uravneniya matematicheskoy fiziki* (Fundamental Differential Equations of Mathematical Physics), Fizmatgiz, Moscow, 1962.
72. KRISHTAL, M. A., DAVYDOV, Yu. I., KORVACHEV, V, D., Local Spectral Analysis for Quantitative Determination of Carbon in Steel, *Zavodskaya laboratoriya*, 1964, No. 8, 950-952.
73. KRISHTAL, M. A., ZHUKOV, A. A., KOKORA, A. N., *Struktura i svoystva splavov, obrabotannykh izlucheniem lazera* (Structure and Properties of Alloys Irradiated by Laser Beam), Metallurgiya, Moscow, 1973.
74. KRYLOV, Yu. K., Motion of Drops and Fine Particles and Determination of Their Temperature at the Moment of Formation, *Trudy LITMO, Voprosy kvantovoy elektroniki* (Problems of Quantum Electronics), is. 65, Leningrad, 1968, 50-57.
75. KUZNETSOV, A. E., ORLOV, A. A., ULYAKOV, P. I., Pulsating Vaporization of Optical Materials by the CO_2 Laser Beam, in *Kvantovaya elektronika* (Quantum Electronics), ed. by Basov, N. G., Sov. Radio, Moscow, 1972, No. 7, 57-60.
76. VEIKO, V. P., LIBENSON, M. N., MELYUCHEV, A. M., *et al*, *Lazernaya tekhnologiya* (Laser Technology), TsNII Elektronika, Moscow, 1970.
77. LANDAU, L. D., LIFSHITS, E. M., *Mekhanika sploshnykh sred* (Mechanics of Continuous Media), Gostekhizdat, Moscow, 1954.
78. LANDAU, L. D., LIFSHITS, E. M., *Teoriya uprugosti* (Theory of Elasticity), Nauka, Moscow, 1965.
79. LIBENSON, M. N., VOLOD'KINA, V. L., KRYLOV, K. I., Heating of Metals in the Oxidizing Atmosphere by the CO_2 Laser Beam, *Fizika i khimiya obrabotki materialov*, 1973, No. 5, 145-146.

80. LIBENSON, M. N., ROMANOV, G. S., IMAS, Ya. A., Effect of Temperature Dependence of Optical Parameters of Metals on the Process of Laser Heating, *Zhurn. tekhn. fiz.*, 1968, 38, 7, 1116-1119.
81. LOKHOV, Yu. N., ROZHNOV, G. N., SHVYRKOVA, I. I., Kinetics of Formation of Liquid Phase and Effect of Heat of Phase Transitions Produced by a Point Heat Source, *Fizika i khimiya obrabotki materialov*, 1972, No. 3., 9-17.
82. LYKOV, A. V., *Teoriya teploprovodnosti* (Theory of Heat Conduction), Vysshaya shkola, Moscow, 1967.
83. MAKAROV, N. I., RYKALIN, N. N., UGLOV, A. A. Theoretical Model for Predicting the Temperature Field in Laser-Welded Plates, *Fizika i khimiya obrabotki materialov*, 1967, No. 3., 9-15.
84. MEZOKH, Z. I., YANUSHKEVICH, V. A., IVANOV, L. I., Development of Point Defects in Ni Produced by Giant Laser Pulses, *Fizika i khimiya obrabotki materialov*, 1971, No. 4, 163-165.
85. KARLOV, N. V., KRYNETSKY, B. B., MISHIN, V. A., *et al*, Metastability of Liquid Phase Under the Conditions of Intense Vaporization of Condensed Media, *ZhETF pis'ma*, 1974, 19, is. 2, 111-114.
86. LIBENSON, M. N., SUSLOV, G. P., KOKORA, A. N., *et al*, *Metody povysheniya tochnosti lazernoy razmernoy obrabotki* (Techniques for Increasing Precision of Laser Machining), Znanie, Leningrad, 1973.
87. MESHKOV, Yu. Ya., Theory of Pearlite-Austenite Transformation with Heating of Steel, in *Fazovye prevrashcheniya v metallakh i splavakh* (Phase Transformations in Metals and Alloys), Naukova Dumka, Kiev, 1965.
88. MIKAELYAN, A. L., TER-MIKAELYAN, M. L., TURKOV, Yu. G., *Opticheskie generatory na tverdom tele* (Solid-State Lasers), Sov. Radio, Moscow, 1967.
89. MIRKIN, L. I., Plastic Deformation of Metals Produced by Laser Pulse with a Duration of 10^{-8} s, *Dokl. AN SSSR*, 1969, 189, No. 3, 528-531.
90. MOGILEVSKY, Yu. V., CHUDNOVSKY, A. F. *Teploprovodnost' poluprovodnikov* (Heat Conduction in Semiconductors), Nauka, Moscow, 1972.
91. BASOV, N. G., BOIKO, V. A., DEMENT'EV, V. A., *et al*, Heating and Ejection of Plasma Produced in Irradiation of a Solid Target with a Giant Laser Pulse, *Zhurn. eksper. i teor. fiz.*, 1966, 51, is. 4(10), 989-999.
92. NOVIKOV, N. P., KHOLODILOV, A A., Disintegration of Thermoplastics by High-Intensity Heat Fluxes, *Mekhanika polimerov*, 1971, No. 1, 122-130.
93. KOKORA, A. N., ZHUKOV, A. A., SHALASHOV, V. A., *et al*, Laser Machining of Steel, *Metallovedenie i termicheskaya obrabotka metallov*, 1966, No. 2, 41-42.
94. GREVTSEV, V. N., MAKAROV, N. I., SKRIPNICHENKO, A. S., *et al*, Effect of Laser Pulse Duration on the Depth of the Molten Pool in Metals, *Fizika i khimiya obrabotki materialov*, 1967, No. 2., 116-120.
95. VEIKO, V. P., LIBENSON, M. N., NIKITIN, M. N., *et al*, Reproducibility of Laser Machining, in *Ispol'zovanie opticheskikh kvantovykh generatorov v sovremennoy tekhnike* (Applications of Lasers in Modern Technology), P. 1, Leningrad, 1969, 17-23.
96. ZUEV, I. V. RAROV, N. N., RYKALIN, N. N., *et al.*, Lifetime of a Narrow Channel in Liquid Phase, *Zhurn. prikl. mekhan. i tekhn. fiz.*, 1974, No. 1, 121-129.
97. KULIKOV, O. F., BRAGIN, O. V., GUR'EV, M. V., *et al*, Mechanism of Disintegration of Benzene with Laser Radiation, *Dokl. AN SSSR*, 1969, 187, 1060-1063.
98. ZHIRYAKOV, B. M., RYKALIN, N. N., UGLOV, A. A., *et al*, Some Features of Disintegration of Metals with Focussed Laser Beam, *Zhurn. tekhn. fiz.*, 1971, No. 5, 1037-1042.
99. ZHIRYAKOV, B. M., RYKALIN, N. N, UGLOV, A. A., *et al*, Some Features of Ejection of Material from the Laser-Irradiated Zone, in *Kvantovaya*

elektronika (Quantum Electronics), ed. by Basov, N. G., Sov. Radio, Moscow, 1973, 1 (13), 119-121.

100. BRAGIN, O. V., KULIKOV, O. F., LIBERMAN, A. L., *et al*, Behaviour of Benzene and Some Other Organic Compounds in Focus of the Laser Beam, *Dokl. AN SSSR*, 1966, 171, No. 3, 616-618.
101. BREKHOVSKIKH, V. F., ZHUKOV, A. A., KOKORA, A. N., *et al*, Local Surface Hardening of Steels by Laser Radiation, *Fizika i khimiya obrabotki materialov*, 1968, No. 4, 35-43.
102. VEIKO, V. P., IMAS, Ya. A., KOKORA, A. N., *et al*, Determination of Metal Temperature in the Laser-Affected Zone, *Zhurn. tekhn. fiz.*, 1967, 37, No. 10, 1920-1923.
103. ORAEVSKY, A. N., PIMENOV, V. P., SHCHEGLOV, V. A., Propagation of Photochemical Waves in Binary Gas Mixtures, *Kratk. soob. po fiz. (FIAN)*, 1971, No. 9, 7-12.
104. ORAEVSKY, A. N., PIMENOV, V. P., SHCHEGLOV, V. A., Photochemical Waves in Gases, *Zhurn. eksper. i teor. fiz.*, 1972, 62, is. 1, 89-99.
105. ORAEVSKY, A. N., SHCHEGLOV, V. A., Effect of Chemical Reactions on Propagation of Photodissociation Waves, *Kratk. soob. po fiz.* (*FIAN*), 1970, No. 5, 3-7.
106. ORAEVSKY, A. N., SHCHEGLOV, V. A., Propagation of Photodissociation Waves in Gases Affected by Chemical Reactions, *Zhurn. eksper. i teoret. fiz.*, 1970, 59, is. 3, 845-856.
107. OREKHOV, M. V., SLAVIN, B. S., TARKHOV, G. N., Experimental Study of Factors Affecting the Development of Laser-Drilled Holes, in *Elektricheskie i elektrokhimicheskie metody obrabotki* (Electrophysical and Electrochemical Processing Techniques), 1969, No. 3, 23-26.
108. UGLOV, A. A., ZHUKOV, A. A., KOKORA, A. N., *et al*, "Shift" of Critical Points of Iron-Carbon Alloys with Laser Heating, *Fizika i khimiya obrabotki materialov*, 1972, No. 2, 3-8.
109. PASHKOV, V. A., ZVEREV, G. M., Disintegration of Ruby and Leucosapphire Crystals with High-Power Laser Radiation, *Zhurn. eksper. i teor. fiz.*, 1966, 51, No. 3, 777-779.
110. GROZIN, B. D., DRAIGOR, D. A., SEMIROG-ORLIK, V. N., *et al*, *Povyshenie ekspluatatsionnoy nadezhnosti detaley mashin* (Enhancement of Performance Reliability of Machine Parts), Mashgiz, Moscow, 1960.
111. GRINBERG, A. A., MEKHTIEV, R. R., RYVKIN, S. M., *et al*, Absorption of Laser Radiation and Disintegration of Semiconductors, *Fiz. tverd. tela*, 1967, 9, No. 5, 1390-1397.
112. PROKHOROV, A. M., Laser Generation with Instantaneous Effect of the Q-Factor, *Radiotekhnika i elektronika*, 1963, 8, No. 6, 1073-1074.
113. ASHKINADZE, B. M., VLADIMIROV, V. I., LIKHACHEV, V. A., *et al*, Disintegration of Transparent Insulators by High-Power Laser Radiation, *Zhurn. eksper. i teoret. fiz.*, 1966, 50, No. 5, 1187-1201.
114. RAROV, N. N., UGLOV, A. A., ZUEV, I. V., Effect of Parameters of Heat Source on Shape of Cavity and Deformation of Liquid Phase Surface, *Dokl. AN SSSR*, 1972, 207, No. 1, 83-85.
115. RIKMAN, E. P., KOKORA, A. N., ZHUKOV, A. A., Distribution of Alloying Elements in the Laser-Affected Zone of Stainless Steel, in *Voprosy metallovedeniya i prochnosti metallov i splavov* (Problems of Physical Metallurgy and Strength of Metals and Alloys), Prioksk. iz., 1968, 182-185.
116. RUBINSHTEIN, L. I., *Problema Stefana* (Stefan's Problem), Zvaigzne, Riga, 1967.
117. RYKALIN, N. N., *Raschety teplovykh protsessov pri svarke lazerom* (Theory of Thermal Processes in Welding), Mashgiz, Moscow, 1951.
118. RYKALIN, N. N., UGLOV, A. A., Heating of Thin Sheets in Laser Welding, *Dokl. AN SSSR*, 1965, 165, No. 2, 319-322.

119. RYKALIN, N. N., UGLOV, A. A., Contribution of Bulk Vaporization in Heating of Metal by Laser Radiation, *Fizika i khimiya obrabotki materialov*, 1970, No. 2, 33-36.
120. RYKALIN, N. N., UGLOV, A. A., Bulk Vaporization in Metals due to Laser Radiation, *Teplofiz. vysok. temper.*, 1971, 9, No. 3, 575—582.
121. RYKALIN, N. N., UGLOV, A. A., Theory of Heating of Metals with the CO_2 Laser Radiation, *Svarochnoe proizvodstvo*, 1973, No. 7, 1-3.
122. RYKALIN, N. N., UGLOV, A. A., Temperature Fields in the Absorbing Media with Local Heat Sources, *Fizika i khimiya obrabotki materialov*, 1967, No. 5, 11-14.
123. RYKALIN, N. N., UGLOV, A. A., Some Features of Heat Sources in Laser Welding, *Svarochnoe proizvodstvo*, 1969, No. 11, 1-5.
124. RYKALIN, N. N., UGLOV, A. A., ZUEV, I. V., Key-Hole Melting of Metals by Electron Beam, in *Elektronnoluchevaya svarka*, Proc. of II All-Union Conf. on Electron-Beam Welding, October, 1970, 19-31.
125. RYKALIN, N. N., UGLOV, A. A., KOKORA, A. N., Laser Effects in Iron Alloys, *Fizika i khimiya obrabotki materialov*, 1972, No. 6, 14-21.
126. RYKALIN, N. N., UGLOV, A. A., KOKORA, A. N., Effects Produced by High-Power CO_2 Laser Radiation on Iron and Its Alloys, *Fizika i khimiya obrabotki materialov*, 1974, No. 1, 3-9.
127. RYKALIN, N. N., UGLOV, A. A., MAKAROV, N. I. Effect of Spike Repetition Rate in Laser Pulse on Laser Heating of Metal Sheets, *Dokl. AN SSSR*, 1967, 174, No. 5, 1068-1071.
128. RYKALIN, N. N., UGLOV, A. A., MAKAROV, N. I., Effect of Non-Ideal Thermal Contact in Laser Welding of Different Materials, *Dokl. AN SSSR*, 1967, 174, No. 4, 824-827.
129. RYKALIN, N. N., UGLOV, A. A., MAKAROV, N. I., Heating of a Two-Layer Plate Welded by Laser Beam, *Dokl. AN SSSR*, 1966, 169, No. 3, 565.
130. SAMOKHIN, A. A., Contribution of Superheating into Advance Vaporization, *Kratk. soob. po fiz.* (*FIAN*), 1973, No. 4, 7-10.
131. SEDOV, L. I. *Metody podobiya i razmernosti v mekhanike* (Similarity Method and Dimensional Method in Mechanics), Nauka, Moscow, 1967.
132. KASATOCHKIN, V. I., KAZAKOV, M. E., SAVRANSKY, V. V., *et al*, Synthesis of a New Allotropic Carbon Modification from Graphite, *Dokl. AN SSSR*, 1971, 201, No. 5, 1104-1105.
133. SOKOLOV, A. V., *Opticheskie svoystva metallov* (Optical Properties of Metals), Fizmatgiz, Moscow, 1961.
134. SOKOLOV, A. K., ZUBAREV, T. N., Temporal Variation in Intensity and Energy of Stimulated Radiation of Ruby Laser with Spherical Mirrors, *Fizika tverd. tela*, 1964, 6, No. 9, 2590-2598.
135. KONYUKHOV, V. K., KULEVSKY, L. A., PROKHOROV, A. M., *et al*, Spectrum of Ruby Laser with External Spherical Mirrors, *Dokl. AN SSSR*, 1964, 158, No. 4, 824-826.
136. BASOV, N. G., MARKIN, E. P., ORAEVSKY, A. N., *et al*, Stimulation of Chemical Reactions by Infrared Laser Radiation, *ZhETF pis'ma*, 1971, 14, is. 4, 251-253.
137. BASOV, N. G., BELENOV, E. M., MARKIN, E. P., *et al*, Stimulation of Chemical Reactions by Laser Radiation, *Zhurn. eksper. i teor. fiz.*, 1973, 64, is. 2, 485-497.
138. BELYANIN, V. A., ZHUKOV, A. A., KOKORA, A. N., *et al*, Structure and Hardness of Surface Layers of Steel Irradiated by Laser Beam, *Fizika i khimiya obrabotki materialov*, 1967, No. 2, 115-116.
139. TAIPOV, R. A. Laser Welding of Circuit Interconnections, *Trudy LITMO*, Problems of Quantum Electronics, 1968, is. 65, 107-113.
140. TAIPOV, R. A., LIBENSON, M. N., STEPANOV, L. N., Laser Welding of Thin Plates Coated with Low-Melting Metal, *Voprosy radioelektroniki, Tekhnologiya proizvodstva i oborudovanie*, 1968, is. 1, 49-55.

141. TAL'ROZE, V. L., BARASHEV, P. P., Chemical Effect of Laser Radiation, *Zhurn. vsesoyuzn. khim. ob. im. D. I. Mendeleeva*, 1973, No. 1, 15-33.
142. VEIKO, V. P., KOTOV, G. A., LIBENSON, M. N., Thermochemical Effect of Laser Radiation, *Dokl. AN SSSR*, 1973, 208, No. 3, 587-590.
143. TIKHONOV, A. N., SAMARSKY, A. A., *Uravneniya matematicheskoy fiziki* (Equations of Mathematical Physics), Nauka, Moscow, 1966.
144. TURYGIN, I. A., *Prikladnaya optika* (Applied Optics), vol. 2, Mashinostroenie, Moscow, 1966.
145. UGLOV, A. A., KOKORA, A. N., KRISHTAL, M. A., Distributions of Some Elements in the Laser-Affected Zones of Alloys, *Fizika i khimiya obrabotki materialov*, 1973, No. 4, 3-7.
146. UGLOV, A. A., KOKORA, A. N., OREKHOV, M. V., Motion of Molten Material in Laser Drilling, *Fizika i khimiya obrabotki materialov*, 1974, No. 2.
147. IOFFE, A. I., MEL'NIKOV, N. A., NAUGOL'NYKH, K. A., *et al*, Shock Wave with Optical Breakthrough in Water, *Zhurn. prikl. mekhan. i tekhn. fiz.*, 1970, No. 3, 125-127.
148. ULYAKOV, P. I., Some Features of Disintegration of Solids with Laser Radiation, *Zhurn. eksper. i teor. fiz.*, 1967, 52, is. 3, 820-831.
149. BASOV, N. G., BOIKO, V. A., KROKHIN, O. N., *et al*, Decrease in Reflectivity of Solid Surface Under Laser Radiation, *Zhurn. tekh. fiz.*, 1968, 38, No. 11, 1973-1977.
150. BASOV, N. G., MARKIN, E. P., ORAEVSKY, A. N., *et al*, Photochemical Effect of Infrared Laser Radiation, *Dokl. AN SSSR*, 1971, 198, No. 5, 1043-1045.
151. FRENKEL', Ya. I., *Sobranie izbrannykh trudov.* (Collected Works), vol. III, AN SSSR, Moscow, 1959.
152. HAUFFE, K., *Reaktionen in und an festen Stoffen*, Berlin u. a., Springer Verl., 1955.
153. HEARD, H., *Laser Parameter Measurements Handbook*, John Wiley & Sons, NY-L, 1968.
154. SCHAWLOW, A. L. Laser and Coherent Light, *American Scientist*, 67, v. 55, No. 3, 197-220.
155. SHLYKOV, Yu. P., GANIN, E. A., *Kontaktny teploobmen* (Contact Heat Exchange), Gosenergoizdat, Moscow, 1963.
156. YAMPOL'SKY, Yu. P., Chemical Effects of Laser Radiation, *Uspekhi khimii*, 1972, v. XLI, is. 6, 1111-1135.
157. ANDERSON, J. E., JACKSON, J. E., An Evaluation of Pulsed Laser Welding in *Laser Welding and Machining* (Seminar on new industrial technologies at the Pennsylvania State University, June 27 to July 2, 1965), 1966, 4-25.
158. ARATA, Y., MIYAMOTO, I. CO_2 Laser Absorption Characteristics of Metal, *Trans. J. Weld. Soc.*, 1972, v. 3, No. 1, Report 2, 1-11.
159. ARATA, Y., MIYAMOTO, I. Metal Heating by Laser Beam, *Trans. J. Weld. Soc.*, 1972, v. 3, No. 1, Report 3, 1-43.
160. ARATA, Y., MIYAMOTO, I., KUBOTA, M., Some Fundamental Properties of High Power CW Laser Beam as a Heat Source, *IIW Doçument*, 14/4/69, March 1969, 1-21.
161. BAN, V. S., KNOX, B. E., Mass-Spectrometric Study of the Laser Induced Vaporization of Compounds of Bismuth with the Elements of Group VIa, *J. Chem. Phys.*, 1970, v. 52, No. 1, 243-247.
162. BAN, V. S., KNOX, B. E., Mass-Spectrometric Study of the Laser Induced Vaporization of Compounds of Arsenic and Antimony with the Elements of Group VIa, *J. Chem. Phys.*, 1970, v. 52, No. 1, 248-253.
163. BARNES, P. A., RIECKHOFF, K. E., Laser Induced Underwater Sparks, *Appl Phys. Lett.*, 1968, v. 13, No. 8, 282-284.
164. BASTON, T. J., BOWDEN, F. P., Localized Damage of Metal Crystals by Laser Irradiation, *Nature*, 1968, v. 218, No. 5137, 150-152.

165. BOC Laser Systems and Applications, *Machinery and Production Eng.*, 1970, 24 June, v. 116, No. 3006, 977-978.
166. BOYER, G., HURIET, J. R., LAMOUROUX, B., Micro-machining by Beams of Coherent Light, *Optics and Laser Technology*, 1970, v. 2, No. 4, 196-199.
167. CARMONE, E. F., CARRIERA, E. M., PROCHASKA, C. J., Photographic Studies of Laser-Induced Pressure Impulses in Liquids, *Appl. Phys. Lett.*, 1967, v. 11, No. 2, 64-66.
168. COLLINS, R. J., NELSON, D. F., SCHAWLOW, A. L., GARRETT, C. G. B. and KAISER, W., Coherence, Narrowing, Directionality, and Relaxation Oscillations in the Light Emission from Ruby, *Phys. Rev. Lett.*, 1960, v. 5, No. 7, 303-305.
169. COOK, J. J., FLOWERS, W. L., ARNOLD, C. B., Measurement of Laser Output by Light Pressure, *Proc. IRE*, 1962, July, v. 50, No. 7, 1963.
170. DAMON, E. K., FLYNN, J. T., A Liquid Calorimeter for High-Energy Lasers, *Appl. Optics*, 1963, v. 2, No. 2, 163-164.
171. EHLER, A. W., Plasma Formed by a Laser Pulse on a Tungsten Target, *J. Appl. Phys.*, 1966, v. 37, No. 13, 4962-4966.
172. *Electron Beam and Laser Beam Technology* (*Advances in Electronics and Electron Physics*, Suppl. No. 4, ed. L. Marton, A. B. El-Karen), Acad. Press, New York, London, 1968.
173. EVTUHOV, V., NEELAND, J. K., Observations Relating to the Transverse and Longitudinal Modes of a Ruby Laser, *Appl. Optics*, 1962, v. 1, 4, 517.
174. FAIRBANKS, R. H., ADAMS, C. M., Laser Beam Fusion Welding, *Weld. J.*, 1964, v. 43, No. 3, 97-S 102-S.
175. GAY, B., Shorter Communication. Comparison of Methods for Solution of the Heat Conduction Equation with a Radiation Boundary Condition, *Intern. J. Heat and Mass Transfer*, 1965, v. 8, No. 3, 507-508.
176. GONSALVES, J. N., DULEY, W. W., Interaction of CO_2 Laser Radiation with Solids and Drilling of Thin Metallic Sheets. I. Drilling of the Metallic Sheets, *Canad. J. Physics*, 1971, v. 49, No. 13, 1708-1713.
177. GROH, G. J., Vacuum Deposition of Thin Films by Means of a CO_2 Laser, *Appl. Phys.*, 1968, v. 39, No. 12, 5804-5805.
178. GUENOT, R., RACINET, J., Heat Conduction Problems in Laser Welding, *Brit. Weld. J.*, 1967, v. 14, No. 8, 427-435.
179. *A Guide to the Laser* (ed. D. Fishlock), Macdonald, London, 1967.
180. HIGUCHI, T., MYAZAWA, T., YOSHIDA, H., OKUDA, T., Application of Laser to Drilling and Cutting, *Mitsubishi Denki*, 1971, v. 45, No. 10, 1298-1303.
181. KARN, F. S., FRIEDEL, R. A., SHARKEY, A. G., Distribution of Gaseous Products from Laser Pyrolysis of Coals of Various Ranks, *Carbon*, 1967, v. 5, 25-32.
182. KARN, F. S., SHARKEY, A. G., Pyrolysis of Coal Materials by Laser Irradiation, *Fuel*, 1968, v. 47, 193-195.
183. LINLOR, W. I., Some Properties of Plasma Produced by Laser Giant Pulse, *Phys. Rev. Lett.*, 1964, v. 12, No. 14, 383-385.
184. LANDAU, H. G. Heat Conduction in a Melting Solid, *Quart. Appl. Math.*, 1950, v. 8, No. 1, 81-94.
185. LOCKE, E., HOAG, E., HELLA, R., Deep Penetration Welding with High Power CO_2 Lasers, *Weld. J.*, 1972, v. 51, No. 5, 245-S 249-S.
186. LUNAU, F. W., PAINE, E. W., CO_2-Laser Cutting, *Weld and Metal Fabric*, 1969, v. 37, No. 1, 9-14.
187. COHEN, M. J., EPERSON, J. Ph., *Application of Laser to Microelectronic Fabrication. Electron Beam and Laser Beam Technology*. Edited by L. Marton and A. B. El-Karen, Acad. Press, New York and London, 1969.
188. MURPHY, R. J., RITTER, G. J., Laser-Induced Damage in Copper Crystals, *J. Appl. Phys. Lett.*, 1966, v. 9, No. 7, 272-273.

189. MURPHY, R. J., RITTER, G. J. Laser Induced Fracture and Graphitization of Diamond, *Industrial Diamond Review*, 1970, v. 30, No. 353, 143-152.
190. NEUMAN, F., Momentum Transfer and Cratering. Effects Produced by Giant Laser Pulses, *Appl. Phys. Lett.*, 1964, v. 4, No. 9, 167-169.
191. *News in Focus. Laser Focus.* 1968, May, 12-19.
192. PHILLIP, H., SMITH. Electronic Applications of the Smith Chart, *Microwaves*, 1970, v. 9, No. 3, 143-144.
193. PLATTE, W. N., SMITH, J. F., Laser Techniques for Metals Joining, *Weld. J.*, 1963, v. 42, 11, 481—S—489—S.
194. READY, J. F., Development of Plume of Material Vaporized by Giant-Pulse Laser, *Appl. Phys. Lett.*, 1963, v. 3, No. 1, 11-13.
195. READY, J. F., Effects Due to Absorption of Laser Radiation, *J. Appl. Phys.*, 1965, v. 36, No. 2, 462-468.
196. READY, J. F., Interaction of High Power Laser Radiation with Absorbing Surfaces, *Proc. Nation. Electronics Conf.*, Chicago, 1964, v. 20, 67-71.
197. ROUSSEAU, D. L., LEROI, G. E., LINK, G. L., Two-Photon Induced Chemical Reaction AgCl, *J. Chem. Phys.*, 1965, v. 42, 4048-4049.
198. PÖSS, D., *Ein optischer rubin — molekularverstarker mit noher modenselektion.* Frequenz, 1963, Bd. 17, No. 2, 61-63.
199. SCHAEFFER, R., PEARSON, R. K., Reaction of Carbon Vapor Produced by Laser Evaporation of Graphite, *J. Am. Chem. Soc.*, 1969, v. 91, No. 8, 2153-2154.
200. SCRASE, T. E., NC Drilling with the Laser, *Metalworking Production* 1970, v. 114, No. 2, 60-61.
201. SHARKEY, A. G., SHULTZ, J. L., FRIEDEL, R. A., Gases from Flash and Laser Irradiation of Coal, *Coal Science*, Advan. Chem. Ser., 1966, v. 55, 643-649.
202. SHULZ, J. L., SHARKEY, A. G., Gases from Laser Irradiation of Coal: Effect of Argon, Nitrogen and Other Atmospheres, *Carbon*, 1967, v. 5, No. 1, 57-59.
203. SPEISER, S., KIMEL, S., Laser-Induced Photolysis of Iodoform, *J. Chem. Phys.*, 1969, v. 51, No. 12, 5614-5620.
204. STIMLER, M., SLAWSKI, Z. I., Measurement of Pulsed Laser Energy by Photon — Momentum Transfer, *Bull. Amer. Phys. Soc.*, 1963, ser. II, v. 8, No. 8, 623.
205. SULLIVAN, A. B. J., HOULDCROFT, P. T., Gas-Jet Laser Cutting, *Brit. Weld. J.*, 1967, August, v. 14, No. 8, 443-445.
206. SWIFT-HOOK, D. T., GICK, A. E. F., Penetration Welding with Laser, *Marchwood Engineering Lab.*, R/M/No. 637, June 1972.
207. TAKI, Ko., KIM, R. H., NAMBA, S., Laser-Induced Chemical Reactions. I. Reactions of Carbon Vapour Produced by a Pulsed Ruby Laser, *Bull. Chem. Soc.*, Japan, 1969, v. 42, No. 3, 823-825.
208. TAKI, Ko., KIM, R. H., NAMBA, S., Laser-Induced Chemical Reactions. IV. Reactions of Carbon Vapour with Hydrogen at Various Pressures, *Bull. Chem. Soc.*, Japan, 1970, v. 43, No. 10, 3278-3280.
209. TIFFANY, W. B., Selective Photochemistry of Bromine Using a Ruby Laser, *J. Chem. Phys.*, 1968, v. 48, No. 7, 3019-3031.
210. VERDIECK, J. F., MAU, A. W. H., Some Laser-Induced Reactions in the Gas Phase, *Chem. Commun.*, 1969, D, No. 5, 226.
211. VOGEL, K., BACKLUND, P., Application of Electron and Optical Microscopy in Studying Laser-Irradiated Metal Surfaces, *J. Appl. Physics*, 1965, v. 36, No. 12, 3697-3701.
212. WEBSTER, J. M., Welding at High Speed with the CO_2 Laser, *Metal Progress*, 1970, v. 98, No. 5, 59-61.
213. ZHUKOV, A., KRISHTAL, M. A., KOKORA, A. N., SNEZNOJ, R. L., La Baikovite une nouvelle structure dans les alliages fer-carbone, *Mem. Sc. Rev. Metallurgie*, 1972, v. 69, No. 3, 211-214.

APPENDIX

1

IMPROVEMENT OF LASER DRILLING TECHNIQUES

Since the laser techniques are widely used for drilling of various materials the ever present problem is to increase the depth and productivity of laser drilling and to improve its quality and accuracy.

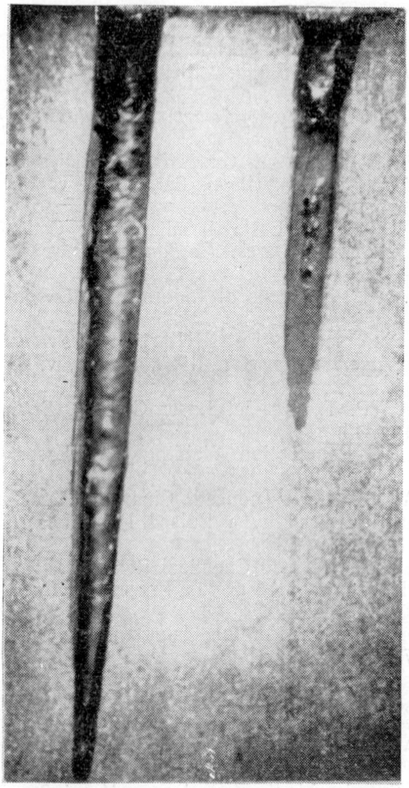

FIG. 1.1. Polished sections of steel 45 (5 mm thick) drilled with *Q*-switched (left) and free-generation (right) laser pulses

As the results of laser drilling are determined by a wide range of factors it is not easy to formulate the optimum parameters of drilling and sometimes, as shown in Chapter 4, such attempts produce contradictory results. For instance, to increase the depth of laser drilling the laser pulse duration must be increased since the laser power density should not be more than 10^8 W/cm^2 as shown in [1]. But an increase in the pulse duration results in a larger volume of the molten material leading to poorer accuracy and so on.

In this case we should determine the optimum ranges of pulse duration and power density at which the volume of the molten material is minimized, and to increase the depth of drilling we should use the multipulse technique described in Chapter 4.

Moreover, the requirements for the laser radiation parameters (for instance, pulse duration) can be too complicated to be realized in laser industrial equipment.

One of the techniques for regulating laser radiation is to use various types of modulation [2, 3]. Here we shall describe a mechanical modulator which increases the efficiency of laser drilling.

The modulator is an opaque disk rotated by an electric motor. There are 7-mm holes with the spacing of 14 mm at the edge of the disk. Basov *et al* [2] were the first to modulate the laser beam with a rotating disk placed in the focus of a telescopic system inside the resonator. In contrast to their modulator, we place the rotating disk inside a spherical resonator which contains only a 7×130 mm neodymium glass active element. The disk rotation rate varies from 12,000 to 16,000 rpm.

With this modulation technique and the optical pumping period of 1.5 ms the laser generates 10 pulses of 40 μs each whose intensity is much higher than the free-generation pulse intensity. It is of interest to note that a modulated laser can generate a pulse when the intensity of the pumping light at the end of the pulse is less than the threshold intensity for free generation. When the active element axis is in the centre of a hole in the stationary disk the free-generation pulse energy is 10 J. The total energy of ten *Q*-switched pulses is 7 J.

Despite a loss of radiation energy modulation of laser pulses results in increased depth of laser drilling. Figure 1.1 presents the profiles of through (left) and blind (right) holes drilled by *Q*-switched and free-generation laser pulses with the same number of pumping light pulses. The hole drilled by *Q*-switched pulses is twice as deep as the hole drilled by free-generation pulses while the radiation energies delivered to the holes for one pumping light pulse are 2.8 and 4 J, respectively (the optical system losses amount to 60%).

Another advantage of the above mechanical modulator is the option to generate pulses with the duration less than 150 μs for long pumping light pulses (0.5-1.5 ms). This is important since it has been shown in reference [4] that the lifetime of the pumping lamp is considerably decreased by generating pumping pulses with the duration less than 150 μs. Moreover, as the laser generates several pulses for one pumping pulse the pumping lamp can be used to produce more holes during its lifetime.

Thus, a simple modulation technique gives rise to a significant improvement of laser drilling efficiency.

REFERENCES

1. KRISHTAL, M. A., ZHUKOV, A. A., KOKORA, A. N., *Struktura i svoistva splavov obrabotannykh izlucheniem lazera* (Structure and Properties of Alloys Treated with Laser Beam), Metallurgiya, Moscow, 1973.
2. BASOV, N. G., ZUEV, V. S., KRYUKOV, P. G., Increase in Power of Pulsed Ruby Laser by *Q*-Switching, *Zhurn. eks. i teor. fiz.*, 1962, 43, is. 1, 353.
3. SOKOLOV, B. M., VAKS, E. D., SOKOLOVA, N. E., BELOVA, G. N., Ultrasound Synchronization of Ruby Laser Radiation, *Elektrofiz. i elektrokhim. metody obrabotki*, 1971, is. 10, 8.
4. ASTAKHOV, A. A., BASHAROV, Yu. P., GUREVICH, E. B., TARKHOV, G. N., FUKSMAN, V. M., SHKLOVSKAYA, T. N., Application of Glass Lasers for Industrial Watch Jewel Machining, *Elektrofiz. i elektrokhim. metody obrabotki*, 1972, is. 6, 23.

APPENDIX 2
EFFECT OF ATMOSPHERE ON LASER MACHINING

Typically, laser machining is done in the air under atmospheric pressure, which can result in economic and technological advantages of this technique compared to electron-beam machining.

However, the oxidizing atmosphere can sometimes be a negative factor in laser machining, especially when the irradiated material can be highly oxidized even during short laser pulses. In such cases a protective neutral-gas atmosphere with a pressure somewhat higher than the normal atmospheric pressure should be used. In Chapter 3 we have discussed some cases of application of the protective atmosphere in laser welding noting that in laser welding of tungsten-rhenium thermocouples the quality of welds is improved by the argon blow through the welding zone.

When the CO_2 laser radiation with the power up to tens of kilowatts is used for welding of metals it can give rise to a cloud of ionized gas near the metal surface, which absorbs a significant part of radiation thus hindering deep-penetration welding. As discussed in Chapter 5 this effect has a radiation power threshold. For instance, for CO_2 lasers with the wavelength of 10.6 μm the threshold power for developing and maintaining the ionized gas cloud is about 2 kW. Physically, the development of the plasma cloud is caused by the breakdown of the air in the metal vapour near the surface, which results in the development of the region with the temperature of $(1\text{-}2)\times 10^4$ K that is higher than the initial temperature of the metal vapour. Raizer [1] has estimated the power parameters of the laser radiation which produces and maintains the plasma cloud.

A plasma cloud can be produced also in gas far from any solid surface; the plasma is first generated by focussing a high-intensity plasma beam ($q \approx 10^{10}$ W/cm^2 and more) and then the plasma cloud is maintained by laser radiation with a lower intensity. Theoretical analysis [1] and experimental results show that it is feasible to develop an optical plasmatron which can be used for machining purposes.

The laser-generated plasma at the surface of the irradiated material has an effect that worsens the precision of machining as it results in the energy transfer through an area larger than the focussed spot.

The authors of references [2-5] have studied the breakdown of the air by the pulsed CO_2 laser radiation at the normal pressure and the breakdown of the gas in the vicinity of metallic surfaces caused by the continuous-wave laser radiation. It should be noted that in the above experiments the radiation power density is considerably lower than the threshold for the breakdown of the dust-free air at the normal pressure far from the material surface. The breakdown threshold for the clean air at the normal pressure for the laser radiation with $\lambda = 10.6$ μm is somewhat lower than that for the radiation with $\lambda = 1.06$ μm, which is about 10^{10} W/cm^2. For both wavelengths the air breakdown threshold in the vicinity of metal surfaces is lower by a few orders of magnitude.

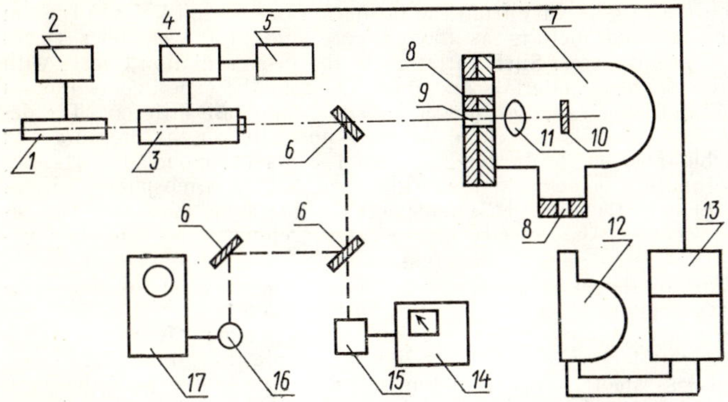

FIG. 2.1. Experimental apparatus for laser irradiation of materials at high pressures

1, 2—helium-neon lasers with control board; *3, 4, 5*—neodymium lasers with control board and power supply; *6*—splitting and deflecting optical elements; *7*—high-pressure chamber; *8*—observation windows; *9*—entrance window for laser beam; *10*—target; *11*—focussing lens; *12, 13*—high-speed photorecorders with control board; *14, 15*—energy-measuring instruments; *16*—photocell; *17*—oscillograph

An increase in the gas pressure generally decreases its laser breakdown threshold. The experiments on the effect of the gas pressure on the breakdown threshold have produced different results owing to the differences in the purity of the gases, the temporal and spatial structures of the laser radiation and the conditions of the experiments. The minimum radiation power threshold for nitrogen is about 5×10^8 W/cm^2 [6] at the gas pressure of about 100 at; the threshold increases with a decrease or increase in the gas pressure.

The breakdown of the gas at an elevated or normal pressure in the vicinity of the laser-irradiated material can be used for changing the local properties of the material such as increasing the microhardness or the content of gases, for carrying out chemical heterogeneous high-temperature reactions and so on.

Let us discuss the results produced by irradiating the materials with pulsed neodymium laser beam ($\lambda = 1.06$ μm) in the atmosphere of argon, nitrogen and helium at the pressures from 1 to 140 at [7-10].

Figure 2.1 presents a schematic diagram of the apparatus used in these experiments. The laser beam passes through a window into a high-pressure chamber and is focussed by a lens on the targets of various materials. The laser radiation parameters are monitored by special instruments. The laser radiation processes are recorded with a high-speed photorecorder using continuous and periodic recording modes. The recordings yield data on the kinetics of the breakdown of the gas near the target surface and the parameters of the process (the geometry of the plasma cloud, the rate of growth of the plasma cloud along the laser beam, variation in these parameters with increasing gas pressure and so on). The targets in the chamber can be moved without loss of air tightness so that the results can be highly reproducible. Studies have been also made of the combined effects of laser radiation, thermal radiation from the plasma cloud and its direct effect on the target.

The experimental results show that the laser radiation effects on targets are determined by the pressure and the properties of gas, the energy parameters of radiation and the properties of the target material. When the laser pulse duration is about 1 ms, the focussed spot radius r_f is between 0.03 and 0.05 cm and the radiation power density is between 10^6 and 10^7 W/cm^2 the radiation effects on the targets vary from the formation of through holes in 1-2-mm plates to heating without melting as the gas pressure in the chamber varies from 1 to tens of atmospheres. Such variation in the laser radiation effects with pressure is found for most materials and gases used in the experiments though the gas pressure and laser radiation power density ranges are different. The development kinetics of the plasma cloud depends not only on the pressure and gas being used but also on the target material. The plasma cloud moves along the laser beam, and its velocity in the argon and helium atmospheres differs noticeably from that in the nitrogen atmosphere for the same materials and laser radiation parameters. The velocity of the plasma cloud along the beam varies during the pulse reaching its maximum at the midpoint of a bell-shaped pulse. The lifetime of the plasma cloud depends significantly on the gas pressure and the target material. As a rule, the plasma lifetime increases with increasing gas pressure. At pressures over 50 at the plasma lifetime is twice the laser pulse duration. For instance, the lifetime of plasma produced by laser radiation of stainless steel 1X18H9T in the nitrogen atmosphere with the pressure of 120 at is about 1.2 ms. Laser radiation of high-temperature ceramics and textolite (laminate plastic) in the high-pressure nitrogen atmosphere produces the plasma cloud with the lifetime of more than 4 ms, that is considerably longer than the laser pulse duration. Such large lifetimes of the plasma clouds produced by laser radiation of textolite and ceramics are suggested to be due to burning of the particles ejected from the irradiated zone of the material [9]. This is caused by the fact that the time τ_b in which the particle reaches the temperature of burning varies as the squared particle radius r_0 and as a logarithm of the plasma temperature (a weak dependence) [11]:

$$\tau_b = \frac{r_0^2}{a k_1^2} \ln \frac{2(T_p - T_0)}{T_p - T_b}$$

where T_b is the burning point for the particle, T_p is the plasma temperature, a is the thermal diffusivity of the material of the particle, $k_1 = \sqrt{3Bi}$, $Bi = \alpha r_0/\lambda$, α is the coefficient of the heat transfer between the particle and the plasma, λ is the heat conductivity of the particle material, $\alpha \approx \lambda_p/r_0$ and λ_p is the heat conductivity of plasma.

The process of "burning-out" of the particles, which depends on their sizes, seems to provide the additional energy for maintaining the plasma cloud when the laser pulse has ended.

When the pressure of the gas atmosphere is high the plasma cloud near the irradiated target develops in tens of microseconds after the beginning of the laser pulse.

The development of the breakdown and subsequent motion of the plasma cloud along the laser beam depend on the properties of the gas and the parameters of the laser radiation. The process of breakdown of the gas near the target caused by the laser radiation with the power density of 10^6-10^7 W/cm^2 takes a few tens of microseconds and depends on the temperature of the target surface and the gas pressure. Laser radiation with development of plasma clouds produces slight melting of the target surface and negligible ejection of the material from such targets as molybdenum and stainless steel. Under the experimental conditions (the distance from the entrance window to the target is 20 cm and the pressure is of tens of atmospheres) the absorption of laser radiation in the unheated gas is shown to be negligible.

The breakdown of gas in the vicinity of the target can be initiated by thermal electrons emitted from the target surface and by the vapour of the target material vaporized when the target surface is heated to about the melting point. It is rather difficult to distinguish beween these two mechanisms producing electrons initiating the plasma cloud near the surface.

Rykalin *et al* [8] have estimated the amount of thermal electrons emitted during the period in which the target surface is heated to the melting point. These estimates show that about 10^{13} electrons are emitted during this period into the region near the surface with the volume equal to the focal volume; this amount is sufficient for the development of electron avalanche. The vaporization of the material naturally only improves the conditions for the development of the avalanche. The contribution of the high pressure to the development of the avalanche is that it confines the electrons to a region of the order of the focal volume since the diffusion coefficient of electrons varies as the reciprocal of pressure. The development of the electron avalanche in the electromagnetic radiation field gives rise to a plasma cloud which moves along the laser beam according to the "slow burning" mechanism [1]. This mechanism is given such a name because of the analogy between it and the process of burning when the velocity of the burning front varies from one to tens of metres per second.

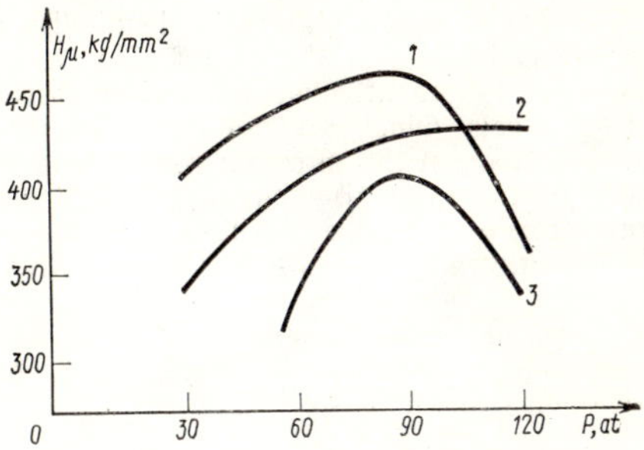

FIG. 2.2. Variation in microhardness of steel (Ст. 3) at different gas (nitrogen) pressures and laser pulse energies

1—8 J; *2*—12 J; *3*—4 J. The pulse duration is 1 ms

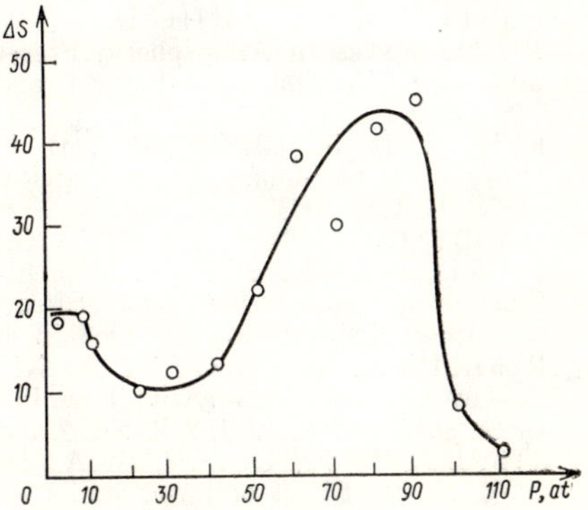

FIG. 2.3. Relative nitrogen content in laser-affected zone as a function of nitrogen atmosphere pressure

The time needed to reach the melting point of the material can be found exactly by solving the nonlinear heat conductivity equations taking into account the temperature dependence of the thermal parameters and the absorption coefficient. Analytical solution of this problem is very difficult therefore the analytical approximation technique has been used in [12]. The approximate results for the nonlinear problem differ significantly (by a factor of about 2) from the time calculated for the linear problem.

In the nitrogen atmosphere the plasma cloud moves along the laser beam owing to a relatively slow process, that is heat transfer from plasma via heat conduction, while in the argon and helium atmospheres where the velocity of the plasma cloud is higher the energy transfer occurs via a faster mechanism [1]. The differences found for the molecular gas (nitrogen) and monoatomic gases (argon and helium) are not limited to the velocities of the plasma clouds but concern also the breakdown thresholds. The ionization potential of helium (24.5 eV) is noticeably higher than that of nitrogen (14.58 eV) or argon (15.7 eV) so that the breakdown of helium occurs at higher pressures and radiation power densities. The temperature of the plasma cloud in helium is higher than that in nitrogen or argon that is why with the same radiation power densities the erosion of the target surface is greater in the helium atmosphere.

The nitrogen plasma is chemically active giving rise to nitrides in the laser-affected zone of the target. This is indicated by the increase in the microhardness in the affected zone of the irradiated materials, including chromium-nickel steels, which are not hardened when they are irradiated by laser beams in the air or in the nitrogen atmosphere at the normal pressure. The microhardness of the laser-affected zones in steels varies with the gas pressure. Laser radiation of steel Cт. 3 (0.3% C) shows that the microhardness as a function of the nitrogen pressure has a saturation point [13] (Fig. 2.2). The saturation point is determined by the compétition between two opposite effects. The nitrogen content in the laser-affected zone increases from the initial level along with variation in the microhardness in this zone. Figure 2.3 presents the plot of the nitrogen content in steel.

REFERENCES

1. RAIZER, Yu. P., *Lazernaya iskra i rasprostranenie razryadov* (Laser Spark and Propagation of Discharges), Nauka, Moscow, 1974.
2. STEGMAN, R. L., SCHRIEMPF, J. T., HETTCHE, L. R., Experimental Studies of Laser-Supported Absorption Waves with 5-ms Pulses of 10.6 μ Radiation, *J. Appl. Phys.*, 1973, 44, No. 8, 3675-3681.
3. FOWLER, M. C., SMITH, D. C., Ignition and Maintenance of Subsonic Plasma Waves in Atmospheric Pressure Air by CW CO_2 Laser Radiation and Their Effect on Laser Beam Propagation, *J. Appl. Phys.*, 1975, 46, No. 1, 138-150.
4. KEEFER, D. R., HENRIKSEN, B. B., BRAERMAN, W. F., Experimental Study of a Stationary Laser-Sustained Air Plasma, *J. Appl. Phys.*, 1975, 46, No. 3, 1080-1083.
5. MOODY, C. D., Maintenance of a Gas Breakdown in Argon Using 10.6-μ CW Radiation, *J. Appl. Phys.*, 1975, 46, No. 6, 2475-2482.
6. *Deistvie lazernogo izlucheniya* (Effect of Laser Radiation), Collected Articles, Mir Publishers, Moscow, 1968, 37-42.
7. RYKALIN, N. N., UGLOV, A. A., NIZAMETDINOV, M. M., Breakdown of Gas by Laser Radiation with Low Power Densities and at High Gas Pressures, *DAN SSSR*, 1974, 218, No. 2, 330-331.
8. RYKALIN, N. N., UGLOV, A. A., DOBROVOL'SKY, I. P., NIZAMETDINOV, M. M., Effect of Laser Radiation on Metals at High Ambient Pressures, *Kvantovaya elektronika*, 1974, 1, No. 9, 1208.
9. RYKALIN, N. N., UGLOV, A. A., NIZAMETDINOV, M. M., Interaction Between Laser Radiation and Metals at High Ambient Pressures, *Zhurn. eksper. i teor. fiz.*, 1975, 69, No. 2(8), 722.
10. RYKALIN, N. N., UGLOV, A. A., NIZAMETDINOV, M. M., Effect of Laser Radiation on Materials in a Wide Range of Argon Pressures, *Kvantovaya elektronika*, 1978, 5. No. 1.
11. LOKHOV, Yu. N., PETRUNICHEV, V. A., UGLOV, A. A., SHVYRKOVA, I. I., Heating and Vaporization of Particles in Low-Temperature Plasma Jet, *Fizika i khimiya obrabotki materialov*, 1974, No. 6, 52-56.
12. RYKALIN, N. N., UGLOV, A. A., NIZAMETDINOV, M. M., Analysis of Laser Heating of Materials Taking into Account Temperature Dependence of Thermal Parameters, *Kvantovaya elektronika*, 1977, 4, No. 7.
13. UGLOV, A. A., NIZAMETDINOV, M. M., Laser Hardening of Steels at High Ambient Pressures, *Fizika i khimiya obrabotki materialov*, 1977, No. 2.

APPENDIX 3
LASER SURFACE HARDENING OF MACHINE PARTS

An integral component of the technological progress is the development of new technologies for improving the reliability, quality and durability of industrial products. One of such new highly promising technologies is the surface hardening of metal parts with laser beams; it opens up a wide range of industrial applications, for instance, in production of tools.

The laser surface hardening is distinguished by the following features:

1. Localized regions (by the depth and the area) of the material can be hardened by the laser beam to a hardness higher (by 15-20%) than that produced by other available thermal hardening techniques while in the rest of the material the properties remain unchanged.

2. Simple optical systems can be used for localized hardening of the difficult-to-treat cavities, holes, etc.

3. Large surfaces can be treated to obtain "spot" hardening which does not produce a continuous brittle surface layer which tends to warp and crack.

4. The microroughness of the hardened surfaces can be produced when necessary.

5. Specified surface properties (physical, mechanical, chemical, etc.) can be obtained by laser alloying of the surface with various elements.

6. The treated workpieces have neither deformations nor buckling owing to the localized character of heat treatment so that practically no finishing is needed.

7. Laser surface hardening can be readily automated even for workpieces of complicated shapes.

The lack of specialized laser equipment limited the wide use of laser hardening. Now such equipment is commercially available [1], and new machines are being developed which make use of powerful continuous-wave CO_2 lasers. Therefore, there is a need now for analysis of specific laser hardening applications and recommendations for optimum treatment regimes. This need has been met by a number of authors (see, for instance, [2-10]).

We shall discuss below the pulsed laser surface hardening of the head bars of the carding machines [10] and the cutting edge of the punching dies [3].

Laser surface hardening of cast-iron workpieces. The machine parts of textile machines, such as the head bars of carding machines, have to meet contradictory requirements. On the one hand, the heads are traditionally cast of low-hardness cast iron to reduce their buckling in operation while on the other hand it is desirable to improve the wear resistance of the bearing surfaces of the head bars which are worn down by dry friction with steel. Thus, the problem is to harden these surfaces in the workpiece of СЧ 18-36 cast iron with the pearlite-ferrite structure.

In this case the high-frequency currents are not suitable for hardening as they result in increased internal stresses in the workpieces with the accompanying buckling tendency. The advantage of the laser surface hardening is that it

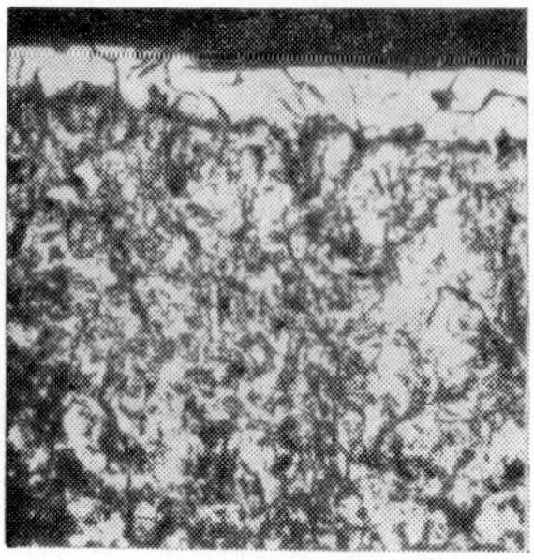

FIG. 3.1. Microstructure of laser-irradiated zone in cast iron (×115)

minimizes the depth of the heated layer [11]. This is convenient for head bars since their permissible wear is 300 μm. Another advantage of laser hardening of grey iron is that a thin molten surface layer freezes as fine-grained ledeburite which has a high wear resistance similar to hardening martensite.

We have experiments with the cast irons containing 3.4-3.6 wt% C, 2.2-2.6 wt% Si, 0.4-0.5 wt % Mn, 0.10-0.25 wt % P and up to 0.14 wt % S. The laser hardening is done with 6-ms laser pulses (50-200 J) and 4-ms pulses (10 J). By changing the conditions of focussing the power density on the workpiece surface is varied from 10^4 to 10^5 W/cm^2. The irradiated "spots" on the surface are located side by side with slight overlapping.

The optimum laser power density (about 8×10^4 W/cm^2) produces a hardened layer 300 μm deep with minimized depth of the molten layer. Owing to the surface tension effects the thin molten layer forms individual protuberances 30 μm high with a very smooth profile (Fig. 3.1). The metal surface has a "paving-stones" structure with wear-resistant ledeburite protuberances which have optimum antifriction, antitear and antiwear properties.

The metallographic analysis of polished specimens shows that directly beneath the cementite layer there is a layer with partially spheroidized graphite inclusions. X-ray diffraction patterns show that in the laser hardened layer the amount of the α-phase decreases, the cementite amount increases (in addition to the pearlite cementite there appears the eutectic cementite due to chilling) and there is a large amount of the γ-phase. The lattice constant of austenite is higher than that of the carbon-free γ-iron; it is 3.60 Å corresponding to 1.2% C in austenite [12, 13].

An unsuccessful attempt has been made to decompose the residual austenite by tempering at 150 and 200 C for 2-4 h. Tempering at 400 C results in the disappearance of austenite but also decreases the hardness of the hardened layer owing to the decomposition of martensite. The initial iron hardness is $HV_5 = 262$ and the hardness in the treated zone is 666-677.

The wear resistance is measured with a test apparatus containing a rotating disk and a stationary block. The steel disk with the thickness of 10 mm and diameter of 50 mm has the hardness $R_C = 45$-48 and the surface finish of class 8. The block is made of the casts for the head bars with the working surface of 0.2×1.0 cm corresponding to the contact area of 0.2 cm^2 after wearing-in. The linear speed of the disk is 0.65 m/s. The pressures are 15, 30, 50, 75 and 100 kgf/cm^2. The duration of the test at 15 and 30 kgf/cm^2 is 40 min and at 50-100 kgf/cm^2, 20 min. The results for two specimens are averaged (see Table 3.1).

TABLE 3.1

State of specimens	Wear, mg/cm^2 m (per 1000 m) at pressures (kg/cm^2)				
	15	30	50	75	100
Without laser hardening	250	283	300	315	338
After laser hardening	52	170	220	288	330

FIG. 3.2. Punching die after laser hardening

The arrow indicates the nonworking surface of the die irradiated with 12 laser pulses (four pulses for each of the three rows) for which the X-ray diffraction pattern was recorded

FIG. 3.3. Microstructure of laser rehardened surface layer (×340)

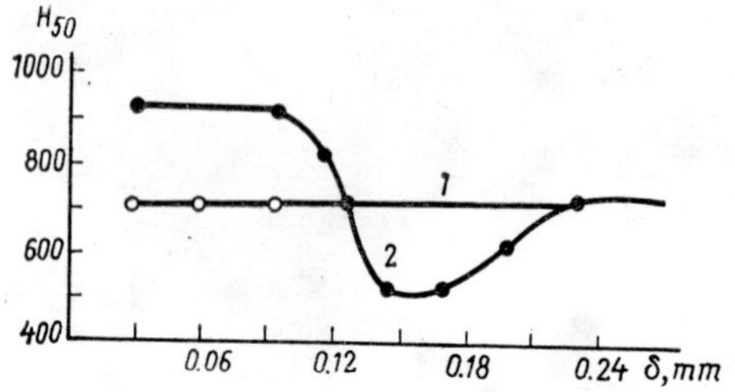

FIG. 3.4. Microhardness of surface layers of die

1—after conventional hardening; 2—after laser rehardening (δ is the layer depth)

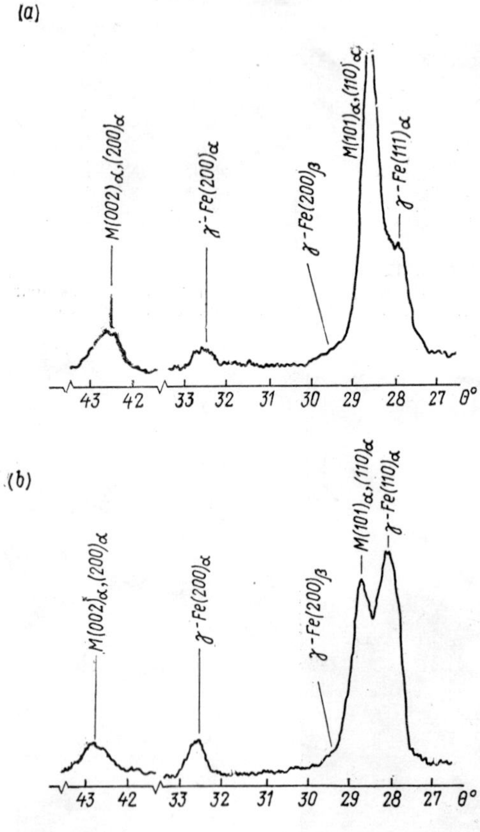

FIG 3.5. X-ray diffraction patterns for working surface of die

(a) after conventional hardening; (b) after laser rehardening

As the pressure increases the wear of the laser-hardened specimens approaches that of the unhardened specimens owing to the removal of the hardened layer soon after the start of the test. At low pressures the difference in wear between the specimens is very considerable.

Since the head bars of the carding machines operate at low pressures the laser hardening of their bearing surfaces proves to be useful. The wear resistance of the laser-hardened head bars is 3-4 times that of the unhardened bars.

Laser hardening of the cutting edge of punching dies. It has been shown [11] that pulsed laser hardening of the previously hardened tool steels increases the hardness of the metal. Here we shall discuss the properties of steel after conventional and laser hardening and the increase in the wear resistance of a tool after such treatment.

A punching die of steel У8 is first hardened in oil at 790 C and tempered at 150 C. The laser hardening of the cutting edge of the die is done with 30-J laser pulses with the duration of 6 ms and the pulse rate of 0.1 Hz. The laser focussed spot on the surface of the die has a size of 4 mm, and the spots are overlapped by about 50%. Figure 3.2 presents a photograph of the working surface of the die after laser hardening. Figure 3.3 shows the microstructure of the laser-hardened layer and the transition zone.

To visualize the structure of the hardened layer a prolonged etching with alcohol nitric acid solution is required. There is a large amount of the residual austenite and fine-needle martensite. Despite a large amount of the residual austenite in the hardened layer its microhardness is significantly higher than the microhardness of the die after conventional hardening (Fig. 3.4). X-ray diffraction studies have also revealed a sharp increase in the residual austenite after laser rehardening of steel (Fig. 3.5).

The paradox of increasing microhardness of the rehardened steel surface despite an increased residual austenite content in it should be analyzed in more detail. The following tentative concepts can be put forward.

1. Martensite produced at high cooling rates (about 10^6 K/s) has an elevated hardness as is well known from the experience with steel hardening by high-frequency currents.

2. Residual austenite produced at high heating rates (about 10^6 K/s) is supersaturated with lattice defects and has an elevated hardness. At the heating rates used there can occur the reverse diffusionless martensite transformation so that austenite inherits the defect structure of martensite with practically no weakening and recrystallization effects during heating.

3. Accumulation of lattice defects in martensite during two direct and one reverse martensite transformations. It is very interesting that the hardened layer contains no defects of hardening such as cracks and microcracks though the hardening conditions are severe without tempering of the hardened zones. This can be attributed to the formation of the tempering zones of two types:

1. A tempered substrate under the hardened layer which separates this layer from the previously hardened base metal. The microhardness of this tempered zone is $H_{50} = 500$ (see Fig. 3.3).

2. The curved tempered zones on the hardened surface of the die which are formed when the focussed spots overlap. These zones amount to not more than 5% of the total volume of the hardened layer. Therefore, they do not make a great contribution to the durability of the die but they have an important useful role as the zone of relaxation of the hardening stresses, that is, as a sort of expansion joint.

The durability of the laser-hardened dies is tested under industrial conditions. The steel of the dies (grade X18H9T) is hardened and tempered to Rc 58. The cutting edges of the dies are hardened using two laser techniques (see Fig. 3.5: *I* — edge hardening, *II* — side hardening). The laser treatment increases the microhardness from 630 to 1200—1350 H_{50}. The durability of the conventionally-hardened dies is 4.5-5 thousand operations to the first grinding. Laser hardening increases the durability to 10,000 (technique I) and 14,000 (technique II).

REFERENCES

1. *Elektronnaya prom.*, 1976, No. 1, 72.
2. MAAS, M. *Elektronika*, 1976, No. 9, 26.
3. ZHUKOV, A. A., KOKORA, A. N., ZARYA, A. N., *et al*, *Fiz. i khim. obrabotki mater.*, 1977, No. 1, 141.
4. VELIKIKH, V. S., GONCHARENKO, V. P., KARTAVTSEV, V. S., *Tekhnologiya i organizatsiya proizvodstva*, 1976, No. 2, 45.
5. VELIKIKH, V. S., *et al*, *Elektronnaya prom.*, 1976, No. 1, 64.
6. *Metal Progress*, 1976, v. 107, No. 5, 61.
7. SEAMAN, F., GNANAMUTHU, D., *Metal Progress*, 1976, Vol. 108, No. 3, 67.
8. *Industrial Heating*, 1973, v. 40, No. 5, 868.
9. *Industrial Heating*, 1974, v. 41, No. 7, 17.
10. ZHUKOV, A. A., KOKORA, A. N., TIMONICH, D. D., *et al*, *Fizika i khim. obrabot. mater.*, 1975, v. II, No. 1, 84.
11. KRISHTAL, M. A., ZHUKOV, A. A., KOKORA, A. N., *Struktura i svoistva splavov obrabotannykh izlucheniem lazera* (Structure and Properties of Laser-Treated Alloys), Metallurgiya, Moscow, 1973.
12. WEVER, F., ROSE, A., PETER, W., *Arch. Eisenhüttenwesen*, 1950, b. 21, 367.
13. ROBERTS, C., *J. Metals*, 1953, v. 5, 203.

APPENDIX 4

CONICAL LENS FOR LASER DRILLING OF LARGE-DIAMETER HOLES

Laser drilling was one of the first industrial uses of lasers. Now the typical holes drilled by single- and multipulse techniques and with the use of lenses and objectives for focussing the beam have the diameters of 0.005-1.0 mm and the depths of 5 mm or more, with the depth-diameter ratio of more than 20. The surface finish of the holes is of class 8 or 10 and the highest accuracy (obtained with the multipulse technique) is 2-3 μm.

The holes with diameters more than 1 mm can be drilled with lens focussing systems only if the laser pulse power is high, which is a factor limiting the industrial use of laser drilling.

A technique has been developed for drilling holes of large size and complex profile in which the workpiece is moved so that the focus of the laser optical system traçes the hole profile. However, this technique has the following significant drawbacks:

1. Motion of the workpiece makes the drilling machine more complicated and impedes automation of drilling.

2. The edges of the drilled hole are irregular to not less than the focussed spot radius. The irregularities may be decreased by significant overlapping focussed spots or decreasing the focussed spot radius. Both methods result in decreased drilling accuracy and productivity.

Another technique for drilling large-diameter holes is to shape the laser beam according to the size and shape of the hole to be drilled.

This can be done by using a conical lens (Fig. 4.1). The optical system shown in Fig. 4.1 consists of laser *1*, a parallel-sided plate *2*, a conical lens *3*, objective *4* with variable focal distance and a workpiece holder *5* which can move along the optical axis of the system.

The laser beam is focussed on the surface of the workpiece by the conical lens *3* and objective *4*, which determines the shape of the beam. The size of the hole is determined by the conical surface *3*; the distance r from the optical axis to a point of the drilling profile in the plane passing through the optical axis is related to the angle α between the generatrix of the conical surface and the optical axis in the same plane:

$$r = F\,[\arcsin(n\cos\alpha) + \alpha - \pi/2] \tag{4.1}$$

where F is the focal distance of the objective and n is the index of refraction of the material of the conical lens.

If the angle α is close to $\pi/2$ Eq. (4.1) can be reduced to

$$r = F\,(\pi/2 - \alpha)\,(n - 1) \tag{4.2}$$

The size of the drilling profile can be varied in a wide range by varying the focal distance of objective *4*. Thus, the optical system can be readily adjusted for drilling holes of various sizes but similar shapes.

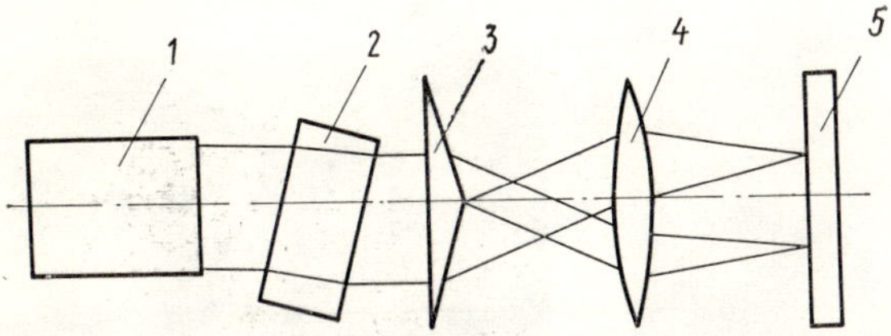

FIG. 4.1. Optical diagram of laser drilling technique using conical lens

By rotating the parallel-sided plate *2* around the optical axis at various angles to it we can gradually vary the distribution of the power density over the hole profile; this feature can be useful if the material has varying thickness, density or other parameters.

For laser drilling of deep cylindrical holes the main parameters of the optical system can be determined in the following way. The deviation of the hole profile from the cylindrical shape increases with increasing angle of convergence of the beam after the conical lens. It can be seen from the geometrical considerations that the deviation of the hole profile from the cylindrical shape decreases with decreasing hole depth and increasing focal distance of the spherical lens.

To increase the accuracy of drilling and the radiation power density the cut width δ ($\delta = \beta F$ where β is the actual divergence of the laser beam) must be as small as possible. Thus we satisfy two contradictory requirements:

1. The focal distance F must be as large as possible to provide for the cylindricity of the hole.

2. The focal distance F must be as small as possible to provide for the minimum width of the drilling profile.

Nevertheless, the focal distance F should be minimized with specified diameter, depth and deviation from cylindricity of the hole to provide for a good drilling accuracy and to increase its efficiency.

Figure 4.2 shows round holes in the 0.3-mm stainless steel specimen drilled using this technique with a single ruby laser pulse. The pulse energy is 50 J and the duration is 1.3 ms. The hole diameter is about 3.0 mm. Figure 4.3*a* shows a round 2.5-mm hole in the 0.3-mm stainless steel specimen drilled by six split pulses from a neodymium laser. The technique of laser pulse splitting with a rotating disk is described in detail in reference [1]. In our experiment the out-

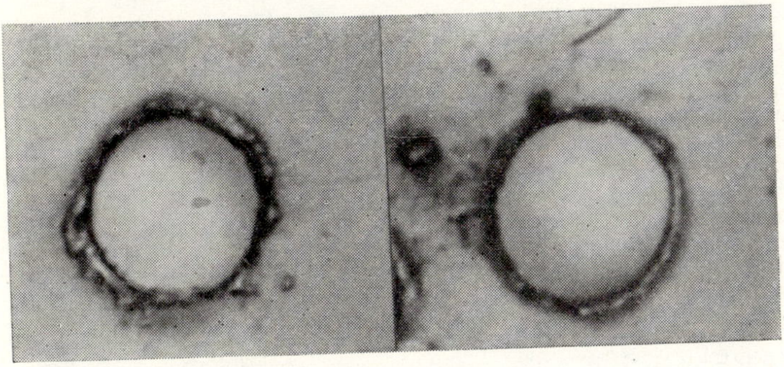

FIG. 4.2. Holes in stainless steel drilled by single ruby laser pulses using conical lens

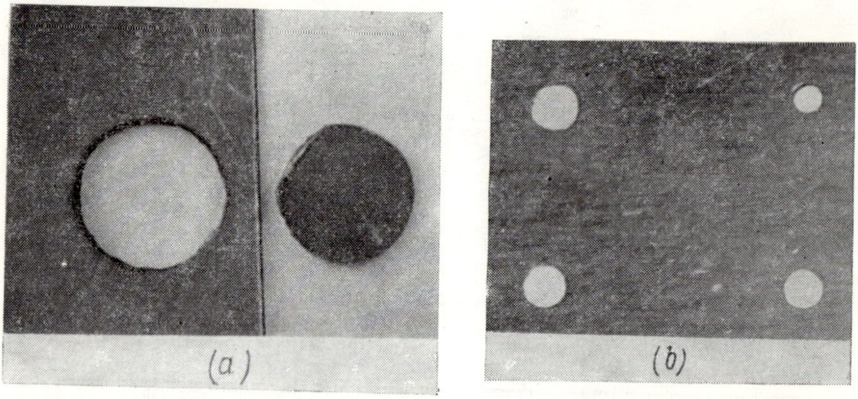

FIG. 4.3. Hole in stainless steel drilled by six split neodymium laser pulses using conical lens (shown beside is the central region of the profile which is not irradiated) (*a*); holes in the same specimen drilled by six split pulses without conical lens (*b*)

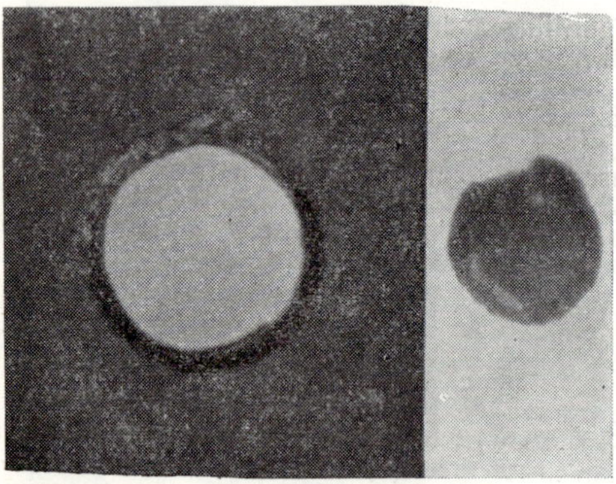

FIG. 4.4. Hole in ferrite drilled by 15 split neodymium laser pulses using conical lens (shown beside is the central region of the profile which is not irradiated)

put pulse energy is 16 J and the pulse duration is 1.0 ms. Each free-generation laser pulse is split into 10 pulses with the duration of 40 μs each. Figure 4.3*b* shows the holes drilled in the 0.3-mm stainless steel specimen with the same laser pulses but without the conical lens. The diameter of these holes is a third of those in Fig. 4.3*a*.

A similar technique is used for drilling 1.2-mm ferrite specimens with 15 split laser pulses producing 2.5-mm holes (Fig. 4.4).

This technique can be used for drilling plain quality holes in various materials including those difficult to work. It can be widely used for drilling systems of large-diameter holes (over 1 mm) in various filters, meshes, etc.

REFERENCES

1. UGLOV, A. A., OREKHOV, M. V., KOKORA, A. N., Improvement of Laser Drilling Efficiency, *Fiz. i khim. obrabotki materialov*, 1976, No. 4, 135.

APPENDIX 5

CUTTING OF NONMETALLIC MATERIALS WITH CO_2 LASER BEAM

Cutting of materials seems to be one of the most promising technical applications of the continuous-wave laser radiation; the most useful laser type for that purpose is the CO_2 laser which has a high efficiency (15-20%). This is confirmed by the fact that the number of patents issued for laser cutting applications throughout the world is higher than for any other laser technological uses.

Moreover, the 10.6-μm laser radiation is readily absorbed by various nonmetallic materials including natural organic materials (leather, wood, linen and cotton fabrics, etc.), man-made materials (artificial leather, synthetic fabrics, etc.) and so on.

The laser cutting has the following advantages compared with other techniques:

(1) narrow cutting zone;

(2) very small size of the heat-affected zone;

(3) small mechanical effects in the material owing to the lack of cutting force;

(4) possibility of cutting the materials along lines of practically any shape in two or more planes;

(5) possibility of automatic cutting at high rate and accuracy.

Meanwhile, the pattern-cutting techniques used now in the light industries are mostly mechanical ones (shearing, punching out, etc.) which have low productivities and necessitate additional labour-consuming operations (preparation of decks, patterning, etc.) [1] while the cutting tools being used hinder automation of the cutting processes. Automation of the pattern-cutting of fabrics depends on the development of a cutting tool which would meet the following requirements: minimization of the mechanical effect on the fabric, simple control of the tool with a remote control option, independence of the width of the cut on the rate of cutting and so on [2].

This is why wide-ranging studies have been carried out both in this country and abroad to develop essentially new nonmechanical non-contact cutting techniques such as the electric spark, microplasma, water jet and other techniques.

Analysis shows that the electric-spark cutting technique has a limited range of cutting rates (1-2.5 cm/s) and can be used for a small number of materials [2, 3].

The microplasma technique seems to be more promising for pattern cutting of various fabrics (see, for instance [2, 4, 5]). When this technique is used for cutting fibre materials the quality of the cut practically does not depend on the cutting rate which is determined to a large extent by the number of layers in the material and its heat conductivity. This technique produces comparatively good results with textiles. Table 5.1 presents the maximum cutting rates for some fabric types and normal operation of the plasmatron [2]. However, this technique does not produce a narrow cut in artificial leather.

TABLE 5.1. Maximum Rates of Cutting of Fabrics with Microplasma Technique

Fabric	Thickness, mm	Cutting rate, cm/s
Rayon	0.19	30
Rayon	0.40	21
Crepe de Chine	0.21	13
Nylon (parallel to the mesh columns)	0.25	30
Nylon (perpendicular to the mesh columns)	0.25	24
Cotton	0.21	16
Wool fabric (for dresses)	0.80	5.5
Wool fabric (for suits)	0.69	3
Terylene-wool fabric	0.54	6

From the reported results (see, for instance [2, 4, 5]) it can be seen that pattern cutting of fabrics with the microplasma technique can be done mostly for one layer of the material, and the cutting rate is comparatively low.

Another cutting technique consists in cutting materials with a high-pressure water jet. The width of the cut made by this technique has been reported to be not more than 0.5 mm [5]. To provide for such a narrow cut the water contains 1.5-2.0% organic additives, and no spraying occurs when the water is ejected from a 0.5-mm nozzle at the pressure of 28,000 kgf/cm^2. It has been reported that for the cut width of about 0.5 mm and the water flow rate of 100 l/h the rate of cutting of 20 layers of artificial leather can be as high as 10 m/s, which is impossible to realize in the design of the cutting machines [5]. Practically, all the types of artificial leather can be cut with this technique but the total thickness of the material cannot be more than 25 mm. Moreover, high pressures used in the machine necessitate using special materials, elements and components to provide for reliable operation.

The laser cutting technique, compared to other cutting techniques, is more efficient; it can be used for pattern cutting of nonmetallic materials (light industry) in single layers with high cutting rates and in multilayer stacks. The laser cutting machines for pattern cutting of single- and multilayer materials have been developed in many countries (USSR, France, USA, West Germany, etc.) [6—11].

However, there are very few reports on the CO_2 laser cutting technique and on the pattern cutting processes in light industry. Bazyuk *et al* [1] have described the single-mode CO_2 laser cutting of various fabrics. However, they did not use blowing with neutral or other gases, they did not cut artificial leather, the power of laser radiation was 60 W and so on.

Here we report the results on the gas-laser cutting of the footwear and leather materials using an automated programmed laser cutting machine (Fig. 5.1) (the detailed description see in [7]).

The installation comprises a two-coordinate pattern cutting machine with automatic feeding of material and removal of the cut elements, the CO_2 laser with the optical system for transmission of the laser beam to the cutting zone and the digital control unit.

The continuous-wave CO_2 laser has the output power instability of not more than 10% (at the output of about 800 W).

Figure 5.1 presents the optical diagram of the programmed laser cutting installation. The laser beam is reflected from the mirror fitted on the cutting tool and focussed by a lens of KCl or other optical material (the focal distance is 240-280 mm) on the material surface.

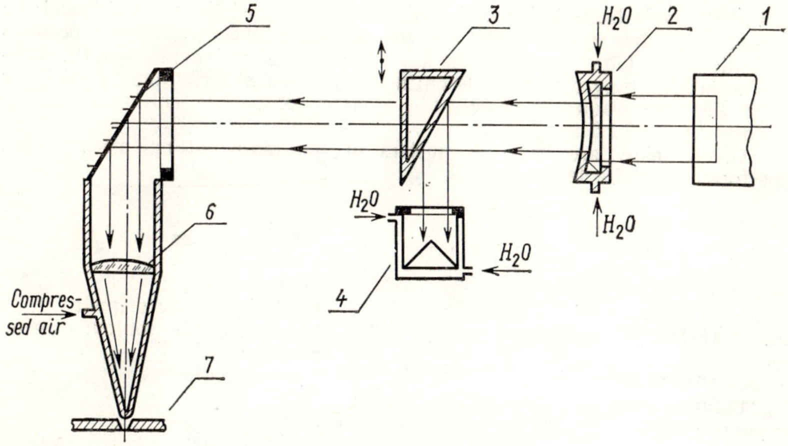

FIG. 5.1. Optical diagram of laser cutting apparatus

1—CO_2 laser; 2—diaphragm; 3—shutter with reflecting plate; 4—radiation absorber; 5—rotating mirror; 6—focussing lens; 7—material

The optical cutting tool of the mirror-lens type has a split casing. A rotating mirror is attached to the internal sloping wall in the upper part of the cutter at an angle of 45° to the beam. The mirror surface is polished to the 0.5 diffraction ring accuracy and coated with a 100-μm copper monoxide layer. The mirror has a reflectivity of about 98% at 10.6 μm and the optical (working) diameter of 80 mm. The mirror coating can withstand the radiation power densities of 10^4 W/cm^2 or more and the temperatures up to 100 C. The compressed air at the pressure of 1-2 at is fed into the cutting zone through the output nozzle of the cutter.

A metal plane shutter with an electromagnetic drive is used for cutting off the laser beam which is reflected from a reflecting plate on the shutter (for instance, of polished aluminium) to an absorber when this is needed for alignment of the optical system or for technological purposes. The stepwise regulation of the laser beam power is done with a diaphragm consisting of a water-cooled metal cylinder and the polished aluminium half-spheres attached to its end with the apertures from 5 to 30 mm. The optical cutter is attached to a carriage sliding along the portal guide with a rack-and-gear drive. Two brackets attach the portal to the machine frame. The table slides along the rails between these brackets. The rails of the table are perpendicular to the guides of the carriage. The cutting table is made of cellular aluminium and driven by a rack-and-gear drive.

The drives of the carriage and the table are identical and comprise hydraulic stepping motors, double-reduction gears of longitudinal and transverse feed and the hydraulic unit.

A vacuum system is used for fixing the material on the table and for removing the thermal decomposition products.

A programmed digital electronic unit controls cutting making it possible to cut complex curvilinear patterns.

Programmed cutting of the material along a given contour is done by simultaneous coordinated motion of the carriage and the table along the coordinate axes X and Z. An important feature of laser cutting technique is the constant speed of the laser beam (that is, the rate of cutting) as it is a major factor providing for a good quality of cutting (small cut width). When the speed of the laser beam is lower than the optimum speed the cut width and the damaged zone size are increased, and the surface of the cutting table becomes superheated. Too high a cutting rate can result in incomplete cutting of the material. The laser cutting rate depends on the kind of the material and its thickness and on the laser radiation power. The accuracy of pattern cutting should be not worse than 0.8-1.0 mm.

Given below are basic parameters of the laser cutting machine and its performance characteristics.

Size of material:	
width	1600 mm, max
length	1500 mm, max
Linear speed of cutting	up to 42 m/min
Motor drive	electrohydraulic stepwise
Control system	two-coordinate contour digital programmed
Interpolation type	linear circular
Programming code	UC^0
Working frequency	2000 Hz
Frequency	8000 Hz, max
Data input	8-channel punched tape with even parity check
Laser type	CO_2 laser
Wavelength	10.6 μm
Divergence of laser beam	2×10^{-3} rad
Generation mode	continuous-wave
Multimode operation output power	800 W
Cooling	forced, liquid
Cooling water flow rate	not more than 5 l/min
Power input of machine	30 kW
Power supply voltage	380 V (three-phase)
Machine mass	2400 kg

The CO_2 laser used in the machine emits the cutting beam with the power of 500 W or more. The KCl lenses in the cutter focus the beam with the divergence of about 2×10^{-3} rad into an 0.5-mm spot.

The effect of the composition and pressure of the gas mixture on the cutting performance. The experiments have been carried out with the artificial leather having a fabric base (Hungary) 1.5 mm thick and with the 4.5-mm thick leather-like rubber.

The depth, width and quality of cutting have been studied as functions of the composition and pressure of the gas mixture. The laser beam power in the cutting zone is 350 W. The optical system with the focal distance of 240 mm focusses the beam on the surface of the material. The cutting speed for the artificial leather is 16.8 m/min and for the rubber 1.5 m/min. When the air pressure is increased from 1 to 4 at and higher the depth of the cut increases to a through cut: 1 at — 1.0 mm, 2 at — 1.2 mm and 3 at — 1.4 mm. The cut width at the depth of $^1/_4$ of the material thickness varies from 0.19 to 0.52 mm increasing with pressure.

When the material is cut in the nitrogen atmosphere the other conditions being the same, the cut width increases from 0.28 mm at 1 at to 0.52 mm at 5 at while the through cut is obtained at a pressure of 2 at or higher. The difference in the cutting depth between the air and nitrogen atmosphere can be attributed to different conditions of radiation screening over the surface of the material.

When rubber is cut in the air the depth of the cut varies from 2.35 mm at 1 at to 2.8 mm at 5 at.

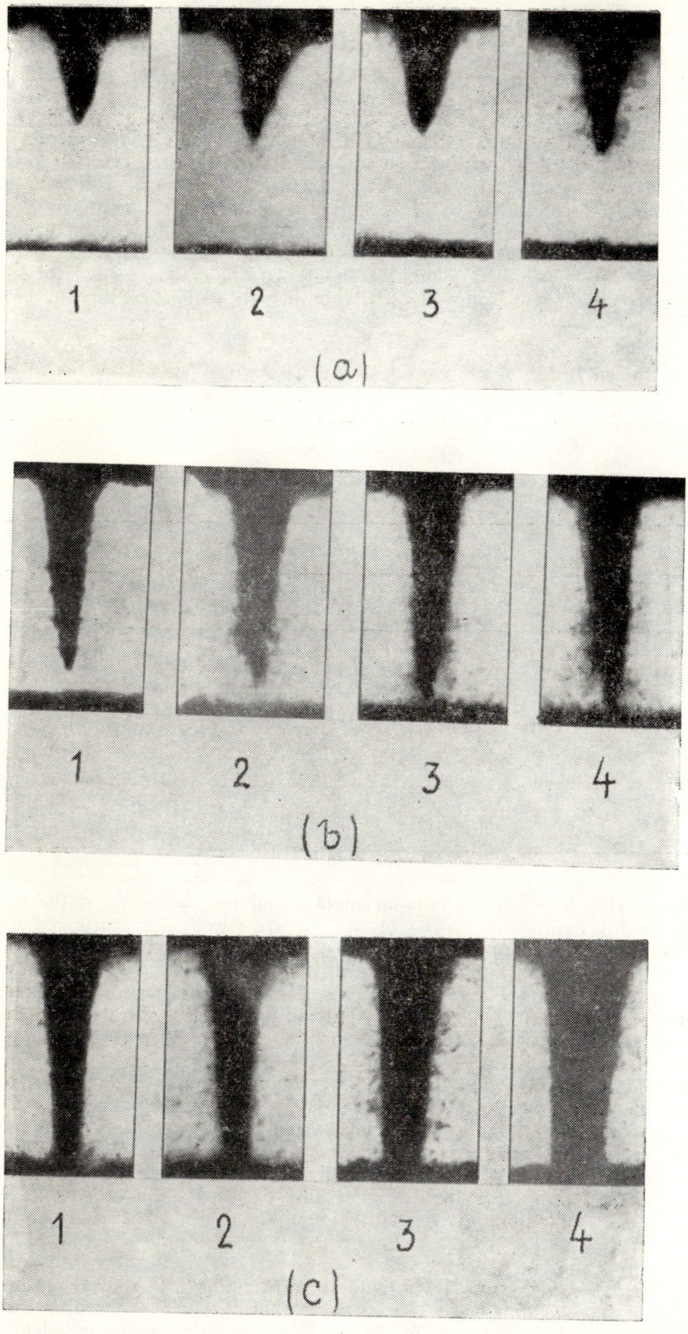

FIG. 5.2. Cross sections of cuts in 2.0-mm carboard made by laser beam in compressed air at various speeds and pressures

Cutting speed: (*a*) 8 m/min; (*b*) 5 m/min, (*c*) 4 m/min; air pressure: *1*—0 at; *2*—1 at; *3*—2 at; *4*—4 at;

We have also studied laser cutting of a 2-mm cardboard in the air at various pressures (0-4 at) and at various cutting speeds (4, 5 and 8 m/min). Figure 5.2 presents the cross sections of the cuts, and Table 5.2 presents their parameters. The width of the cuts (both from front and back sides) increases with the air pressure, and the increase is higher at lower cutting speeds. For instance, at the cutting speed of 8 m/min the width of the front cut increases from 0.42 mm at zero pressure to 0.5 mm at 4 at while at 4 m/min the respective cut widths are 0.54 and 0.79 mm. It should be noted that there is practically no heat-affected zone along the cut and that the edges of the cut are smooth.

TABLE 5.2. Variation in Parameters of Cuts in Cardboard with Air Pressure and Cutting Speed *

Air pressure, at	Cutting speed, m/min							
	8			5			4	
	cut width, mm		depth, mm	cut width, mm		depth, mm	cut width, mm	
	front	back		front	back		front	back
0	0.42	The material	0.85	0.48	The material	1.76	0.54	0.24
1	0.48	is not cut	0.97	0.54	is not cut	1.82	0.6	0.3
2	0.48	completely	0.97	0.54	completely	1.94	0.67	0.36
4	0.50		1.1	0.54	0.09	× **	0.79	0.48

* The laser power is 350 W, the beam is focussed with a 280-mm KCl lens on the surface of the material.

** The material is cut completely, the depth of the cut is 2.0 mm.

The relationship between the cut parameters and the focussing of the laser beam. The width, depth and profile of the cut are significantly affected by the focussing of the laser beam, that is, by the focal distance of the optical system and by the position of its focal plane with respect to the surface of the material. In the experiments the focal plane is positioned both above the surface of the material (6 mm) and inside it (10 mm) from the surface. The laser power is 500 W, the air pressure is 2 at and the cutting speed is 10 m/min. The material is the leather-like rubber.

Figure 5.3 presents various focussing arrangements and the profiles of the respective cuts. It may be seen that when the focal plane is above the surface of the material the width of the cuts is considerably larger than when the focal plane is below the surface (1.17 mm and 0.82 mm, respectively). When the focal plane is shifted deeper into the material the cut width increases to 1.1 mm (here $\Delta F = 10$ mm, ΔF being the difference between the actual distance and the focal one). The minimum cut width is obtained when the focal plane is below the surface ($\Delta F = +3$ mm), and when the focal plane is shifted in any of the directions the cut width increases since the size of the focussed spot on the material surface increases. The maximum cut depth (2.48 mm) is found when the focal plane is 3 mm below the surface, that is, for the minimum cut width. When the focal plane is below or above the optimum level the cut depth decreases to 1.03 mm ($\Delta F = -6$ mm) and 1.45 mm ($\Delta F = +10$ mm). It is of interest that the profiles of the cuts are similar irrespective of the focussing arrangements.

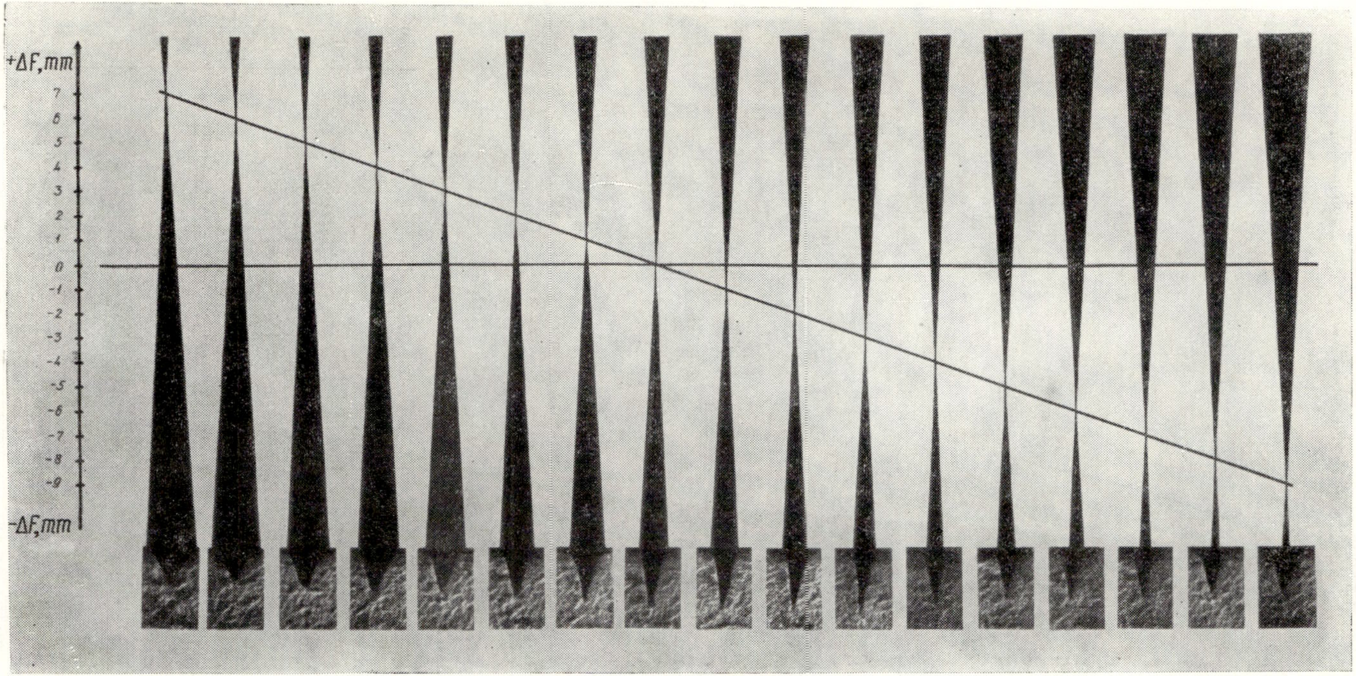

FIG. 5.3. Width and depth of cuts in rubber as a function of laser beam focussing
Laser power=500 W; cutting speed=10 m/min

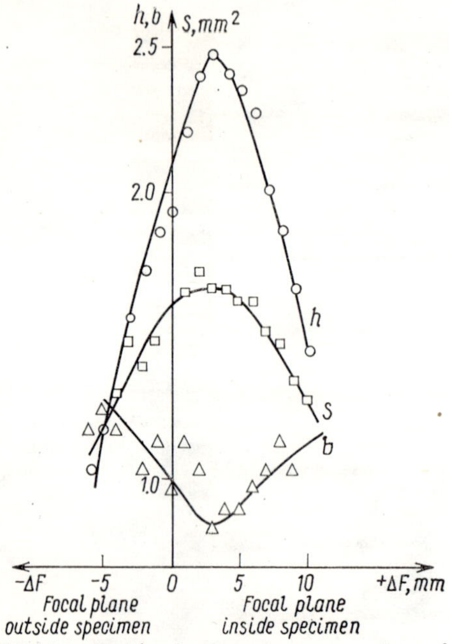

FIG. 5.4. Width, depth and cross section of cuts in rubber as a function of laser beam focussing

Table 5.3 presents the parameters of the cuts (width, depth and cross section) as functions of ΔF, and the respective plots are given in Fig. 5.4.

Thus, the experiments show that the maximum depth and the minimum width of the cut are obtained when the focal plane of the focussing system is 3 mm below the material surface (for a 280-mm lens).

The effect of the laser power on the cut width and the cutting speed. Ten materials used in the footwear and leather industries have been studied. The air pressure is 2 at, the laser power is 500 W (240-mm KCl lens) and the focal plane of the optical system is below the material surface by about 50% of its thickness.

The experimental results are given in Table 5.4. Practically all the materials studied (with the exception of asbestos and kersey) are cut completely at the cutting speed of up to 24 m/min. The maximum cut width is 0.8 mm (only for the asbestos it is as high as 1.25 mm).

TABLE 5.3. Width, Depth and Cross Section of Cut as Functions of Focussing Conditions

ΔF, mm	Width b, mm	Depth h, mm	Cross section, mm²
+10	1.103	1.448	1.276
+9	1.03	1.655	1.343
+8	1.103	1.862	1.482
+7	1.03	2.0	1.515
+6	0.966	2.276	1.621
+5	0.897	2.345	1.621
+4	0.897	2.414	1.355
+3	0.828	2.483	1.655
+2	1.03	2.414	1.722
+1	1.03	2.207	1.655
0	0.966	1.931	1.449
−1	1.103	1.862	1.482
−2	1.03	1.724	1.377
−3	1.379	1.586	1.483
−4	1.310	1.310	1.310
−5	1.241	1.172	1.206
−6	1.172	1.03	1.101

TABLE 5.4. Parameters of Cuts Produced by Laser Beam in Various Footwear and Leather Materials

Material	Thickness, mm	Cutting speed, m/min	Cut width, mm	Cut
Fabric-base artificial leather (Hungary)	1.5	15	0.75	Complete
		20	0.75	Complete
		21	0.625	Complete
		22	0.575	Complete
		23	0.5	Complete
		24	0.5	
Vinyl artificial leather	1.5	15	0.625	Complete
		20	5.25	Complete
		21	5.0	Complete
		22	5.0	Complete
		23	5.0	Complete
		24	0.325	
Footwear plastic	1.2	15	0.5	Complete
		20	0.175	Complete
		21	0.25	Complete
		22	0.25	Complete
		23	0.25	Complete
		24	0.125	
Porous-monolithic plastic film	1.8	15	0.45	Complete
		20	0.3	Complete
		21	0.3	Complete
		22	0.275	Complete
		23	0.275	Complete
		24	0.295	Complete
Reinforced plastic film	1.0	15	0.75	Complete
		20	0.625	Complete
		21	0.5	Complete
		22	0.45	Complete
		23	0.5	Complete
		24	0.375	Complete
Chlorinated cloth	1.3	15	0.8	Complete
		20	0.6	Complete
		21	0.5	Complete
		22	0.45	Complete
		23	0.5	Complete
		24	0.5	Complete
Artificial leather	1.4	10	0.425	Complete
		11	0.3	Complete
		12	0.25	Complete
		13	0.25	Complete
		14	0.25	Complete
		15	0.2	Complete
		16	0.2	Complete
		17	—	Incomplete

Table 5.4 (cont.)

Material	Thickness, mm	Cutting speed, m/min	Cut width, mm	Cut
Asbestos fabric	2	10	0.125	Complete
		2	1.0	Complete
		3	0.875	Complete
		4	0.75	Complete
		5	0.625	Complete
		6	—	Incomplete
Kersey (impregnated multi-layer fabric)	1.5	10	0.25	Complete
		11	0.225	Complete
		12	0.15	Complete
		13	0.125	Complete
		14	0.125	Complete
		15	0.75	Complete
		16	—	Incomplete
Artificial leather with non-woven fabric base	1.8	15	0.8	Complete
		20	0.75	Complete
		21	0.675	Complete
		22	0.55	Complete
		23	0.55	Complete
		24	0.5	Complete

Figure 5.5. presents the cut width as a function of the cutting speed for some materials. The cut width can be seen to decrease with the increasing cutting speed. Figure 5.6 presents the profiles and widths of the cuts in cardboard according to the cutting speeds. Figure 5.7 shows the laser power required for cutting the vinyl leather as a function of the cutting speed. The air pressure is

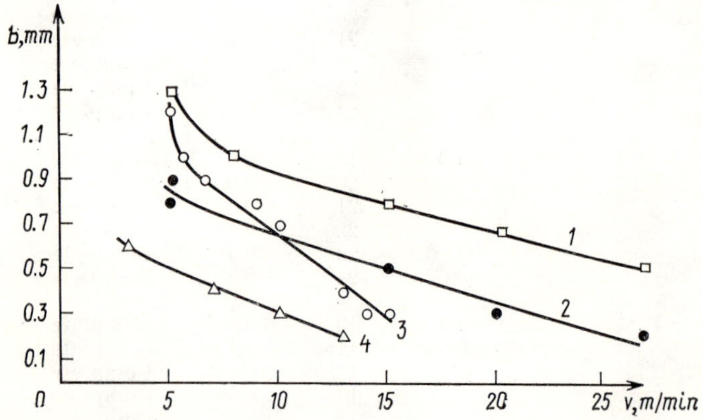

FIG. 5.5. Cut width as a function of cutting speed in various materials
1—leatherette; *2*—glove leather; *3*—plastic film; *4*—kersey. Laser power=500 W; air pressure=2 at; 240-mm KCl lens

b,mm	0.75	0.625	0.425	0.25	0.2
v,m/min	I	2	4	6	8

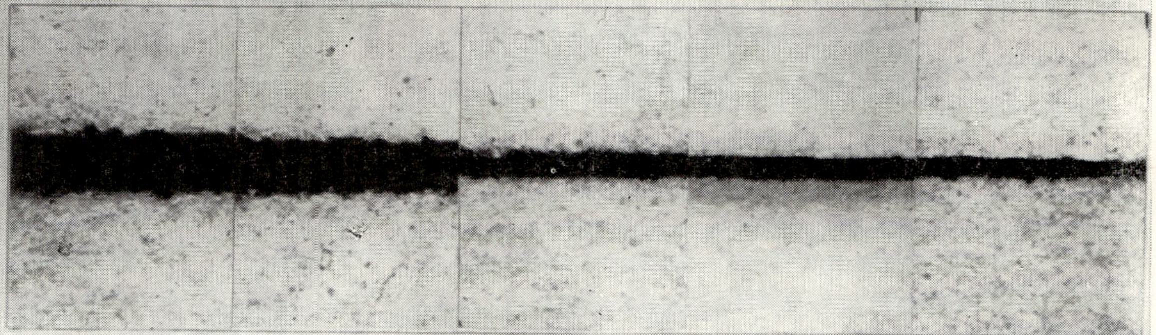

FIG. 5.6. Cuts in cardboard made at various cutting speeds (top view)

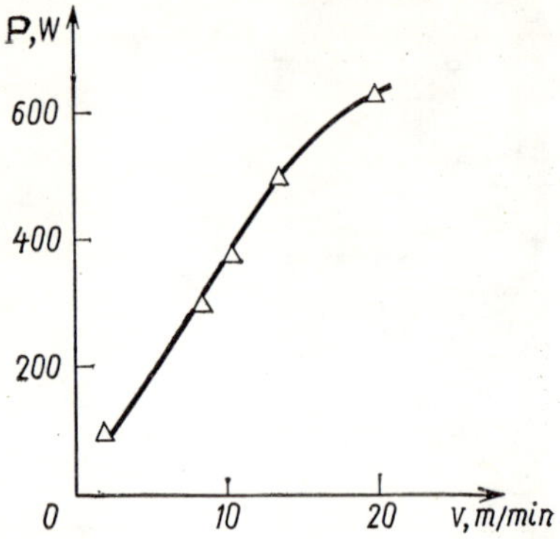

FIG. 5.7. Laser beam power required for cutting vinyl leather as a function of cutting speed

2 at, the beam power is varied by its diaphragming. It can be seen that the cutting beam power is practically a linear funciion of the cutting speed.

The focussed spot in the experiments is not more than 0.8 mm in diameter with the maximum beam power (240-mm lens). The cut width is typically less than 0.8 mm. This difference can be attributed to the fact that in the focussed spot the diameter of that part of the beam where the power density is higher than the threshold is smaller than the diameter of the focussed spot.

Figure 5.8 shows the cuts made in some materials. The cuts are practically clean, smooth and have a molten outer edge. As the materials have a low temperature conductivity and the cutting speeds are high the width of the heat-affected zone is not more than 0.05-0.1 mm. However, the surface of the materials is sometimes damaged (burned, etc.) along the cut owing to the defects in the cutter design, in the systems of gas blow and removal of the decomposition products, etc.

Evaluation of the laser cutting parameters. Let us discuss briefly some physical features of laser cutting for the above materials.

The starting process of laser cutting is the absorption of radiation by the material. In contrast to metals, in dielectrics radiation is absorbed in the bulk of the material, the penetration of radiation depending on the physical properties of the material. There can occur also the resonance absorption of radiation. In a very short time absorption of radiation results in the development of a heat source in the material, which produces thermal decomposition of the material after a certain temperature T_0 has been reached. The decomposition products in the cutting zone are removed by the gas flow into a removal system. We can make some qualitative estimates using the same models as for laser decomposition of the metals. The minimum power required for disintegration of the material with the thickness h can be estimated as

$$P_{\min} = S_0 h \rho v \sqrt{\frac{\pi}{k}} \approx S_0 h \rho v r_f \sqrt{\pi} \qquad (5.1)$$

where $S_0 = cT_0$, c is the specific heat, T_0 is the temperature of decomposition, ρ is the material density, v is the cutting speed, and r_f is the radius of the focussed spot on the material surface.

Equation (5.1) shows that $P_{\min}$ is proportional to v as confirmed by the experimental results on cutting of the vinyl leather with the laser power varying from 100 to 500 W and the cutting speed varying from 3 to 12 m/min. An increase in the focussed spot size also results in higher $P_{\min}$.

Note that Eq. (5.1) does not take into account the presence of the gas atmosphere, its pressure and the type of gas. However, Eq. (5.1) indirectly takes into account the gas pressure. It is assumed in Eq. (5.1) that at T_0 the complete cutting (melting or decomposition) of the material occurs. Therefore, the difference in the properties of materials can be reflected in the interaction between the laser beam and the evaporated material and in different degrees of ionization of the plasma cloud above the material.

For most materials discussed here it is difficult to use Eq. (5.1) because their thermal properties are not known. Qualitative estimates can be derived for

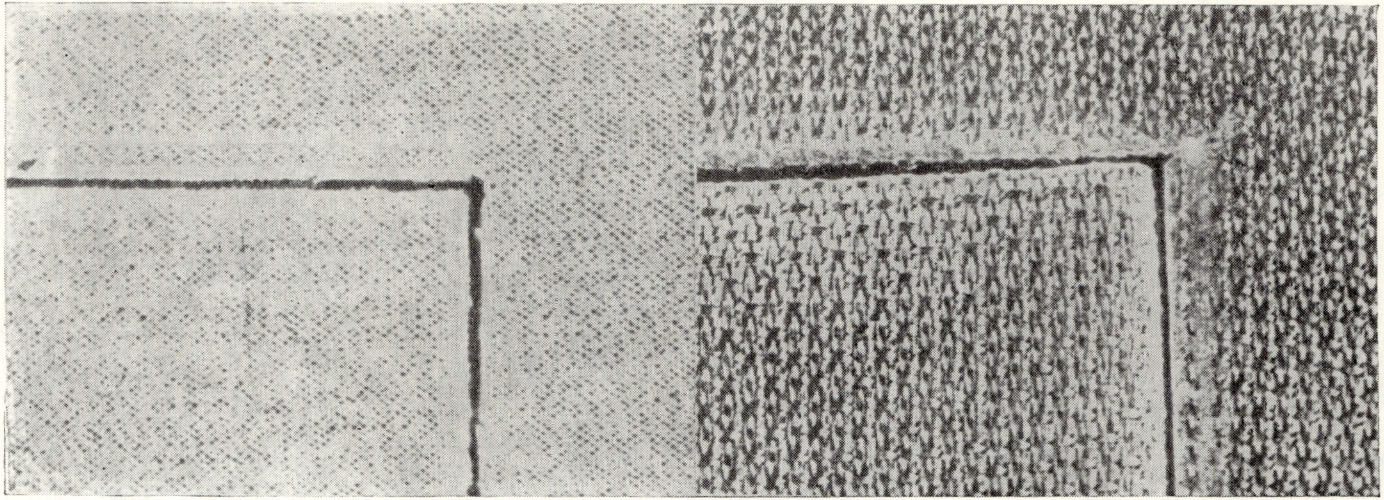

FIG. 5.8. Cuts made by laser beam in various materials (×3)
Laser power = 500 W; cutting speed = 24 m/min; air pressure = 2 at

vinyl leather for which $S_0 = 1.8$ kJ/g. Substituting into Eq. (5.1) the experimental results ($h = 1.5 \times 10^{-1}$ cm, $\rho = 1.5$ g/cm^3, $r_f = 0.1$ cm, $v = 25$ cm/s) we obtain $P \approx 1.7$ kW. This result does not agree with the experimental data ($P \approx 0.5$ kW). This can be due to an overestimated value of S_0 and ρ used in the calculations.

The width of the heat-affected zone can be estimated from a simple relation for the temperature field produced by a normal-distribution band heat source [12]:

$$T(y) = \frac{P}{vh\sqrt{4\pi c\rho\lambda t_0}} \exp\left(-\frac{y^2}{4at_0}\right) \tag{5.2}$$

where $T(y)$ is the temperature distribution along the coordinate perpendicular to the cut, λ is the heat conductivity, $t_0 = {}^1/_4\, ak$, a is the thermal diffusivity and P is the power of the source.

If the heat-affected zone is defined as a zone of the width y_0, and at a distance of y_0 from the cut the temperature is a certain fraction of the decomposition temperature T_0 Eq. (5.2) can be used to find y_0:

$$y_0 = 2\sqrt{at_0 \ln A/T(y_0)} \tag{5.3}$$

Here $A = P/vh\sqrt{4\pi c\rho\lambda t_0}$

Since $at_0 = a/4ak = r_f^2/4$, we obtain

$$y_0 = r_f \ln^{1/2} A/T(y_0) \tag{5.4}$$

that is, the width of the heat-affected zone increases with the radius of the focussed spot, and its dependence on variation in the beam power is weak.

The quantitative estimates from Eq. (5.4) for the typical parameters of the studied materials agree with the experimental data.

REFERENCES

1. BAZYUK, G. P., BARCHUKOV, A. I., KAPUSTIN, I. I., KORNILOV, V. P., CO_2 Laser Cutting of Garment Industry Materials, *Shveinaya prom.*, 1968, No. 3, 12-15.
2. ZRAZHEVSKY, V. A., PAVLENKO, Yu. S., Cutting of Fabrics with a Microplasma Arc., *Shveinaya prom.*, 1971, No. 2, 14-18.
3. BAZYUK, G. P., KAPUSTIN, I. I., KORNILOV, V. P., Electric Spark Technique for Pattern Cutting of Garment Industry Materials, *Shveinaya prom.*, 1968, No. 2, 18-21.
4. PALEY, Ya. S. *Avtomaticheskaya vyrubka detaley.* (Automatic Punching of Machine Parts), Legkaya industriya, Moscow, 1969.
5. CHIEL, F., Modern Techniques of Pattern Cutting and Machining of Elements in Footwear and Leather Industries, Report at the Intern. Exhibition Inlegmash-76, Moscow, 1976.
6. TYCHINSKY, V. P., *et al*, Programmed Laser Cutting Machine for Sheet Materials, *Svarochnoe proizv.*, 1972, No. 4, 52.
7. RAEVICH, V. K., ZAIDENBERG, G. A., KOKORA, A. N., *et al*, Automated Laser Cutting Machine for Nonmetallic Materials, *Kvantovaya elektronika*, 1976, v. 3, No. 7, 1543.
8. *Techniques de l'habillement*, 1971, No. 196, 47.
9. *Bekleidung and Wasche*, 1972, No. 2.
10. *The Shoe and Leather*, 1971, 220, No. 2877(5), 10.
11. MAAS, M. A., The Industrial Laser — a Special Tool for Special Needs, *Electronics*, 1976, No. 9.
12. RYKALIN, N. N., *Raschety teplovykh protsessov pri svarke* (Theory of Thermal Processes in Welding), Mashgiz, Moscow, 1951.

INDEX

Printed in the Union of Soviet Socialist Republics